Foundations of Mathematics: A Preparatory Course

Guido Walz • Frank Zeilfelder •
Thomas Rießinger

Foundations of Mathematics: A Preparatory Course

Guido Walz
Mannheim, Germany

Frank Zeilfelder
Mannheim, Baden-Württemberg
Germany

Thomas Rießinger
Bensheim, Germany

ISBN 978-3-662-67808-4 ISBN 978-3-662-67809-1 (eBook)
https://doi.org/10.1007/978-3-662-67809-1

This book is a translation of the original German edition „Brückenkurs Mathematik" by Walz, Guido, published by Springer-Verlag GmbH, DE in 2014. The translation was done with the help of artificial intelligence (machine translation by the service DeepL.com). A subsequent human revision was done primarily in terms of content, so that the book will read stylistically differently from a conventional translation. Springer Nature works continuously to further the development of tools for the production of books and on the related technologies to support the authors.

This Springer imprint is published by the registered company Springer-Verlag GmbH, DE, part of Springer Nature.
The registered company address is: Heidelberger Platz 3, 14197 Berlin, Germany

Paper in this product is recyclable.

Introduction and Foreword

Mathematics as a subject is so serious that no opportunity should be missed to make this subject more entertaining. (Blaise Pascal)

Even today, it is said to happen from time to time that someone picks up a book in a quiet hour to browse through it; however, in the rarest cases this will be a mathematics book, but this is not so much due to the mathematics itself, but to the way it is presented in many books. If terms like "dry", "brittle" or "incomprehensible" come to mind here, we are already well on the way to a common assessment.

Unfortunately, mathematics has a high "yuck" factor in society, and this is often due to traumatic school experiences and bad textbooks for students. With such a school career, however, one is usually afraid of the mathematical parts of studies, and quite in the spirit of a self-fulfilling prophecy, one then really gets into trouble there.

With this book, we try to save you from these entry problems by building a bridge in an entertaining way that gently guides you over all shoals into the interior of college mathematics. This bridge starts on one side with simple number crunching, as you probably encountered it in middle school, and takes you across to the basics of linear algebra, differential calculus, and probability, which will be the main content of your first semesters. You will always face this content there, and when dealing with it, you can then say with confidence, "I know it already!"

The book is aimed at students of all disciplines who come into contact with mathematics in the course of their studies or even their professional lives. One of our main concerns is to prove that it is possible to write a mathematics book that is easy to read from cover to cover without getting lost in formalism or humourless dryness, but which, after reading, has nevertheless given you the necessary knowledge and technical confidence for a successful mathematics part of your studies.

A large number of exercises, the solutions of which you will find at the end of the book, will help you to further deepen what you have read. We have deliberately avoided proofs for the most part – you will have to deal with this often enough in the course of your studies – rather, we simply want to whet your appetite for mathematics, even if this may seem contradictory to you at the moment.

A major annoyance in our eyes is the “nurse plural” displayed by many textbook authors (“How are we doing today?”, “So tomorrow we’ll have our livers out!”, “Now let’s integrate a function!”). Horrible! (Mind you, this refers only to this formulation bad habit, *not to* the profession of nursing, a group whose representatives each individually probably do more for humanity than all the mathematics professors in the world put together). In order to avoid this bad habit, we have decided to address you personally in each chapter in first-person style – each of us has written those chapters whose contents he can convey competently and (hopefully) instructively, and this should also be expressed by the writing style.

And now (and we mean it!): Have fun studying the following pages!

Mannheim, Germany — Guido Walz
Mannheim, Germany — Frank Zeilfelder
Bensheim, Germany — Thomas Rießinger
Summer 2005

Preface to the Second Edition

The gratifyingly good acceptance of our book has already made a second edition necessary. We have taken the opportunity to correct some minor errors in the text that were pointed out to us by attentive readers.

We have also included a new chapter on the basics of complex numbers. We hope you enjoy this as much as the rest of the text.

Mannheim, Germany — Guido Walz
Mannheim, Germany — Frank Zeilfelder
Bensheim, Germany — Thomas Rießinger
Spring 2007

Preface to the Third Edition

Despite all efforts to be error-free – a state that probably no book in the world will ever reach – there were still a few printing errors in the second edition, which we have now corrected.

This third edition includes a new take-out formulary that provides a condensed version of the most important formulas in the book, which you can carry with you for any occasion.

Mannheim, Germany — Guido Walz
Mannheim, Germany — Frank Zeilfelder
Bensheim, Germany — Thomas Rießinger
Spring 2011

Preface to the Fourth Edition

The unbroken strong demand for our book has again made a new edition necessary, which we are very pleased about. We take this opportunity to eliminate some printing errors in the last edition and thank all readers who brought them to our attention.

As in previous editions, we have added a new chapter on descriptive statistics to the book.

Mannheim, Germany	Guido Walz
Mannheim, Germany	Frank Zeilfelder
Bensheim, Germany	Thomas Rießinger
Summer 2014	

Contents

Elementary Calculation Methods

Before one ascends into the higher spheres of mathematics, one should first be sure that the handling of the elementary arithmetic procedures is no problem. In this first chapter, I will therefore first of all recall the basic arithmetic operations, especially fractions, and practice dealing with terms, brackets and other mathematical symbols and notations.

1.1 Basic Arithmetic Operations

Believe it or not, I am going to start with some very simple number crunching – although first it needs to be made clear what kind of "numbers" we are talking about here in the first place.

The simplest numbers are certainly the numbers 1, 2, 3, …, which you already encounter as a small child when you count your building blocks or the cookies lying on the table in front of you (Personally, I never got very far here, as I preferred to eat the cookies rather than count them, but that does not belong here). These numbers are called **natural numbers**, and the symbol $\mathbb{N}$ is used to denote the set of all these numbers. A **set** in mathematics is always a summary of objects, we also say **elements**, and the elements that make up a set are written in curly brackets {and}, we also say **set brackets**.

With this notation regulation is therefore

$$\mathbb{N} = \{1, 2, 3, \ldots\}. \tag{1.1}$$

On the basis of these simplest numbers we can already introduce the first laws of arithmetic, whose transfer to more general number ranges is no problem later.

G. Walz et al., *Foundations of Mathematics: A Preparatory Course*,
https://doi.org/10.1007/978-3-662-67809-1_1

By the way, it is disputed within the mathematical community whether the set $\mathbb{N}$ still includes the zero or not. I will not comment on these discussions here, but simply ask you to adopt the definition given in line (1.1). For the set of natural numbers including the zero, I will use the symbol $\mathbb{N}_0$ if necessary, i.e. $\mathbb{N}_0 = \{0,1,2,3, \ldots\}$.

I really do not want to get into the definition of the sum or product of two natural numbers, otherwise you would be bored and put this book aside, and I do not want that. Instead, let us look at simple properties of these arithmetic operations.

As you may know, it is completely irrelevant whether you (on a humid and merry evening) first drink two and soon after again four glasses of red wine or whether you down four glasses right at the beginning and later again two glasses: the result, i.e. the headache on the following day, is identical. This is simply because $2 + 4 = 4 + 2$, and of course this is true not only for these particular numbers, but for all of them; in mathematics, such a thing is called a regularity, or a law for short, and this one is called the commutative law. This also applies to multiplication, and I am going to formulate it in general terms, while I am also going to tell you about a related law, the associative law:

Commutative Law and Associative Law

An arithmetic operation * on a set M is called **commutative** if for all a and b from M holds:

$$a^*b = b^*a.$$

In this case, we also say that the operation * satisfies the **commutative law**.

An arithmetic operation * on a set M is called **associative** if for all a, b and c from M holds:

$$a^*(b^*c) = (a^*b) * c.$$

In this case, we also say that the operation * satisfies the **associative law**.

Do you have any idea why I just put these definitions here? Well, that is why:

Addition and Multiplication Are Commutative and Associative

Addition + as well as multiplication · are both commutative and associative on the set $\mathbb{N}$.

With which we would have the first two laws of arithmetic for these operations already under roof. For example, $2 + (3 + 4) = (2 + 3) + 4$ (=9), and it is also clear to the toddler mentioned above that he has eaten nine biscuits at the end of the day, regardless of whether he eats two biscuits in the morning and three and four biscuits in the afternoon, or whether he eats two biscuits in the morning and three more and is content with four in the afternoon; with which we have also seen an example of the associative law.

Exercise 1.1 Show that for three natural numbers a, b, c holds:

$$a + b + c = c + b + a. \quad \blacksquare$$

Also the next property, which I want to introduce, namely distributivity, can be formulated for general arithmetic operations, but as here always two different operations have to be combined, I want to write it down concretely, in order not to confuse you right at the beginning.

Distributive Law

With multiplication · and addition +, the set $\mathbb{N}$ satisfies the **distributive law**, that is, it holds for any three natural numbers a, b, c:

$$a \cdot (b + c) = a \cdot b + a \cdot c. \tag{1.2}$$

By the way, the parentheses in this expression are not for embellishment, but have the effect that the calculation $b + c$ must be carried out first, and only then is multiplied by a. This is the most elementary rule of the *parenthesis*, which we will look at in more detail below.

As an example of (1.2), I calculate $3 \cdot (2 + 5) = 3 \cdot 7 = 21$, because this is also what you get if you calculate according to the right-hand side of (1.2): $3 \cdot 2 + 3 \cdot 5 = 6 + 15 = 21$.

You may have wondered why I have not yet mentioned subtraction, which is generally regarded as the second basic arithmetic operation. This is because subtraction is unfortunately not possible within the set of natural numbers alone: Although it is clear that $5 - 2 = 3$, what should $2 - 5$ be? Obviously, it must be a number to which you have to add 3 to get 0. This number is called -3 and is sometimes also called the opposite number of 3. Every natural number has such an opposite number, the totality of all these opposite numbers of the natural numbers is called the **negative numbers**, and these together with the natural **numbers** and zero form the set of **integers**, denoted by $\mathbb{Z}$:

$$\mathbb{Z} = \{\ldots, -3, -2, -1, 0, 1, 2, 3, \ldots\}.$$

By the way, these newly introduced negative numbers are not an artificial invention of unworldly mathematicians, but are quite relevant to practice; think, for example, of the outside temperatures on a cold winter's day or your bank balance at the end of the month.

Within the number range $\mathbb{Z}$, you can now happily add, subtract and multiply without falling out of the set.

However, subtraction, because of which we introduced the integers in the end, has a big disadvantage: It does not fulfil the simplest laws of arithmetic, because it is neither commutative nor associative. I have already shown the lack of commutativity in the first example ($5-2 \neq 2-5$), and the following example shows that subtraction is also not associative:

$$7-(3-1)=7-2=5,$$

but

$$(7-3)-1=4-1=3,$$

and 3 is not equal to 5. Of course, this is chosen rather arbitrarily, there are countless other numerical examples that would have done here as well. But that is just the essence of an important proof principle in mathematics, which we learned here in passing, so to speak: To *disprove* an assumption (in this case, the assumption that subtraction would be associative), *a single* counterexample suffices.

In order to be able to calculate with integers, especially negative ones, we still have to clarify what an expression like $-(-a)$ means. Well, this is supposed to be the opposite number of $-a$, i.e. the one that results in zero when added to $-a$gerade. But there we would already have a candidate, namely a. So we have shown:

$$-(-a)=a. \tag{1.3}$$

This is, so to speak, the prototype of all rules of arithmetic which produce among all of us such associations as "minus times minus equals plus", for it follows from (1.3) that for arbitrary integers a and b hold:

Products of Negative Numbers

$$(-a)b=a(-b)=-ab$$

and

$$(-a)(-b)=ab.$$

All further rules, which we will get to know in the section on bracket calculus, among others, follow from these rather simple rules.

Exercise 1.2 Calculate the following numbers:

(a) $(3 - 6)(-2-3)$;
(b) $-(8-3)(-2 + 5)$;
(c) $(9-3 - 8)(3-1 - 7)(-2 + 1)$.

1.2 Fractions and Rational Numbers

There is, of course, a fourth basic arithmetic operation that you have surely been eagerly waiting for: division. To deal with it, however, we must leave the whole numbers and enter the world of fractions. Courage, I am with you!

Let us start again with an example, and since I, like Obelix, always think about food, here again something from this area: Imagine you are sitting on a nice summer evening with three people having dinner and in front of you on the grill are six juicy steaks. Since everyone is equally hungry, these six steaks are divided fairly among the three people, you can also say "divided" or "divided", and everyone gets $6 : 3 = 2$ steaks. So far, so good. The next day is bad weather, barbecue falls flat and you have the popular frozen pizza. Unfortunately, you forgot to go shopping because of all the maths learning and there are only two pizzas (this plural of pizza really exists, I looked it up in the dictionary) left for three people. The problem is clear: With a fair division, no one gets a whole pizza now, and that is just because 2:3 is not a whole number.

What to do? With the pizza, a sharp knife helps; with division, what mathematicians always do when a problem initially has no solution: they simply define something new. In this case, you expand the range of integers, in which we could move all the time, to the set of **rational numbers** or **fractions**, symbolized by the sign $\mathbb{Q}$: It is

$$\mathbb{Q} = \left\{ \frac{p}{q} \text{ with } p, q \in \mathbb{Z} \text{ and } q \neq 0 \right\}.$$

This needs some explanation, because this is the first time in this book that I have not simply enumerated a set, but defined it. This definition means that the set $\mathbb{Q}$ contains all expressions (fractions) of the form p/q, where p and q may be any integers, but q may not be zero. We call p the **numerator and** q the **denominator of** the fraction.

Fractions whose numerator is equal to one are also called **root fractions**. Conversely, of course, fractions whose denominator is equal to one are always equal to the numerator, a whole number. Therefore, the integers are included as a subset in the set of rational numbers.

A fraction is, so to speak, a notation for a division "not yet done", because p/q means in principle the same as $p:q$; every fraction can be transformed into a decimal number by performing this division, which either has only finitely many decimal places or whose decimal places repeat periodically.

Examples are

$$\frac{1}{4} = 0.25$$
$$\frac{2}{13} = 0.153846153846153846153846153846\ldots$$
$$\frac{35}{7} = 5.0.$$

Our pizza problem above is quite easy to solve with this new notation, because everyone now gets 2/3 pizzas.

However, the nice experience of being able to divide any (whole) number is somewhat spoiled by the unpleasant fact that you now have to deal with the ever-popular **fraction. N. B.: There** are a few rules for this, and I am going to tell you about them now, whether you like it or not. The first concerns what is known as **extending** a fraction:

Extending Fractions

Multiplying the numerator and denominator of a fraction by the same non-zero number does not change its value; in formulas:

$$\frac{p}{q} = \frac{p \cdot a}{q \cdot a} \quad \text{for all } a \neq 0.$$

This is more accurately called extending the fraction by the factor a. Practical example? Here you go: Whether you eat half a pizza or two quarters of a pizza is the same both nutritionally and mathematically, because

$$\frac{1}{2} = \frac{1 \cdot 2}{2 \cdot 2} = \frac{2}{4}.$$

The inverse of expanding is **shortening** a fraction:

Shortening Fractions
If the numerator and denominator of a fraction contain the same nonzero factor, you can divide both by that factor without changing the value of the fraction, in formulas:

$$\frac{p \cdot a}{q \cdot a} = \frac{p}{q} \quad \text{for all } a \neq 0.$$

This is used to avoid unnecessarily large numbers in the numerator and denominator of a fraction. Of course, the factor a is rarely explicit; rather, it must first be identified as a common factor of the numerator and denominator. For example, you can see relatively quickly that the common factor 4 is contained in the fraction 8/12, and you calculate

$$\frac{8}{12} = \frac{2 \cdot 4}{3 \cdot 4} = \frac{2}{3}.$$

It gets more difficult with things like 104/152. Some authors now advise to make a decomposition into prime factors, in order to recognize the common factor (here it is 8) and to do it with one stroke. Personally, I do not think it is a bad idea to first cut out small, easily recognizable factors and then possibly try your luck again; note that in the above definition, no one asked for a to be the *largest possible* factor in both terms.

In the example, you can immediately see that the numerator and denominator are both even, i.e. contain the factor 2, and determine

$$\frac{104}{152} = \frac{52}{76}.$$

With a little luck, you will see that the remaining fraction can even be shortened by 4 (if you are not so lucky, you will have to shorten it twice by 2) and you will find the following as the final result

$$\frac{104}{152} = \frac{52}{76} = \frac{13}{19}.$$

Since 13 and 19 are prime numbers, no further truncation is possible.

Exercise 1.3 Shorten the following fractions as much as possible:

$$\frac{231}{22}, \quad \frac{52}{28}, \quad \frac{16}{64}. \quad \blacksquare$$

So far, we have not really calculated with fractions, but only transformed them. The first type of arithmetic that I will now introduce is a real challenge: the **addition** (and synchronous **subtraction) of fractions**; I am sure that the mention of this topic will bring one or two traumatic school experiences to your mind. That is because, unfortunately, you are *not* allowed to add fractions by the numerator and denominator, but must first make sure they have the same denominator. Namely, it is true that:

Sum of Two Fractions with the Same Denominator

The sum of two fractions of the form p_1/q and p_2/q is

$$\frac{p_1}{q} + \frac{p_2}{q} = \frac{p_1 + p_2}{q}, \tag{1.4}$$

which is in line with life experience that one quarter of a pizza plus two quarters of the same pizza just add up to three quarters of that pizza.

But this really *only* works if the initial fractions have the *same denominator*. Since you will rarely be so lucky, in the general case you have to first equalize the denominators by expanding them. One method that always works without much thought is to expand with the denominator of the other fraction. Is that how you understood it? I do not think so, and this is not *your* fault but mine, sometimes a formulaic representation is really more appropriate. So:

If one wants to determine the sum of the two fractions p_1/q_1 and p_2/q_2, one can first expand the first fraction with q_2, the second with q_1 and obtain – using (1.4) – the calculation rule

$$\frac{p_1}{q_1} + \frac{p_2}{q_2} = \frac{p_1 q_2}{q_1 q_2} + \frac{p_2 q_1}{q_1 q_2} = \frac{p_1 q_2 + p_2 q_1}{q_1 q_2}$$

and also of course the rule for subtraction

$$\frac{p_1}{q_1} - \frac{p_2}{q_2} = \frac{p_1 q_2}{q_1 q_2} - \frac{p_2 q_1}{q_1 q_2} = \frac{p_1 q_2 - p_2 q_1}{q_1 q_2}.$$

One calculates herewith for example

$$\frac{2}{3}+\frac{3}{7}=\frac{14}{21}+\frac{9}{21}=\frac{23}{21}.$$

Since I have never seen three sevenths of a pizza in my life, I am afraid I cannot think of an example here, but I am sure you can verify it that way by doing the math.

This method of adding or subtracting two fractions always works in principle, but is sometimes not useful: For example, if you want to add or subtract the sum of

$$\frac{1}{5000}+\frac{1}{10,000}$$

you would have to bring both fractions to the denominator 50,000,000, but this does not make sense, because you can see with the naked eye that 1/5000 = 2/10,000, so the sum.

$$\frac{1}{5000}+\frac{1}{10,000}=\frac{3}{10,000}$$

results.

In general, one can make the addition of two fractions more efficient by taking the **least common multiple (lcm)of** the two denominators q_1 and q_2 as the target of the expansion. The lcm, as the name implies, is the smallest of all numbers that are multiples of both q_1 and q_2. In the above example, this lcm is equal to 10,000, because it is a multiple of both 10,000 itself (as a one-fold) and 5000. Of course, it is not always quite that simple; in the worst case, the lcm is equal to the product of the two numbers involved; this happens, for example, when both are prime numbers.

The following formula line should not scare you too much; I just want to write down the sum of any two fractions as a closed formula; sometimes I just need something like that:

Sum of Two Fractions

$$\frac{p_1}{q_1}+\frac{p_2}{q_2}=\frac{p_1\cdot\frac{lcm}{q_1}+p_2\cdot\frac{lcm}{q_2}}{lcm}.$$

When searching for the lcm – there are also sophisticated algorithms for this, but I do not want to do them to you and me here – you can start, for example, with the product as the first candidate and then see whether there are still smaller common multiples. For example, you find that the lcm of 6 and 10 equals 30: the first candidate, 60, is still too big, half will do, and even smaller common multiples than 30 do not seem to exist either. Therefore, for example

$$\frac{1}{6}+\frac{3}{10}=\frac{5}{30}+\frac{9}{30}=\frac{14}{30}=\frac{7}{15},$$

where I shortened the final result again.

Do not worry, we have got through the worst of it, because **multiplication of fractions** really does happen the way every student imagines it in their mathematical pipe dreams: by numerator and denominator.

Product of Two Fractions
The **product of two fractions** is calculated as follows:

$$\frac{p_1}{q_1}\cdot\frac{p_2}{q_2}=\frac{p_1\cdot p_2}{q_1\cdot q_2}.$$

Believe it or not, I am coming at you again with a pizza example: remember that rainy evening where you had to make do with 2/3 pizza? Well, let us further assume that at this point your big brother shows up and demands 3/4 of your slice for himself (it is a *very big* brother!). How much do you have to cede? According to the rules you just learned

$$\frac{3}{4}\cdot\frac{2}{3}=\frac{3\cdot 2}{4\cdot 3}=\frac{2}{4}=\frac{1}{2},$$

so half a pizza! What you then still remain, I do not calculate now, it is too depressing.

Exercise 1.4 Calculate the following products:

$$\frac{9}{4}\cdot\frac{6}{3},\quad\frac{13}{7}\cdot\frac{21}{26},\quad\frac{52}{76}\cdot\frac{19}{13}.\quad\blacksquare$$

Since any integer can be interpreted as a fraction with a denominator of 1, it is now clear how to calculate the product of a fraction and a number: multiply the numerator by the integer and leave the denominator unchanged:

Product of a Fraction and an Integer
For $a \in \mathbb{Z}$ and $p/q \in \mathbb{Q}$ is

(continued)

$$a \cdot \frac{p}{q} = \frac{a \cdot p}{q}.$$

Pizza? Alright: If you ate a quarter of a pizza four times, that makes a total of

$$4 \cdot \frac{1}{4} = \frac{4 \cdot 1}{4} = 1,$$

so it is a pizza. Mathematics can be this simple and true to life!

The last basic arithmetic operation is division, but it is not much more difficult than multiplication. To be able to formulate the division rule for fractions, I need the notion of the reciprocal fraction (some also say reciprocal) of a fraction: The **reciprocal fraction of** the fraction *p/q* is the fraction *q/p*, where of course p must not be zero. In particular, the reciprocal of an integer a is the **unit fraction** 1/a.

So you get the reciprocal of a fraction by swapping the numerator and denominator. Now we can move on to division:

Dividing Fractions

You divide by a fraction by multiplying by its reciprocal, in formulas:

$$\frac{p_1}{q_1} : \frac{p_2}{q_2} = \frac{p_1}{q_1} \cdot \frac{q_2}{p_2} = \frac{p_1 \cdot q_2}{q_1 \cdot p_2}.$$

The result of the division is called the **quotient of** the two initial fractions.

Without much preamble, two numerical examples; it is

$$\frac{13}{15} : \frac{2}{7} = \frac{13 \cdot 7}{15 \cdot 2} = \frac{91}{30}$$

and

$$\frac{1}{5} : \frac{3}{5} = \frac{1 \cdot 5}{5 \cdot 3} = \frac{1}{3}.$$

Again, you can formulate a practical life example. Since too much pizza is not good for your body, now once something healthy: Potato chips! Assume you want to watch a movie that lasts 150 min, that is 5/2 h. (In daily life, you would probably say "two and a half hours" here, but in arithmetic you should avoid "mixed numbers" like 2 1/2, as this can all too easily be confused with the product 2 1/2 (=1)). Unfortunately, you only have three-quarters of a bag of chips left, and you want to distribute them as evenly as possible over the running time of the movie. The necessary calculation is

$$\frac{3}{4} : \frac{5}{2} = \frac{3 \cdot 2}{4 \cdot 5} = \frac{3}{10},$$

... and I have shortened the result immediately. So you should eat three tenths bags per hour. Have fun measuring!

Exercise 1.5 Calculate the following quotients:

$$\frac{3}{4} : \frac{7}{8}, \quad \frac{52}{76} : \frac{13}{19}, \quad \frac{143}{11} : \frac{130}{22}. \quad \blacksquare$$

It only remains to emphasize that all the laws of arithmetic presented at the beginning for the natural numbers and the integers also remain valid for the rational numbers.

At the end of this section, I will now let you in on the secrets of the rule of three – if this is at all necessary, because the rule of three is of course the subject of school mathematics. But just as naturally, most people have a certain awe of the actually quite harmless rule of three, and if you also belong to this group, I now want to take this away from you.

The best way to start is with an example: A man buys five loaves of bread and has to pay 6 Euros for them. The next day (it is a large family) he buys seven loaves. How much do these seven loaves cost?

"Actually" one should now first calculate what a loaf of bread costs, viz.

$$\frac{6 \text{ Euro}}{5 \text{ Breads}} = 1.20 \text{ Euro/Bread},$$

and then determine the price of seven loaves by multiplication:

$$7 \text{ Breads} \cdot 1.20 \text{ Euro/Bread} = 8.40 \text{ Euro}.$$

The rule of three gets around this back-calculation to a unit (here: the price of a loaf of bread) by exploiting the fact that the ratio of 6 Euros to five loaves is the same as that of the unknown price p to seven loaves. In formulas this reads:

$$\frac{6 \text{ Euro}}{5 \text{ Bread}} = \frac{p \text{ Euro}}{7 \text{ Bread}}.$$

If we now resolve this to p, we get

$$p \text{ Euro} = \frac{6 \text{ Euro}}{5 \text{ Breads}} \cdot 7 \text{ Breads} = \frac{42}{5} \text{ Euro} = 8.40 \text{ Euro}.$$

This is the essence of the rule of three, that is, the task of directly calculating a fourth, unknown, quantity from three given quantities:

Rule of Three
In the rule of three, the unknown quantity is put in proportion to a total, and this ratio must be equal to a known ratio. By solving this equation for the unknown, you can then calculate it directly.

No, you do not have to understand that right away. I would rather make one or two more examples: A crew of 15 workers takes 9 h to dig a pit. The next day, due to illness, only 6 workers show up for the same job on an identical construction site. Question: How long do they need?

The rule of three

$$\frac{x \text{ hours}}{15 \text{ workers}} = \frac{9 \text{ hours}}{6 \text{ workers}}$$

now makes it possible to bypass the conversion to "man-hours" and to use the solution

$$x \text{ hours} = \frac{9 \text{ hours}}{6 \text{ workers}} \cdot 15 \text{ workers} = 22.5 \text{ hours}$$

directly.

This example can also be used to illustrate very nicely the limits of mathematical modelling: The same rule of three results in the fact that the desired work can be done by one million construction workers in considerably less than 1 s, which certainly does not quite correspond to reality for various reasons.

You already know I have a thing for pizza, so why not another example from that genre? Now then: It takes 750 g of dough to make three pizzas. How much dough is needed to prepare a party with 36 pizzas? To solve this, use the rule of three:

$$\frac{x \text{ Gramm}}{36 \text{ Pizzas}} = \frac{750 \text{ Gramm}}{3 \text{ Pizzas}}.$$

Solving for x gives

$$x \text{ Gramm} = \frac{750 \text{ Gramm}}{3 \text{ Pizzas}} \cdot 36 \text{ Pizzas} = 9000 \text{ Gramm}$$

So you need to prepare 9 kg of dough for your pizza party. Your problem!

1.3 Bracket Calculation

In the course so far, I have already been unable to dispense with the use of brackets in a few places, for example in the formulation of the distributive law or in the properties of subtraction. In both cases, the brackets served less to embellish the expressions than to determine the order in which the calculations were to be performed. In general:

Bracket Rule
Arithmetic operations that are enclosed by brackets must always be executed first.

We have already intuitively got this right with the distributive law, because $4 \cdot (2 + 5)$ just means that the addition in the parentheses must be carried out *before* the multiplication; the result is 28. The expression $4 \cdot 2 + 5$ written without parentheses, on the other hand, would mean that you must calculate $4 \cdot 2$, and only when you have done that may you add 5; the result is now 13. The reason is that at some point a long time ago it was established that multiplication and division must always be carried out *before* addition and subtraction ("dot arithmetic before dash arithmetic", remember?).

Once you have done the math in the parenthesis, you no longer write the parenthesis, it has, so to speak, vanished into thin air, and this is why this process is also called „unraveling"parentheses; e.g., is

$$-(3-5) \cdot (-2+4) = -(-2) \cdot 2 = 2 \cdot 2 = 4.$$

Parentheses often appear in connection with subtraction because it is not associative; for example, 11–5 – 3 is just *not* the same as 11–(5–3). (In case you are too tired to do the math yourself: The first expression gives 3, the second 9).

You think that calculations like 11 – (5–3) cannot occur in extra-mathematical life? Well, imagine you are watching a football match of your favourite team. It is a rather rough match and suddenly five players of your team are lying injured on the sidelines. After a short treatment three of them can play again, two have to go to hospital. How many players of your team are still on the pitch? (Since it is a *very* rough match, the substitute players have all fled already). Now, the calculation you need to do for this is exactly the one given at the beginning of the paragraph, and you can see that there are nine players left. If you left off the bracket, there would only be three left and the referee would stop the game.

Back to the math; parentheses can also occur in a pack, in which case:

Nested Brackets
If brackets are nested, the innermost one must be resolved first.

Two small examples of this. It is

$$-(18-(2+3\cdot(4+1)))=-(18-(2+3\cdot 5))=-(18-(2+15))=-(18-17)$$
$$=-1$$

and

$$1-(1-(1-(1-(1-1))))=1-(1-(1-(1-0)))=1-(1-(1-1))$$
$$=1-(1-0)=1-1=0.$$

That was a fairly short and probably not too difficult section. Maybe you think it is superfluous (which is a good sign, because then you are already familiar with it), but when we are calculating with terms and variables below, you will be glad you have looked at bracket arithmetic here before.

Exercise 1.6 Calculate the following expressions:

(a) $(((2-3)\cdot 4)-2)\cdot(-2)$
(b) $2-((1-4)\cdot(3-2)+4)$ ▪

1.4 Powers and Roots

Even without my having covered it here before, I am sure you remember how to figure out the contents of a square with side length 2: You simply calculate $2 \cdot 2$ (=4). If you now add a third dimension, i.e. consider a cube of edge length 2, its content is calculated to be $2 \cdot 2 \cdot 2$ (=8). This is already the end of the line for us simple human beings, our living space has no more than three dimensions, but let us assume that Captain Kirk would encounter a four-dimensional space, where one can then calculate four-dimensional cubes of edge length 2; their content would then be $2 \cdot 2 \cdot 2 \cdot 2$ (=16). And then, when even Captain Kirk is at a loss, Luke Skywalker comes along and finds a ten-dimensional cube! No, do not worry, I am not going to write down a product of ten twos, because mathematics has found an abbreviation for this, the *power notation*; our ten-dimensional cube would then have the content 2^{10} (=1024), and it is generally defined like this:

Exponentiation with Natural Numbers

If a *is* an arbitrary number and n is a natural number, then a^n is defined as the *n-fold* product of a with itself, i.e.

$$a^n = \underbrace{a \cdot a \cdots a}_{n\text{-times}}.$$

Call a the **base** and n the **exponent** of a^n and call a^n itself the n **-th power of** a.

Furthermore one defines $a^0 = 1$, but then a must not be zero.

For example

$$3^4 = 3 \cdot 3 \cdot 3 \cdot 3 = 81$$

and

$$\left(\frac{7}{2}\right)^3 = \frac{7}{2} \cdot \frac{7}{2} \cdot \frac{7}{2} = \frac{343}{8}.$$

I do not want to keep us here with further numerical examples, I do not think that is necessary, and after all there is still a lot to do. For example, formulate the first laws of calculation:

Power Laws

For all nonzero numbers a and b and all natural numbers m and n, the following equations hold:

$$a^m \cdot a^n = a^{m+n} \text{(P1)}$$

$$a^m \cdot b^m = (a \cdot b)^m \text{(P2)}$$

$$(a^m)^n = a^{m \cdot n} \text{(P3)}$$

Simple numerical examples of this: It is $2^3 \cdot 2^2 = 2^5$, because $2^3 = 8$, $2^2 = 4$, so $2^3 \cdot 2^2 = 32$, which agrees with 2^5. This illustrates (P1).

To explain (P2), I calculate $3^2 \cdot 5^2$, which gives $9 \cdot 25 = 225$. By rule (P2), one can just as well compute $(3 \cdot 5)^2$, which is 15^2, and indeed $15^2 = 225$.

Finally, we look at what 3^{24} gives: It is $9^4 = 6561$, and this is also obtained as the result of $3^{2\,\cdot\,4} = 3^8$, which was to be expected according to rule (P3).

Now that you have mastered calculating with powers as long as the exponents are natural numbers, let us venture further by allowing all integers as exponents, including negative ones. Only, what should be for example a^{-1}? In order for rule (P1) to apply in this case as well, it must be a number that has the property

$$a^{-1} \cdot a = a^{-1+1} = a^0 = 1$$

which, multiplied by a, leads to 1. But there is only one real candidate, namely the number 1/a. So we have seen:

$$a^{-1} = \frac{1}{a}$$

and in general:

Powers with Negative Exponents
For all integers n holds:

$$a^{-n} = \frac{1}{a^n}. \quad (P4)$$

The power rules (P1) to (P3) formulated above also apply to non-positive exponents m and n.

Example 1.1 A few examples here; according to the first rule, in the term $5^4 \cdot 5^{-5} \cdot 5^3$, one can simply add the exponents and get

$$5^4 \cdot 5^{-5} \cdot 5^3 = 5^2 = 25.$$

Even an expression of the form $(x^{-2} \cdot y^{-3})^{-3} \cdot x$ need not frighten us, for with the help of the power rules we find that

$$\left(x^{-2} \cdot y^{-3}\right)^{-3} \cdot x = \left(x^{(-2)\cdot(-3)} \cdot y^{(-3)\cdot(-3)}\right) \cdot x = x^6 \cdot y^9 \cdot x = x^7 \cdot y^9$$

is. After all, the unpleasant looking fracture can be

$$\frac{\left(17^2\right)^{-6}}{17^9 \cdot 17^{-3}}$$

transform into

$$\frac{\left(17^2\right)^{-6}}{17^9 \cdot 17^{-3}} = \frac{17^{-12}}{17^6} = \frac{1}{17^{12} \cdot 17^6} = \frac{1}{17^{18}}.$$

Not a pure pleasure either, but at least summarized compactly.

Exercise 1.7 Calculate the following terms:

(a) $(1/2)^3 \cdot (2/3)^5 \cdot (3/4)^2$
(b) $(a^2b^3c^{-1})^2 \cdot (c^2)^2$

So far, so good. But now that this is going so well with integer exponents, let us get foolhardy and try taking rational numbers in the exponents. But what on earth, for example

$$17^{\frac{173}{229}}$$

may be? Well, you do not have to start that complicated, let us rather begin with the simplest rational numbers, the root fractions, and here again with the simplest (if I leave out the trivial root fraction 1/1), namely 1/2.

Before I go any further: I have just heard the word "trivial", which is popular with mathematics lecturers and authors, flowing from my pen or keyboard; perhaps I should explain it. In mathematics, a *trivial case* is one that is so simple or obviously correct that one does not need to elaborate on it, much less prove anything. (Malicious tongues claim, however, that an author also likes to write "trivial" when he himself has no idea how to approach the matter and prefers to leave the reader alone with the problem).

Back to the topic. So what is $a^{1/2}$, for example concretely $3^{1/2}$. Well, according to rule (P3) this must be a number for which holds:

$$\left(3^{\frac{1}{2}}\right)^2 = 3^{\frac{1}{2}\cdot 2} = 3,$$

But there is no such number in the entire set of rational numbers (which, strictly speaking, you would have to prove, but a certain Euclid of Alexandria kindly did that for us a few thousand years ago), and if something does not exist, then you just define it, as you already know. And since this obviously does not have anything to do with the special number 3, you do it in general, as follows:

Powers with Exponent 1/2 – Square Roots
For each positive number a, $a^{1/2}$ is defined as the positive number which, when multiplied by itself, gives a. One designates $a^{1/2}$ as **square root** or simply as **root from** a and writes for it also the symbol $\sqrt{a}$, thus

$$a^{\frac{1}{2}} = \sqrt{a}.$$

For some integers, also **called squares**, the root is also an integer, for example $\sqrt{16} = 4$, because $4^2 = 16$, and so is $\sqrt{36} = 6$. In the vast majority of cases, however, the root is not an integer, not even a rational one. What else is it supposed to be? Please be patient for a few more lines, then I will tell you. Before that, I am afraid I have to remind you that we have come only minimally closer to our goal of allowing arbitrary rational numbers as exponents, because we have so far clarified the matter for only one rational number, 1/2.

But do not worry, that breaks the spell and the rest is easy. First, define powers with any root fraction in the exponent in the way proven with 1/2:

Powers with a Root Break in the Exponent – *nth* Roots
For every positive number a and every natural number n, $a^{1/n}$ *is* defined as that positive number which, *when* multiplied by itself n *times*, gives a. One designates $a^{1/n}$ as **nth root of** a and writes for it also the symbol $\sqrt[n]{a}$, thus

$$a^{\frac{1}{n}} = \sqrt[n]{a}$$

Again, there are some selected integers whose *nth* roots are integers again, for example $\sqrt[3]{64} = 4$, because $4^3 = 64$, or $\sqrt[4]{16} = 2$, because $2^4 = 16$. For $n = 3$, such numbers are called **cubes**, for larger values of n they no longer have their own designation.

Boring? Here you go:

Exercise 1.8 Calculate the following root expressions:

(a)

$$\sqrt[4]{81}$$

(b)

$$\sqrt[3]{2^2 \cdot 2}$$

(c)

$$\sqrt[5]{(a^5)^2}$$

If you are now thinking, "Now get on with it, we still have an infinite number of rational exponents ahead of us!" then I can assure you, we have just about got it. For if you find a number a with any rational number p/q as its exponent, you can use the rule (P3) to perform the following transformation:

$$a^{\frac{p}{q}} = a^{\frac{1}{q} \cdot p} = \left(a^{\frac{1}{q}}\right)^p. \tag{1.5}$$

But with this everything is clear, because we have just learned the exponentiation with the root fraction 1/q and we can also do the exponentiation with the *integer p*. In (not very nice, but perhaps necessary) words this means as follows:

Exponentiation with Any Rational Number
For any positive number a, the expression $a^{p/q}$ denotes the *pth* power of the *qth* root of a, in formulas:

$$a^{\frac{p}{q}} = \left(a^{\frac{1}{q}}\right)^p = \left(\sqrt[q]{a}\right)^p.$$

And it gets even better: because of the commutative law of multiplication (remember that?), instead of doing the transformation in (1.5), you can do the following:

$$a^{\frac{p}{q}} = a^{p \cdot \frac{1}{q}} = (a^p)^{\frac{1}{q}}. \tag{1.6}$$

This means that $a^{p/q}$ is also equal to the *qth* root of the *pth* power of a. This is worth a text box in the summary:

Identities for $a^{p/q}$

For all positive numbers a and all rational exponents p/q holds:

$$a^{\frac{p}{q}} = \left(a^{\frac{1}{q}}\right)^p = \left(\sqrt[q]{a}\right)^p = (a^p)^{\frac{1}{q}} = \sqrt[q]{a^p}.$$

For example, after each of the identities given here, you get that $8^{2/3} = 4$, because

$$8^{\frac{2}{3}} = \left(8^{\frac{1}{3}}\right)^2 = 2^2 = 4,$$

but also

$$8^{\frac{2}{3}} = (8^2)^{\frac{1}{3}} = 64^{\frac{1}{3}} = 4,$$

because $4^3 = 64$.

So you can calculate powers with rational exponents. The power calculation rules apply unchanged also for rational exponents, for example

$$3^{\frac{1}{2}} \cdot 3^{\frac{3}{2}} = 3^{\frac{1}{2}+\frac{3}{2}} = 3^2 = 9.$$

This should certainly take some practice, so here are a few suggestions for self-study:

Exercise 1.9 Calculate the following numbers:

(a) $4^{1/3} \cdot 2^{1/3} \cdot 16^{1/4}$
(b) $120^{1/2} \cdot 900^{1/4}$
(c) $\sqrt{0.16}$

I mentioned above that roots are rarely rational numbers. But what are they then? Well, first of all, you give them and their fellow sufferers a name: Numbers that are not rational are called **irrational numbers**.

But naming alone is not enough, you also want to know how big the irrational number is. And you can do that by approximating its decimal representation. I will show you this using the irrational number $\sqrt{2}$ as an example.

I stalk the decimal representation of this number step by step by determining one decimal place after the other. To get a first orientation about the number $\sqrt{2}$, I find out that $1^2 = 1$ is smaller than 2 and $2^2 = 4$ is larger than 2. This makes it clear that $\sqrt{2}$ is between 1 and 2, so the decimal representation starts with 1, something. To examine "something" more closely, I now focus on the first decimal place and calculate

$$1.4^2 = 1.96 \quad \text{and} \quad 1.5^2 = 2.25.$$

So one square is smaller than 2, the other larger than 2, so $\sqrt{2}$ is between 1.4 and 1.5, and is therefore of the form 1.4 + something. You already know how I examine the new "something": I calculate

$$1.41^2 = 1.9881 \quad \text{and} \quad 1.42^2 = 2.0164.$$

Thus we know that $\sqrt{2}$ is between 1.41 and 1.42 and thus of the form 1.41 + something.

You guessed it: there are an infinite number of "somethings", because the decimal representation of the irrational number $\sqrt{2}$ never breaks off. However, you can calculate it arbitrarily precisely and you find out:

$$\sqrt{2} = 1.4142135623730950488016887242209\ldots$$

You must admit that this has succeeded in giving an "approximate" estimate of the size of this figure.

So or similarly one can deal with any irrational number, each of them can be sized arbitrarily exactly and thus sorted into the existing scale of rational numbers. One can even show, that between any two rational numbers there are always an infinite number of irrational ones; these numbers lie, as one also says, arbitrarily close to each other. If you mark them as points lying next to each other, then these points form a closed line, which is also called a **number line**. The set of these numbers has a special name:

Real Numbers

The combination of all rational numbers with all irrational numbers is called the set of **real numbers**, denoted by the symbol $\mathbb{R}$.

The set $\mathbb{R}$ of real numbers now actually includes all numbers you can imagine, of course first of all the integers and the rational numbers, but also famous irrational numbers such as the circle number π or the Eulerian number e, which you will encounter later, and of course also an infinite number of unspectacular irrational numbers such as $\sqrt{17}$, which usually go unnoticed. The important thing is, that you can calculate with all real numbers in the same way as I worked out for natural, integer and rational numbers. Finally I want to summarize this as a quotable rule:

Arithmetic Rules for Real Numbers
All the rules of calculation that we have found out so far for natural, whole and rational numbers apply unchanged to the real numbers as well.

1.5 Special Expressions and Notations

Mathematics is teeming with special notations and formula symbols. To a certain extent (some authors overdo it a little, I think) this is right, because it supports the precision and brevity necessary in mathematics. For example, the expression

$$\sqrt{16} = 4 \tag{1.7}$$

is much shorter than

,,That positive number which – multiplied by itself – gives 16 is 4.

The only problem with this is that many students are put off by this unfamiliar formula Chinese, especially at the beginning, and do not understand connections that are actually quite easy to grasp, because they simply cannot "translate" the formula symbols; for example, the brevity and precision of the expression in (1.7) is of no use to you if you do not know what the symbol "$\sqrt{42}$" means. It has probably happened to many a lecturer that he has written on a blackboard full of the most elegant statements and connections without noticing that half of his audience has already dropped out at the beginning at the first sum symbol.

To avoid this in your case, I will explain below some notations and conventions commonly used in mathematics.

Although I have tried to work as concretely as possible with numbers in the course of this first chapter so far, I still had to "calculate with letters" at one point or another, as my then young son once said. Well, this "calculating with letters" is the basis of all mathematics, without it formulas, solution methods, functions, systems of equations and other beautiful things would not be formulable. Letters, with which one calculates, one calls **variables**, depending upon situation also **unknowns** or **parameters**. They serve as placeholders in a mathematical expression and are later replaced by elements of a given base set, usually numbers. A mathematical expression that contains variables is called a **term**. By the way, this has nothing to do with thermals or thermal baths, which is why it is written without an "h".

These terms should not frighten you, you know such a thing for a long time; for example, is

$$2 \cdot (a + b)$$

a simple term. If you call the two side lengths of a rectangle a and b, this term gives the perimeter of the rectangle. The advantage (still very small here) is that you can use it to calculate the perimeter of any rectangle in the world (including a rectangular pizza, by the way) without having to add the four side lengths each time.

Later, in the section on functions, we will practice in detail how to deal with terms containing roots, powers or similar nasty things, so I would not dwell on that here. Instead, I will now introduce you to the handling of terms by means of parentheses, which I introduced for numbers in Sect. 1.3.

In the following let a, b, c and x, y, z be variables representing any real number; later we will simply say for this purpose: Let a, b, c, d and x, y, z be any real numbers.

With the simple dissolution of parentheses it is now no longer so easy, because in a term like $3a \cdot (b + 2\ c)$ I can not shorten $b + 2\ ci$, that must already remain so. However, you can simplify such terms by multiplying them out, because the **distributive law** also applies to terms. Thus

$$3a \cdot (b + 2c) = 3ab + 6ac.$$

Of course, two or more parenthesized terms can also be multiplied; for two, it looks like this:

Multiplying Out Brackets

For all real numbers a, b, x and y is:

$$(a + b) \cdot (x + y) = ax + bx + ay + by. \tag{1.8}$$

To justify it, you only have to apply the distributive law twice in a row:

$$\begin{aligned}(a + b) \cdot (x + y) &= (a + b) \cdot x + (a + b) \cdot y \\ &= ax + bx + ay + by.\end{aligned}$$

Of course, one must also observe the sign rules when applying (1.8), for example

$$(a - b) \cdot (x + y) = ax - bx + ay - by.$$

Example 1.2 Two easy examples of this: It is

$$(3x - 5z) \cdot (-b + 3c) = -3bx + 5bz + 9cx - 15cz$$

and

$$-(xy + c) \cdot (ab - 2xz) = -(abxy + abc - 2x^2yz - 2cxz) = -abxy - abc + 2x^2yz + 2cxz. \quad \blacksquare$$

Here I have made use of the convention that when you have a product of variables, you write down the individual factors in alphabetical order. This is of course (because of the commutativity of multiplication) mathematically the same as any other order, but helps you keep track of long expressions; or would you have seen offhand that

$$abcxyz - zaxcby + yxbazc - byaxcz = 0$$

is? Probably not, at least not me. But if you put each of the four products in the correct alphabetical order, you will immediately see that there are four $a\ b\ c\ x\ y\ z$.

Special cases of the rule (1.8) are the **binomial** formulas, a real crown of school mathematics:

Binomial Formulas
For all real numbers a and b holds:

$$(a + b)^2 = a^2 + 2ab + b^2$$
$$(a - b)^2 = a^2 - 2ab + b^2$$
$$(a + b) \cdot (a - b) = a^2 - b^2$$

By the way, these are not, as I once read in a textbook, named after the Italian mathematician Allesandro Binomi (simply because he does not exist), rather the name comes from the Greek and says that "two things" (namely a and b) are treated here. The binomial formulas, which are often referred to as the first, second and third binomial formulas in the order given above, are simply proved by applying (1.8) and then summing them up. Thus one obtains, for example

$$(a+b)\cdot(a-b)=a^2-ab+ab-b^2=a^2-b^2,$$

which proves the third formula.

The summarizing of terms with the same name I have put under you here quasi incidentally, it should of course always be done after the multiplying out, to avoid too long expressions. Let us hold on to that for a moment:

Summarizing Terms with the Same Name
After multiplying out parentheses, you should always combine terms with the same name (that is, terms that contain the same variables).

Since you probably feel sick at the mere thought of pizza and have run out of potato chips, I am going to give you a serious apples and oranges lesson, the classic of elementary arithmetic par excellence: If you get the result of a multiplication, which I cannot imagine to be that exact either:

$$5\,\text{apples} + 6\text{ pears-3 apples-2 pears},$$

you will certainly not leave it at that, but summarize it to

$$2\,\text{apples} + 4\text{ pears}.$$

This is exactly what is meant by the above-mentioned "combining of terms with the same name", whereby these terms are not called "apples and pears" in mathematics, but a b, x^2 y or similar. The following is a rather mathematical example in this sense, in which the multiplication according to (1.8) is practiced again:

$$\begin{aligned}(2x+3)\cdot(4xy+y) &= 8x^2y+12xy+2xy+3y\\ &= 8x^2y+14xy+3y.\end{aligned}$$

Of course, more than two parentheses or even more than two expressions in one parenthesis can occur in such expressions. Consistent multiple application of the rules just learned leads to the goal here as well, for example

$$\begin{aligned}(a+1)\cdot(4a+3)\cdot(1-a) &= (a+1)\cdot\left(4a+3-4a^2-3a\right)\\ &= (a+1)\cdot\left(a+3-4a^2\right)\\ &= a^2+a+3a+3-4a^3-4a^2\\ &= -4a^3-3a^2+4a+3\end{aligned}$$

or

$$\begin{aligned}(1+x+y-xy)\cdot(x-y) &= x-y+x^2-xy+xy-y^2-x^2y+xy^2\\ &= x-y+x^2-y^2-x^2y+xy^2.\end{aligned}$$

I cannot see any other simplifications, so I will leave it at that.

The combination of these two generalizations, that is, both more than two brackets and more than two terms in the brackets, is too complicated for me; fortunately, I just realized that I have not done an exercise problem in a long time, so that would come in handy now, do not you think?

Exercise 1.10 Simplify the following terms as much as possible:

(a) $(a^2 + b - c)(a + b\ c)(a^2\ b)$
(b) $(1 + x + x^2 + x^3 + x^4)(1 - x)$ $\square$

Just as in simple arithmetic, **nested brackets** can of course occur when dealing with terms; there is nothing mysterious about this now, the rule applies here too that in this case the innermost bracket must always be resolved first.

It is not very useful to give a formula for this, I think one or two examples will clarify the situation better.

Example 1.3 First, I consider the term

$$T = 3b - (2b \cdot (3 - (2a + 3b) \cdot (3b \cdot (1 - a)))).$$

True to the motto "from the inside out", the innermost bracket is resolved first, which provides

$$T = 3b - (2b \cdot (3 - (2a + 3b) \cdot (3b - 3ab))).$$

Now the two inner brackets are multiplied out, after which T *has* the following form:

$$T = 3b - \left(2b \cdot \left(3 - \left(6ab + 9b^2 - 6a^2b - 9ab^2\right)\right)\right).$$

Dissolving the now innermost bracket now simply means to drag in the minus sign in front of it; this results in

$$T = 3b - \left(2b \cdot \left(3 - 6ab - 9b^2 + 6a^2b + 9ab^2\right)\right).$$

After multiplying the now innermost bracket we have

$$T = 3b - \left(6b - 12ab^2 - 18b^3 + 12a^2b^2 + 18ab^3\right),$$

and finally as a final result

$$T = -3b + 12ab^2 + 18b^3 - 12a^2b^2 - 18ab^3. \quad ■$$

In the preface it was said that this book should make you want to read mathematics texts, and I am not *quite* sure whether I have succeeded in doing that with this example; I actually think rather not. Therefore, I will now dispense with the second example originally planned here and move on to the next topic.

In a sense, I am now treating the inverse of multiplying out parentheses: factoring out a common factor of the individual addends of a sum. Any more questions, Watson? I think so, because that was not very clear. For once, perhaps a formulaic representation is clearer than a verbal one here:

Factor Out
Transforming a sum of the form $a\,x + a\,y$*into* the form $a \cdot (x + y)$ is called factoring **out** a from the sum.

In this topic, working with terms is for once easier than working with numbers, because with terms you can often see the common factors of the individual summands with the naked eye. An example: The summands in the sum

$$abx^2y + axy^2$$

obviously both contain the factor axy, thus one can factor out as follows:

$$abx^2y + axy^2 = axy \cdot (bx + y).$$

The crux of efficient factoring out is obviously finding the **greatest common divisor of** the summands, which is usually referred to as the gcd. This term is actually self-explanatory: The gcd is the largest term that divides all involved summands, i.e. the largest possible factor that can be extracted. When finding this factor, especially for sums with more than two summands, one pragmatically proceeds as follows: One takes any – usually the first –

summand as a candidate for the gcd and truncates it by checking the other summands one by one until the largest *common* factor remains at the end.

Example? Of course, you are right:

Example 1.4 Let us consider the term

$$6abx^2 - 6a^2x + 3ax + 12ab.$$

Since I am not too creative at this time of day, I simply take the first summand, $6abx^2$, as the first candidate for the greatest common factor. Already when testing the second summand, $-6a^2x$, the candidate has to give up feathers, because this second summand does not contain b as a factor, and x also only occurs in the first power, so $6ax$ remains as a candidate. In the third summand, $3ax$, this almost fits, only the prefactor becomes 3, so the candidate becomes $3ax$. Unfortunately, there is no x in the last summand, so the factor $3a$ is identified as the ggT. Thus:

$$6abx^2 - 6a^2x + 3ax + 12ab = 3a \cdot \left(2bx^2 - 2ax + x + 4b\right). \tag{1.9}$$

By the way, it would have been faster if I had started with the third summand $3ax$, but one does not only make happy decisions in life.

Exercise 1.11 Factor out the largest possible factor:

(a) $34\,x\,y\,z^2 - 17\,x^3\,y\,z + 51\,y^4$
(b) $2\,a^2\,b\,c - 4\,a\,b^2\,c + 8\,a\,b\,c^2$

An important application of factoring out common factors is the **shortening of fractions.** To do this, one has to identify a common factor in the numerator and in the denominator, and if numerator and/or denominator consist of a sum of terms, this means factoring out. A first simple example is the fraction

$$\frac{4ax - 2ay}{ay}.$$

Here we can see with the naked eye that in the numerator the factor $2a$ can be factored out, in the denominator there is nothing to do. With this we can transform:

$$\frac{4ax - 2ay}{ay} = \frac{2a \cdot (2x - y)}{ay}$$

and truncate (by a):

$$\frac{2a \cdot (2x - y)}{ay} = \frac{2 \cdot (2x - y)}{y}.$$

More is not possible here, as tempting as the factor 2 in the numerator or even the y in the parenthesis may look. Speaking of "tempting": I thought about it for a long time, but I just could not think of a pizza application for this topic; if you have a good idea, please write to me, you will then also be mentioned with praise in the next edition of the work, I promise!

Somehow the dry mathematician seems to be deeper in me than I thought, because unlike pizza examples, I can think of masses of formulaic examples right now; we will do one more of those, alright? Thank you:

Example 1.5 So let us take a look at the following fraction:

$$\frac{6abx^2 - 6a^2x + 3ax + 12ab}{6axy^2 - 12ax^2 + 6a^2x}.$$

You asked for it! But it is not quite that bad, because we have already done the groundwork above and have already completely taken care of the numerator in formula (1.9). If we look at the denominator, i.e. the term $6axy^2 - 12ax^2 + 6a^2x$, we see that all three addends contain the factor $6ax$, and a closer inspection shows that this is also the largest common factor. Thus

$$6axy^2 - 12ax^2 + 6a^2x = 6ax \cdot \left(y^2 - 2x + a\right),$$

and the entire fraction becomes

$$\frac{6abx^2 - 6a^2x + 3ax + 12ab}{6axy^2 - 12ax^2 + 6a^2x} = \frac{3a \cdot \left(2bx^2 - 2ax + x + 4b\right)}{6ax \cdot (y^2 - 2x + a)},$$

where I used the result in (1.9). Shortening (by $3a$) is now a pure pleasure and gives the result

$$\frac{6abx^2 - 6a^2x + 3ax + 12ab}{6axy^2 - 12ax^2 + 6a^2x} = \frac{3a \cdot \left(2bx^2 - 2ax + x + 4b\right)}{6ax \cdot (y^2 - 2x + a)} = \frac{2bx^2 - 2ax + x + 4b}{2x \cdot (y^2 - 2x + a)}. \qquad \blacksquare$$

That is all you can do here, but since inactivity is quite bad in mathematics as well as in life, here is something for you:

Exercise 1.12 Shorten as much as possible:

(a) $\frac{4a^3(bc)^2}{(2abc-4ab)a^2}$

(b) $\frac{(3x^2yz)^3}{9(xy^2z)^3}$?

With that, we were pretty much done with the topic of parentheses, but there is one special thing from the dirty tricks department I want to show you: If you read the third binomial formula backwards, it says.

$$a^2 - b^2 = (a + b) \cdot (a - b),$$

and this is now very commonly used to "pull apart" expressions of the form $a^2 - b^2$. For example, for all a in the world

$$a^4 - a^2 = (a^2 + a) \cdot (a^2 - a) \tag{1.10}$$

and above all the relationship

$$1 - x^2 = (1 + x) \cdot (1 - x)$$

is always a popular choice. (If you got puzzled by the transformation in (1.10), look again very briefly into the section on power calculation, I wait here so long). I am almost sure that you will encounter something like this in one of your first lectures or even practice or exam assignments, and then you will be grateful!

In the previous calculations we had to deal with only a handful of numbers or variables, all of which could be written down explicitly. This will not always be the case; for example, if you want to calculate the sum of, say, 16 numbers $a_1, a_2, a_3, \ldots, a_{16}$, you will no longer want to write them all down; you can, of course, use the dot notation

$$a_1 + a_2 + \cdots + a_{16} \tag{1.11}$$

in the first place. But this is not only very space-consuming, but sometimes also not exact enough, because the expression in (1.11) can also mean the sum $a_1 + a_2 + a + a_{48} + a_{16}$, the law of formation of the indices to be summed is not unique. Therefore, the summation sign $\sum$ has been invented. This is a sigma, i.e. a Greek "S", and it stands for "sum". What is to be summed is written after the sum character, in our example this would be $\sum a_i$, the index i, over which the sum is to be calculated, was set here as a variable. The range through which this index is to pass is written above and below the sum sign, the first value below, the last value above; the sum of the above-mentioned 16 numbers is thus written in the exact notation 16 numbers is thus in exact notation:

$$\sum_{i=1}^{16} a_i,$$

and if you have to pronounce it once, you say: "Sum over a_i for i from 1 to 16." Note that here there is no longer any uncertainty about the running range of the index, because by convention i here runs through *all the* integers between the lower and upper bounds. If you do not want to have all the numbers in the sum, you have to do a little trickery; for example, is

$$a_1 + a_2 + a_4 + a_8 + a_{16} = \sum_{i=0}^{4} a_{2^i}.$$

(You would better check, you never know with me!) Of course, the upper and/or lower bounds can themselves be variable quantities, and the index need not be called i (common, but not mandatory, letters here are i, j, k, m, n, sometimes the Greek letters μ (mu) and ν (nu)).

Some examples of this: The sum of the first 100 even numbers is simple

$$\sum_{i=1}^{100} 2i,$$

and

$$\sum_{j=5}^{n} \frac{1}{j}.$$

is the sum of the root fractions between 1/5 and 1/n, where n denotes any natural number greater than or equal to 5. Finally

$$\sum_{k=-3}^{2} 2^{k+1} = 2^{-2} + 2^{-1} + 2^0 + 2^1 + 2^2 + 2^3 = \frac{1}{4} + \frac{1}{2} + 1 + 2 + 4 + 8,$$

whatever that is supposed to be good for. Before I leave you alone with some exercises, one more remark: Especially with variable sum limits, it can happen that the starting value of the index (which is below the sum) is already larger than the final value above. Since there is obviously nothing to sum here, you agree that the sum value is zero in this case:

Empty Sums

$$\sum_{i=n}^{m} a_i = 0, \quad \textit{if } n > m.$$

Now you:

Exercise 1.13 Represent using the sum sign:

(a) the sum of all numbers divisible by 3 below 100;
(b) the sum of the first 11 odd numbers.

Sums can also occur nested, for example means

$$\sum_{i=1}^{m} \sum_{j=1}^{n} a_{i,j}$$

the sum of all numbers $a_{i,j}$ (which of course must be given), where the first index runs between 1 and m, the second between 1 and n. This sum therefore has a total of $m \cdot n$ summands.

When it comes right down to it, the summation limit of the inner sum still depends on the running index of the outer one; just look at this cream puff:

$$\sum_{i=1}^{3} \sum_{j=-i}^{i-2} a_{i,j}.$$

Written out, it reads:

$$\begin{aligned}\sum_{i=1}^{3} \sum_{j=-i}^{i-2} a_{i,j} &= \sum_{j=-1}^{-1} a_{1,j} + \sum_{j=-2}^{0} a_{2,j} + \sum_{j=-3}^{1} a_{3,j} \\ &= a_{1,-1} + (a_{2,-2} + a_{2,-1} + a_{2,0}) + (a_{3,-3} + a_{3,-2} + a_{3,-1} + a_{3,0} + a_{3,1}),\end{aligned}$$

where I actually put unnecessary parentheses for clarification.

Exercise 1.14 Calculate the sum

$$\sum_{i=-1}^{1} \sum_{j=-2i}^{i} i^2 \cdot j. \blacksquare$$

Just as with sums, you want to use the dot notation for multiple products such as

$$a_1 \cdot a_2 \cdot a_3 \cdots a_n \tag{1.12}$$

and just as with sums, the same procedure is followed here: The abbreviation for "product" is "P" and the Greek letter P is the $\prod$ (Pi). Thus (1.12) becomes

$$\prod_{i=1}^{n} a_i.$$

All the comments I made above on the totals carry over verbatim to the products, which is why I actually want to spare myself and you them here.

In probability and statistics, we often deal with the product of several consecutive natural numbers, usually starting with 1. We already have a notation for this, the product of the first n *natural* numbers is just $\prod_{i=1}^{n} i$, but in the long run this is a bit unwieldy. Therefore, the abbreviated designation $n!$ (pronounced: "n factorial") has been introduced:

Factorial

For any natural number n

$$n! = \prod_{i=1}^{n} i.$$

In addition, one sets $0! = 1$.

The fatal thing is that the exclamation mark "!" used here, in contrast to the sum and product symbol, is also used in everyday language and writing, and therefore in mathematical texts you sometimes have to look very carefully to see what is meant. For example, if you ask the question "Calculate 8!" in an exam, you might get the answer: "First of all, I don't need to calculate 8, because it's already there, and secondly, I find the exclamation mark at the end of the question rather rude." Well.

With the introduction of the factorial sign, this section is almost at an end, because you will learn in the course of your studies what you can calculate and represent with it. I would just like to show you how you can use it to represent an arbitrary product of natural numbers (i.e. not starting at 1) quite simply: If k and n *are natural* numbers and if $k \leq n$, then

$$\prod_{i=k+1}^{n} i = \frac{n!}{k!}. \tag{1.13}$$

So slowly you have to get used to the fact that in mathematics you have to prove everything in principle. However, the proof of this statement is not too difficult, you only have to use the definition of the factorial, whereby I exceptionally, since there can be no doubt about the running range of the variable here, once again fall back on the dot notation: It is

$$\frac{n!}{k!} = \frac{1 \cdot 2 \cdot 3 \cdots (k-1) \cdot k \cdot (k+1) \cdots (n-1) \cdot n}{1 \cdot 2 \cdot 3 \cdots (k-1) \cdot k},$$

where I have written in some of the inner factors to clarify the situation in the numerator. Since k is not greater than n, I can now truncate the complete denominator and get

$$\frac{n!}{k!} = (k+1) \cdot (k+2) \cdots (n-1) \cdot n,$$

which agrees with the left-hand side of (1.13).

Exercise 1.15

(a) Calculate $8!$

(b) Represent the product $9 \cdot 10 \cdot 11$ using factorials.

And so – exhausted, tired and hungry (did someone say “pizza!”?) – we have reached the end of this introductory chapter. I am sure much of it was already familiar to you, or at least you were able to retrieve it from the depths of your memory with the help of these lines, and maybe you even asked yourself from time to time, “Do we have to?” Yes, I think so, because for the following chapters and even more so for the entire study you need a basis to build on, and that has now been laid; and if you have perhaps been asking yourself all along, “All this number crunching and notational lyricism is all well and good, but when is *real mathematics* finally coming?” then I can answer, “Right now, read on!”

Basic Information About Functions

2

For most people, "real" mathematics begins with the first appearance of functions: As soon as one can draw a diagram of a function and determine the course of a function, their mathematical heart beats. I must confess that I am not entirely free from this way of thinking either, although I want to emphasize that all that was the subject of the first chapter is mathematics to be taken quite seriously. But still, I will gladly join the chorus of those who say, "Here we go!"

The problem is, however, that no sooner has the joy over the appearance of functions faded than the jitters begin: What is a function, what important properties can it have, what special functions do you need to know, etc.?

But these fears are really not justified, functions something "completely normal". To give you a feel for it, I will start with a very simple example; imagine driving your car for a few hours at the constant speed of 120 km/h (I know that is hard to imagine these days, but just mentally put yourself in the interior of Australia, you can do that). To calculate the distance of s km covered after a time of t h, you can then use the formula

$$s = 120 \cdot t$$

whereby I generously omitted the units in the best mathematician's manner. In applications, I would like to warn you right away, you should always include the physical units, either explicitly on both sides or in square brackets for the whole equation.

Using the above formula, you can calculate that after 1 h you have travelled 120 km, after 2 h 240 km and, for example, after 5 h 600 km. But it is equally possible to calculate how far you have travelled after half an hour (namely $120 \cdot 1/2 = 60$, i.e. 60 km), how far after 1 min (2 km), and if you feel like it also the distance travelled after 1 sec or even after 273/334 h. So you see that you can use *any* positive value for t, and depending on this

G. Walz et al., *Foundations of Mathematics: A Preparatory Course*,
https://doi.org/10.1007/978-3-662-67809-1_2

value, the formula returns the corresponding distance s. To emphasize this dependence more strongly, one usually writes

$$s(t) = 120 \cdot t, \tag{2.1}$$

and we have already seen – secretly, quietly – the first example of a function. Since t here depends on nothing at all, it is also called the **independent variable**, correspondingly s the variable dependent (on t); for any input value of t you feed the function with, it will give you a uniquely determined output value s. This is exactly the essence of a function, and I will first write this down precisely in the next section.

2.1 Definition Range, Value Set and Image Set

Since in life there are not only cars and driven times (what you could believe with some people though), the independent variable in mathematics is usually called x (and not t) and the function is called f (and not s).

Function, Domain and Set of Values

Let D and W be sets which should be non-empty; usually, but not necessarily, they will be sets of numbers. A **function** f is a rule (mapping rule, formula, ...) which assigns to each element x *of* the set D an element $y = f(x)$ of W *in* a uniquely determined way.

One calls D the **definition range of** f (because f *is* to be *defined* for all values x from D) and W the **value set of** f, because y is called the **function value of** x and W gathers all possible function values of f in itself.

Often one writes f: $D \to W$ *for* short and thus has defined the domain of definition and the set of values of f *in* one fell swoop. In the following I will do this occasionally, but not always.

I suspect you need a strong coffee right now to digest what you have read. But maybe I can help you with a few examples:

Example 2.1 For convenience, I first choose the set of real numbers as both the definition set and the set of values, that is, $D = W = \mathbb{R}$. The first mapping rule I want to test is

$$f : \mathbb{R} \to \mathbb{R}, \quad f(x) = 2x.$$

Now this is a function of purest water, because it assigns to each real number x a uniquely determined value, namely the double of x, and this value is also in the admissible range of values, namely the set of real numbers.

Originally I had planned to calculate some function values of f in an exercise, but that seems too childish to me now.

Example 2.2 Instead, a second example, the mapping rule this time being

$$f : \mathbb{R} \to \mathbb{R}, \quad f(x) = x^2.$$

This is also a function, because each real input value x is assigned a unique number, namely just its square. Nevertheless, this example shows a novelty: True to the rule "minus times minus results in plus", negative numbers never appear here as function values, although they would be allowed in principle.▪

So it is important to emphasize that the set of values W contains the set of all numbers allowed as function values, but these do not necessarily have to be actually "hit" by the function. This should be noted:

The Value Pool Does Not Have to Be Exhausted

The value set W of a function contains the set of all *potential* function values; it may well contain numbers that never occur as function values.

I call the set of all elements of W that actually appear as function values the **image set of** D and symbolize it by $f(D)$.

So the image set of the function $f(x) = x^2$ is just the set of non-negative real numbers.

I admit that the subtle distinction between value set and image set is not everybody's favourite topic, but it is important nevertheless; even worse: The notation is not uniform in the literature, and it may well happen that later on a lecturer or textbook author will prefer to call what I call value set a value domain, the image set in turn may be called value set, and so on. I cannot change it, you just have to always look carefully what the respective author defines at the beginning of the text.

Another rather monotonous but correct example of a function that does not exhaust its set of values is the following: Let $D = \mathbb{R}$, $W = \{0,1,2\}$ and $f(x) = 1$.

What is going on here? Well, this is certainly a function, because it assigns a value to every real number x in an unambiguous way, namely the number 1. The store of values of this function is already not too large, it comprises only three numbers, and yet it does not exhaust it, because the numbers 0 and 2 never appear as function values. So here the image set is just the one-element set $\{1\}$.

Very often the domain of definition of a function is an interval, and this notion as well as the related notation I will introduce right here:

Interval

If a and b are real numbers where a *is* said to be less than or equal to b, that is, $a \leq b$, then the set of all real numbers that lie between a and b is called the **interval** with boundaries a and b.

If a and b themselves also belong to this interval, this is called a **closed interval** and is denoted by square brackets:

$$[a, b] = \{x \in \mathbb{R} \ \text{ with } a \leq x \leq b\}.$$

If a and b *do* not belong to this interval, this is called an **open interval** and is denoted by round brackets:

$$(a, b) = \{x \in \mathbb{R} \ \text{ with } a < x < b\}.$$

Also the mixed forms

$$(a, b] = \{x \in \mathbb{R} \ \text{ with } a < x \leq b\}$$

and

$$[a, b) = \{x \in \mathbb{R} \ \text{ with } a \leq x < b\}$$

exist, they are called **half-open intervals.**

With the help of this notation, for example, a function whose domain is to be the set of all real numbers between 2 and 3 (including these two) can be written briefly in the form

$$f : [2, 3] \to \mathbb{R}$$

How about a little *mental snack in between* at this point?

Exercise 2.1 Determine the exact image set of the function

$$f : [-1, 2] \to \mathbb{R}, \quad f(x) = \frac{1}{x - 3}. \quad \blacksquare$$

Now if you are so slow to think that pretty much everything under the sun that starts with "*f*(*x*) =" is a function, I have to show you the following:

Let $D = \mathbb{R}$, $W = \{-1, 1\}$ and for all $x \in D$

$$f(x) = 1 \text{or} - 1, \text{depending on how I feel.}$$

Now this is certainly not an unambiguous rule for determining the value of $f(x)$, because how are you supposed to know how I feel, I often do not know myself; thus f is *not a* function.

But also the following two examples, again rather mathematically inclined, do not represent functions – for different reasons:

For example, let us put

$$g : \mathbb{N} \to \mathbb{N}, \quad g(x) = \sqrt{x},$$

it turns out relatively quickly that, for example, the value $g(2)$, i.e. $\sqrt{2}$, is not in the given set of values of g, because $\sqrt{2}$ is not a natural number (not even a rational one, but that is not important anymore).

But even if I extend the domain of definition and the set of values of g to $D = W = \mathbb{R}$, I do a disservice to the would-be function, because now it is not even defined on the entire set D, since real roots from negative numbers are not explained.

So you see, you always have to check the function rule in connection with the domain of definition and the set of values in order to be able to say for sure whether it really is a function. And – to lapse into the nurse plural popular with many authors – let us practice this a bit right away:

Exercise 2.2 Which of the following represents a function?

(a) f: $\mathbb{N} \to \mathbb{N}, f(x) = x^2$
(b) g: $[-1,1] \to \mathbb{R}, g(x) = 1/x$
(c) h: $\mathbb{Q} \to \mathbb{Q}, h(x) = 2\,x^4 - 5/3\,x^3 + 1/2\,x$

If both the domain of definition and the set of values of a function lie in the real numbers – one then also says in short that it is a **real function** –, then one can also represent its course pictorially. Although you are probably already familiar with this from various occasions, I will briefly write it down here for the sake of completeness:

You start by drawing a **coordinate system**, i.e. two axes perpendicular to each other, which usually intersect at the respective zero point. On these, a division is made by marking at least one number. The horizontal axis pointing to the right is usually called the *x-axis,* since the *x-values* are marked on it. The vertical axis pointing upwards is usually called the y-axis, since at least in earlier times the function values marked on this axis were not called $f(x)$, but simply *y. The x-axis is called* the *y-axis.*

After the great philosopher and mathematician René Descartes, whose name was Latinized to Renatus Cartesius, such a coordinate system is also called **Cartesian coordinate system**.

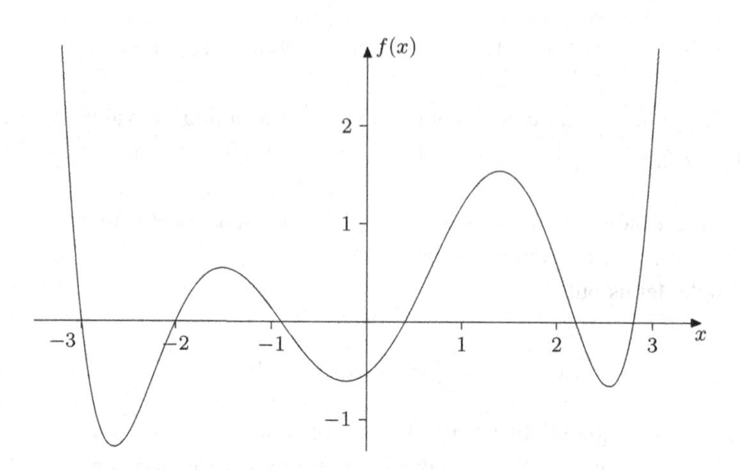

Fig. 2.1 Graph of a function

In order to get a picture of the course of a function in the truest sense of the word, one interprets the number pair $(x, f(x))$ as coordinates of a point in this system just drawn for each x from the definition range of the function and draws this point. For example, for the function $f(x) = 3\,x$ the pairs (0,0), (1,3), (−2, −6) and (100,300) result.

If this is done not only for four pairs, but for many pairs lying very close to each other, the points drawn in this way result in a continuous line for not too crazy functions, which is called the **graph of** the function. A rather arbitrary example is shown in Fig. 2.1.

2.2 Concatenation of Functions; Monotonicity and Invertibility

Let us say you are the happy owner of a small company that has a monopoly on a number of high-demand products, and you want to recalculate the prices of your products. Since you want to make a tidy profit, you proceed as follows: First, raise the old price by 100, and then triple it again (If this seems outrageous to you, I agree with you, but, for example, the oil companies do the same thing when they recalculate their petrol prices).

Since you have secretly been reading a math book, describe each of these two operations by a function that is said to be defined on the set of positive numbers: If x is the old price, the function to be applied first is

$$g(x) = x + 100,$$

and the second is

$$f(y) = 3 \cdot y.$$

For example, for $x = 100$, you calculate that $g(x) = 200$ and $f(200) = 600$; a product that previously cost 100 Euros now costs 600 Euros.

So you can do this bit by bit with all product prices. For example, a product that initially cost $x = 10$ Euros, after your small recalculation comes to $g(x) = 110$ and finally $f(110) = 330$ Euros.

However, you will probably notice pretty soon that you can actually omit the intermediate determination of $g(x)$ and rather put the output value of g back into f right away. What you are doing is called concatenation **of** the functions g and f *in* mathematics. I call the new function created by the concatenation h for the moment; I calculate it explicitly by plugging $g(x)$ into f and calculating it:

$$h(x) = f(g(x)) = f(x + 100) = 3 \cdot (x + 100) = 3x + 300.$$

So it is

$$h(x) = 3x + 300,$$

and also this function is defined on the set of all positive numbers. In this representation you can now calculate a little faster than above that $h(100) = 600$ and $h(10) = 330$.

In principle, this works for all functions in the world, but it must be guaranteed that the function f knows what to do with the output value of g, i.e. that the image set of g lies in the domain of definition of f.

I really should write that down precisely now:

Concatenation of Functions

Let $f: F \to W$ and $g: D \to E$ be two functions with the property that the domain of definition F of f lies in the image domain $g(D)$ of g.

Then the function is called $f \circ g: D \to W$, defined by

$$f \circ g : D \to W, \quad f \circ g(x) = f(g(x))$$

for all $x \in D$, the **concatenation** of f and g; occasionally we also say **successive execution** or **composition**.

Note the sequence: g is applied *first* and *then* f.

I have no idea how to pronounce the $\circ$ sign, I have only ever written it down in my life. Some people say "squiggle" to it, some people say "linked to", and some people just say "after", so "$f \circ g$" in this case is pronounced as "f after g". This is not as bad as it might seem at first glance, because it again clarifies the order in which the two functions are to be performed in succession: First g and then f. People often get this wrong, especially at the beginning, because they are used to reading from left to right (possibly Japanese people make fewer mistakes here, I do not know). The notation $(f\circ g)(x)$ is supposed to suggest that g is applied to x first, because it is right next to it, and only then does f take its turn.

Before I get lost in philosophical considerations about advantages and disadvantages of this notation, I would like to give some examples; in order to write them more compactly, I would like to introduce the symbol $\mathbb{R}^+$ for the set of non-negative real numbers, i.e.

$$\mathbb{R}^+ = \{x \in \mathbb{R}; x \geq 0\}.$$

Then we were good to go:

Example 2.3 Let us consider the functions

$$g : \mathbb{R} \to \mathbb{R}, \quad g(x) = (x+9)^2$$

and

$$f : \mathbb{R}^+ \to \mathbb{R}, \quad f(y) = \sqrt{y} - 8.$$

Since the image set of g is just the non-negative real numbers ("minus times minus…"!), I can safely concatenate f with g and get:

$$(f\circ g) : \mathbb{R} \to \mathbb{R}, \quad (f\circ g)(x) = f\left((x+9)^2\right) = \sqrt{(x+9)^2} - 8 = x + 9 - 8 = x + 1,$$

so $(f\circ g)(x) = x + 1$. So the concatenation just causes x to increase by 1. You have to admit that we saved a lot of work here compared to evaluating f and g separately!▪

Example 2.4 Now let

$$g : \mathbb{R}^+ \to \mathbb{R}, \quad g(x) = \sqrt{\frac{x+2}{x+1}}$$

and

$$f : \mathbb{R}\backslash\{0\} \to \mathbb{R}, \quad f(y) = \frac{1}{y^2}.$$

The function f digests as input value every real number except zero, and since this does not occur as value of g, we do not need to worry about the matching of the ranges; the calculation or simplification of the concatenation I will now do in single steps: If we insert $g(x)$ into f, $g(x)$ is squared and packed into the denominator; but squaring just cancels the root formation, so that the denominator of f is now $\frac{x+2}{x+1}$. And since you divide by a fraction by multiplying by its reciprocal, we get as a result:

$$(f \circ g) : \mathbb{R}^+ \to \mathbb{R}, \quad (f \circ g)(x) = \frac{x+1}{x+2}. \quad \blacksquare$$

Perhaps you have been wondering all along what happens if you swap the order of the concatenated functions in the examples above when you concatenate them? Too late, I was faster and ask *you*:

Exercise 2.3 For the pairs of functions discussed in the examples above, check whether the concatenation $g \circ f$ is also possible; if so, state the concatenated function as simply as possible.

A certain special role in the concatenation of functions is played by the so-called inverse function of a function. This is the function that reverses a given function by concatenation; I will define this in more detail in a moment, but first a few examples:

If you concatenate the function

$$g : \mathbb{R} \to \mathbb{R}, \quad g(x) = x - 28$$

with the function $f(y) = y + 28$, also defined on the whole $\mathbb{R}$, we get

$$(f \circ g) : \mathbb{R} \to \mathbb{R}, \quad (f \circ g)(x) = (x - 28) + 28 = x$$

and likewise

$$(g \circ f) : \mathbb{R} \to \mathbb{R}, \quad (g \circ f)(x) = (x + 28) - 28 = x.$$

Thus, concatenating the two functions leaves each x unchanged, regardless of the order, as if no function had been applied at all. In other words: The function applied first in each case is reversed by the application of the second. This is, of course, completely independent of the special number 28 and, incidentally, corresponds completely to everyday life experience: if, for example, you win a certain amount in roulette and lose it completely in the very next game, nothing has changed overall in your financial situation, true to the old adage: "Easy come, easy go."

In the next example, the inversion still works quite well: To use the function

$$g : \mathbb{R} \to \mathbb{R}, \quad g(x) = -2x$$

again, one only has to concatenate it with the function $f(y) = -1/2y$, because then for all $x \in \mathbb{R}$:

$$(f \circ g) : \mathbb{R} \to \mathbb{R}, \quad (f \circ g)(x) = -\frac{1}{2} \cdot (-2x) = x$$

and likewise

$$(g \circ f) : \mathbb{R} \to \mathbb{R}, \quad (g \circ f)(x) = -2 \cdot \left(-\frac{1}{2}x\right) = x.$$

All well and good, you will say, what is the problem? The problems will come, do not worry. But before that, it is high time to define the term inverse function exactly. Of course, the examples just considered should serve as a guide.

Reverse Function

Let f be a function with domain D and image set $f(D)$. A function, let us call it f^{-1}, whose domain of definition is $f(D)$ and which has the property

$$(f^{-1} \circ f)(x) = x$$

for all $x \in D$ is called an inverse **function** of f. If an inverse function exists for a function f, then f itself is called **invertible**.

Not every function is invertible.

Perhaps the question has been on the tip of your tongue for quite a while, why on earth one is actually interested in the inverse function, one could just as well leave the execution of the function and thereby also save the execution of the inverse function. That is correct in itself, but mostly one is not interested in the invalidation of a function, but one wants to draw conclusions about the value x, from which the function value originates, on the basis of a presented function value $f(x)$. And this is what the inverse function is for.

Any more questions my dear Watson? Well, I think so, and it is probably due to my muddled expression. I would better make an example. Think back to the example formulated at the very beginning of this chapter about driving a car. There I had – in formula (2.1) – the function

$$s(t) = 120 \cdot t$$

which calculates the distance travelled $s(t)$ for a car travelling at speed 120 at any time t.

Now, however, you might also want to know, in the opposite question, how long it takes you to cover a certain distance if you drive at a constant 120. In other words, you are asking to which value of time t a distance s selected by you belongs. But this is nothing else than asking for the inverse function of the function $s(t)$.

With a little thought, you find that this inverse function s^{-1} is given by

$$s^{-1}(y) = \frac{y}{120},$$

because

$$s^{-1}(s(t)) = \frac{s(t)}{120} = \frac{120t}{120} = t$$

for all t. So, for example, if you want to know how long it takes you to travel 300 km, just calculate

$$s^{-1}(300) = \frac{300}{120} = \frac{5}{2}.$$

So you need 5/2, that is 2.5 h. Life can be that simple with a little math!

It is the last remark, somewhat coyly placed at the end of the box above, that leads us to the problems already mentioned above. As an example I show you the function

$$f : \mathbb{R} \to \mathbb{R}, \quad f(x) = x^2,$$

which – or rather whose graph of the function – is also called a **normal parabola.** It is a special case of the so-called polynomials, which I will look at with you below; for now, we have enough to do with this one function. We already considered above that the image set of this function is just the set of non-negative real numbers.

Thus, if f is to be reversible on the entire domain of definition $\mathbb{R}$, then one must find a function that is defined on all of $\mathbb{R}^+$ and reverses the action of f everywhere there.

You might be thinking, "Well, come on, tell me it's the root function, and let me go get a cup of coffee." Sorry, coffee is cancelled for now, for the following reason: First of all, things look quite good, because the root function, i.e. the function $w(x) = \sqrt{x}$, which assigns its (square) root to every number, is in fact defined for every non-negative integer, in other words for $\mathbb{R}.^+$

Even if you calculate, for example, $w(4) = 2$ you are still in good spirits, because since $f(2) = 2^2 = 4$, the root actually reverses this process, since

$$w(f(2)) = w(4) = 2,$$

and you can do that not only for 2, but for any positive number.

But here it comes: Try to invert the function $f(x) = x^2$ at the point $x = -2$! You get the intermediate result $f(-2) = 4$, and if you now apply the square root to it, you get the value 2, as just seen, and *not* -2. If both +2 and -2 are in the domain of definition, the function is not invertible, and neither is it in any other negative value, if the corresponding positive value is in the domain of definition. So all in all, it turns out:

The function $f(x) = x^2$ defined on all $\mathbb{R}$ is not invertible.

Why is that? Well, you have to think about what "inversion" of a function f actually means: It is about "throwing back" each function value $f(x)$ of f to its initial value – one also says its **archetype** – x, as if no mapping had taken place at all. For this purpose, however, it is obviously necessary to be able to say unambiguously for each function value where it came from, i.e., from which *x-value* it originated. But this is not possible with the normal parabola $f(x) = x^2$, because already with my example 4 as function value nobody can say where this value came from, whether from −2 or from +2.

But why is it now no longer possible with this actually quite harmless function, which still worked quite well with the two examples mentioned above? Well, if you look at the diagrams of the three functions in Fig. 2.2, you will notice that the first two, i.e. $f_1(x) = x - 28$ and $f_2(x) = -2x$, run wonderfully straight from the bottom to the top or vice versa, without making a detour on the way and going back a bit to the top or bottom. The normal parabola, on the other hand, first arrives with a lot of verve from the top to the bottom, but then suddenly turns around at the zero point and goes straight back up again.

But this means that every positive value occurs twice as a function value, and this is exactly where reversibility fails.

I am afraid I have strayed a little bit from the noble goal of writing an entertaining mathematics book, as stated in the preface: Admit it, you are bored. Well, I tried for a really long time to include a pizza example here, but I just have no idea how to square a pizza. After all, with a frozen pizza, that would be like squaring the circle! (O.K., forget it, was one of my worse jokes).

Therefore, contrary to my usual habit, I will now head straight for the goal without a long preface: It will soon become clear that the property of not making any "swerves", one can also say of running "monotonously", is exactly the right prerequisite for reversibility. But since, as far as I know, the term "swerve" does not appear in mathematical

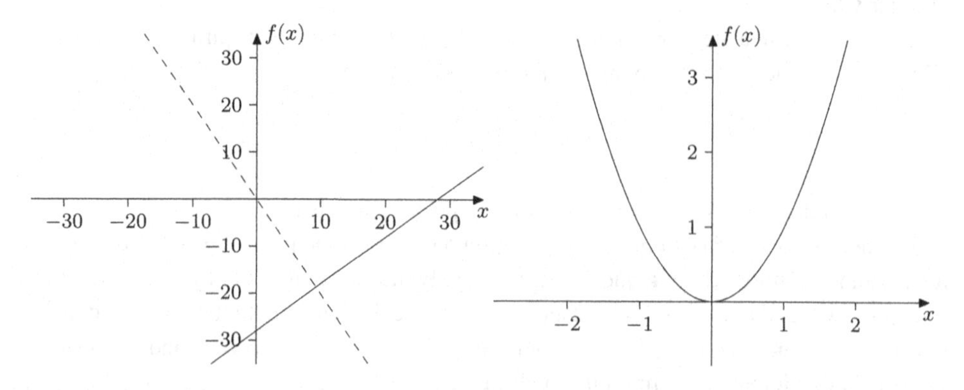

Fig. 2.2 The functions $f_1(x) = x - 28$ (*left, black*), $f_2(x) = -2x$ (*left, dashed*), and $f_3(x) = x^2$ (*right*)

terminology, I am afraid I will first have to say again precisely what a monotonic function is supposed to be:

Monotone Function

Let f be a function and I a subset of the domain D of definition of f; this may or may not be the whole domain of definition.

One calls f **monotonically increasing** on I if holds that whenever x_2 is greater than x_1 (i.e., lies "to the right" of x_1), then $f(x_2)$ is greater than or equal to $f(x_1)$. In brief:

$$\text{If } x_1 < x_2, \text{then } f(x_1) \leq f(x_2).$$

Analogously, a function f is called **monotonically decreasing** on I if holds that whenever x_2 is greater than x_1 (i.e., lies "to the right" of x_1), then $f(x_2)$ is less than or equal to $f(x_1)$. In brief:

$$\text{If } x_1 < x_2, \text{then } f(x_1) \geq f(x_2).$$

A function is called **monotonic** if it is monotonically increasing or monotonically decreasing.

We have already seen examples of both cases above: The function

$$f(x) = x - 28$$

is monotonically increasing everywhere in $\mathbb{R}$; this can be seen not only in the figure (which would not be a proof), but can also be calculated: If two input values x_1 and x_2 are such that x_2 is greater than x_1, then x_2–28 is also greater than x_1–28, so $f(x_2)$ is greater than $f(x_1)$, and since "greater" of course means "greater than or equal to" a fortiori, the function f is monotonically increasing; in short:

$$\text{If } x_1 < x_2, \text{then } x_1 - 28 < x_2 - 28, \text{so } f(x_1) < f(x_2).$$

Similarly, you may consider that the function $f(x) = -2x$ *is* monotonically decreasing *everywhere* in $\mathbb{R}$; I think by now you have become so accustomed to the short form that I can expect you to take it right away without much preamble; if not, please write to me or turn confidently to your doctor or pharmacist – if the man has learned mathematics. But now:

$$\text{if } x_1 < x_2, \text{then} -2x_1 > -2x_2, \text{so } f(x_1) > f(x_2).$$

If the "so is" step seems strange to you, I can only offer you an example for now and otherwise put you off until the chapter on inequalities. As an example, I will just take $x_1 = 1$ and $x_2 = 2$. Surely 1 is smaller than 2; but if I apply the function f to these two values, I get

$$f(1) = -2 \text{ and } f(2) = -4,$$

so $f(1)$ is larger than $f(2)$, because it lies further to the right on the number line.

If monotonicity, as said at the beginning, has something to do with reversibility, then the normal parabolic function $f(x) = x^2$, which is defined for the whole of $\mathbb{R}$ and of which we know in the meantime that it is not reversible, should actually also have its problems with respect to monotonicity. And indeed, it turns out to be *nonmonotonic*, neither increasing nor decreasing, when considered over all of $\mathbb{R}$. To prove this computationally, I first pick the two points $x_1 = -1$ and $x_2 = 0$. Certainly $x_1 < x_2$, and since $f(-1) = 1$ is *not* less than or equal to $f(0) = 0$, the function is not monotonically increasing everywhere. But it is also not monotonically decreasing everywhere, as you can (and should) easily test, for example, using the new pair of points $x_1 = 0$ and $x_2 = 1$.

Notice what happened here, because it is a general principle: if one wants to disprove a conjecture (here the conjecture that the function is monotonically increasing or monotonically decreasing on all of $\mathbb{R}$), *a single* counterexample suffices.

Exercise 2.4 Check whether the following functions are monotonic:

$$\begin{aligned} f &: \mathbb{R} \to \mathbb{R}, \quad f(x) = \frac{x}{10} - 17 \\ g &: \mathbb{R} \to \mathbb{R}, \quad g(x) = -\frac{x}{17} + 10 \\ h &: [1, 17] \to \mathbb{R}, \quad h(x) = \frac{1}{x} \end{aligned}$$

Perhaps you have already wondered that in the definition of monotonicity I also allowed the equality of the function values, i.e. I did not demand that, for example, in the case of a monotonically increasing function from $x_1 < x_2$ it follows: $f(x_1)$ is really smaller than $f(x_2)$.

The reason for this is that one also wants to call such functions monotonic, which increase nicely and bravely almost over their entire definition range and only rest once in a while very briefly and remain constant for a bit; you can see an example of such a function in Fig. 2.3.

Now, unfortunately, there are functions which shamelessly exploit this generosity on my part (and I am not alone in this within the mathematical community) and do not move from the spot at all; probably the most impudent representative of this genus is the function

$$f : \mathbb{R} \to \mathbb{R}, \quad f(x) = 0,$$

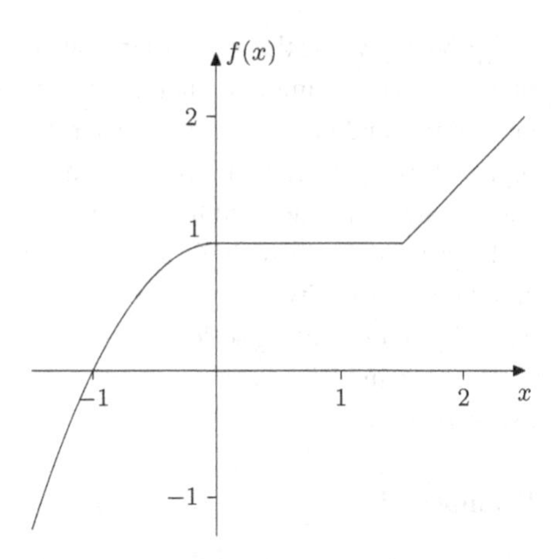

Fig. 2.3 A piecewise constant monotone function

which maps everything and anything from the set of real numbers to the value zero, or figuratively speaking, which does not move a millimeter down or up during its entire life. The stupid thing now is that this function is *both monotonically increasing and monotonically decreasing* according to the above definition, because for every two real numbers x_1 and x_2 $f(x_1) = f(x_2)$ and thus $f(x_1)$ is both less than or equal to and greater than or equal to $f(x)._2$.

In order to exclude such pathological cases, one tightens the requirements a little and defines the concept of *strict* monotony:

Strictly Monotonic Function

Let f be a function and I a subset of the domain D of definition of f; this may or may not be the whole domain of definition.

One calls f **strictly monotonically increasing** on I if holds that whenever x_2 is greater than x_1 (i.e., lies "to the right" of x_1), then $f(x_2)$ is greater than $f(x_1)$. In a nutshell:

$$\text{If } x_1 < x_2, \text{then } f(x_1) < f(x_2).$$

Analogously, a function f is called **strictly monotonically decreasing** on I if holds that whenever x_2 is greater than x_1 (i.e., lies "to the right" of x_1), then $f(x_2)$ is less than $f(x_1)$. In a nutshell:

$$\text{If } x_1 < x_2, \text{then } f(x_1) > f(x_2).$$

A function is called **strictly monotonic** if it is strictly monotonically increasing or strictly monotonically decreasing.

I probably wrote this box faster than you read it, because I was able to copy and paste the one above about simple monotonicity almost word for word, only the word "strict" had to be inserted, and in two places "greater than or equal to" or "less than or equal to" had to be replaced by "greater than" or "less than"; which would also have pointed out what you should pay more attention to when reading this box.

I can actually do without new examples at this point, because the above examples of simple monotonicity are even examples of strict monotonicity; please look at them again carefully, I am waiting so long here.

Back again? Well, you are right, if you are a little dissatisfied, I should bring one new example; how about this:

Example 2.5

$$f:\mathbb{R}\to\mathbb{R},\quad f(x)=\begin{cases}-x^2, & \text{if } x<0 \text{ is,}\\ x^2, & \text{if } x\geq 0 \text{ is.}\end{cases}$$

Before I now examine this function for (strict) monotonicity, I should perhaps explain the function of the curly bracket to you again: This is used here to make a **case distinction.** For each input value *x, it* must be checked whether it is less than zero or greater than or equal to zero; in the first case, the upper line behind the curly bracket is used, and the number $-x^2$ is assigned to this x as the function value, with the minus sign in front of the x and thus not squared. For example, the function value of $x = -3$:

$$f(-3) = -(-3)^2 = -9.$$

If the current input value of x is greater than or equal to zero, you have to look at the bottom line of the right-hand side to find the function value; you can see that the function value should be just the square of the input value; so, for example, $f(4) = 4^2 = 16$.

So far, so good, now we can handle the function a bit, but is it monotonic?

Yes, it is, even strictly monotonic on all of $\mathbb{R}$, and increasing. To prove this, I again take two points x_1 and x_2 with

$$x_1 < x_2$$

and compare their function values. For this purpose, I distinguish three cases:

Case 1: x_1 and x_2 are both negative, so $x_1 < x_2 < 0$. To find the function values, I have to square the *x values* in both cases and then put a minus sign in front of them. But if $x_1 < x_2 < 0$, then x_1^2 is a larger number than x_2^2; now if I put a minus sign in front of both, then $-x_1^2$ is to the left of $-x_2^2$ and is therefore smaller. Not so clear? Admittedly, comparing these negative numbers is always a bit counterintuitive, we had already seen that above. So let us make an example: If $x_1 = -5$ and $x_2 = -3$, then undoubtedly

$x_1 < x_2 < 0$. x_1^2 is then 25 and x_2^2 is 9, so x_1^2 is the larger number. However, after putting the minus sign in front and thus completing the function value formation, $-25 < -9$ holds; thus, $f(x_1) < f(x_2)$.

Case 2: x_1 and x_2 are both non-negative, so $0 \leq x_1 < x_2$. Well, this means that to find the function values both numbers have to be squared evenly, and if already x_1 is smaller than x_2, a fortiori x_1^2 is smaller than x_2^2. So also in this case the strict monotonicity is proved.

Case 3: Admit it, you thought for at least a little moment that there was not another case! (If not, I hereby formally apologize to you and ask you to immediately start majoring in mathematics). But there is one more case not considered so far, namely that x_1 is negative, but x_2 is not negative. But this is now quite easy: if x_1 is negative, then $f(x_1) = -x_1^2$ is also negative, while $f(x_2) = x_2^2$ is non-negative. So, in particular, $f(x_1)$ is smaller than $f(x_2)$.

In each of the three cases, it thus follows that from $x_1 < x_2$ it always follows: $f(x_1) < f(x_2)$. Thus, the function f is strictly monotonically increasing.▪

All the work must have been worthwhile, too, and the reward comes in the form of the following theorem, which traces the question of whether a function is invertible back to monotonicity, which is quite easy to verify:

Strict Monotonicity and Reversibility

Let $f: D \rightarrow W$ be a strictly monotone function on all of D with image set $f(D)$. Then there exists an inverse function f^{-1} of f defined on $f(D)$, so f is invertible.

If f is strictly monotonically increasing, then f^{-1} is also strictly monotonically increasing; if f is strictly monotonically decreasing, then f^{-1}.

With this beautiful result (mathematicians sometimes have their own taste in beauty, you will have to get used to that) I will end this section and introduce you to some concrete functions in the next ones; before that, though, you should practice what you have learned a bit, and as the name suggests, this is what the exercise problems were invented for:

Exercise 2.5 State the inverse functions of the functions from Exercise 2.4.

2.3 Power and Root Functions

Above, I had already given you the normal parabolic function $f(x) = x^2$ and the (square) root function $f(x) = \sqrt{x}$ in passing. Both are special cases of the power and root functions, respectively, which I want to introduce to you in this section.

Let us start with the power functions:

Power Functions
For any natural number n, the function is called

$$p_n : \mathbb{R} \to \mathbb{R}, \quad p_n(x) = x^n$$

power function with exponent n or simply *nth* power function.

In the light of things, this is actually nothing new for you: Already in the first chapter, we practiced the exponentiation of real numbers with arbitrary exponents, and therefore you know that you can calculate the function value of the *nth* power function by multiplying the current value of x n *times by* itself; so, for example, is

$$p_3(-2) = -8 \quad \text{and} \quad p_4(3) = 81,$$

because $(-2)^3 = -8$ and $3^4 = 81$. The new aspect here is that I now want to look at the whole thing as a *function* and focus on properties of this function such as monotonicity and invertibility.

The simplest power function, which actually does not really deserve this name, is $p_1(x) = x$. So here we are not really exponentiating. The graph of this function is the bisector of the coordinate system (see Fig. 2.4), a rather boring matter.

By the way, another word for "boring" is "monotonic", and since this graph is even *very* boring, I can also say "strictly monotonic". In fact, the first power function is strictly monotonically increasing on all of $\mathbb{R}$ and thus, as you already know, is everywhere invertible. Worse yet, it is its own inverse function, because.

$$p_1(p_1(x)) = p_1(x) = x$$

for any real number x. I will come back to this invertibility in a moment, do not forget!

The "next" power function is $p_2(x) = x^2$. We had already examined this – under the name normal parabola – in the last section and had seen in particular that it is *not* invertible, at least not if one wants to invert it on its whole domain of definition $\mathbb{R}$.

Figure 2.5 shows an example of the graphs of some power functions. It can be seen that there seem to be two types of graphs: If n is even, then the function graph moves only within the nonnegative region and is symmetric about the *y-axis*. But if n is odd, then obviously every real number occurs as a function value and the function graph is symmetric about the zero point.

It must now be shown that this holds not only in the examples shown, but in fact for all $n \in \mathbb{N}$.

First, I consider the case where n is an even number. This means that n *is* divisible by 2 without remainder, i.e. contains the factor 2. Because mathematicians cannot simply

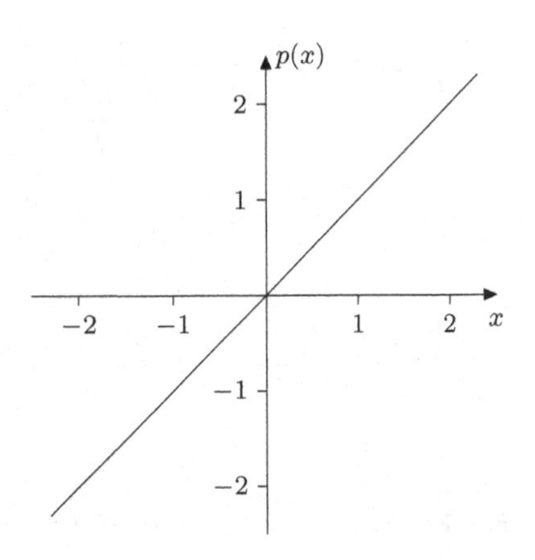

Fig. 2.4 The function $p_1(x) = x$

leave something so simple to formulate, but must immediately translate it into their formula language (which, however, later enables them to deal with such properties briefly and concisely), they usually formulate it as follows:

If n is even, then it can be written in the form $n = 2\ m$ *with* a natural number m. For example, $14 = 2 \cdot 7$ and $30 = 2 \cdot 15$.

Let us now consider the *nth* power function $p_n(x) = x^n$. If n is *even,* which I currently assume, then I can replace n by $2\ m$, and thus the power function can be represented as

$$p_n(x) = x^n = x^{2m}.$$

Now comes the point, where the trouble with the power rules in the first chapter finally pays off. It was shown there that $x^{2\,m} = (x^{2})^m$. This means that the evaluation of the *nth* power function for *even n can* be thought of as a concatenation of two functions as follows: First, the function x^2 is applied. This function is known to produce no negative values, and apart from zero, each function value x^2 occurs exactly twice, as an image of +x and of −x, so the function graph is symmetric about the *y-axis*. The subsequent application of the *m-th* power (where m, as I said, is half of n) does not change these two facts, of course, because the application of the *m-th* power does not produce any negative numbers from positive input values, and since it is presented with the input value x^2 twice (once from +x and once from −x), it also dutifully calculates the same output value twice.

Overall, then, it follows that x^n for even n has a graph that is symmetric about the *y-axis* and runs in the non-negative region; qualitatively, it looks like that of the normal parabola. In particular, this function is not invertible on all $\mathbb{R}$.

Since you have already come to know and love this way of reasoning, I will be a little more brief in the remaining case where n is odd (do I hear you breathing a sigh of relief?).

If n is odd, then $n - 1$, the natural number preceding n, is even, hence of the form $2\ m$ as shown above. Thus $n - 1 = 2\ m$ and thus n is of the form

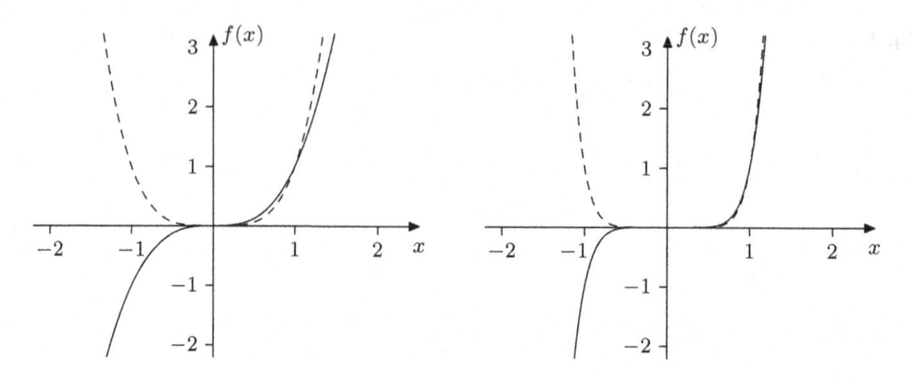

Fig. 2.5 Some power functions: x^3 *(left, black)*, x^4 *(left, dashed)*, x^7 *(right, black)*, x^8 *(right, dashed)*

$$n = 2m + 1.$$

Thus, the *nth* power function can be written as follows for odd *n*:

$$p_n(x) = x^n = x^{2m+1} = x^{2m} \cdot x,$$

where in the last step I again secretly and quietly used a power law.

Given this representation of the function $p_n(x)$, I can now argue for odd n as follows: We have just painstakingly determined what the output of the function x^{2m} looks like. Now, if $n = 2m + 1$, this output is multiplied by x again; but what happens in the process is pretty clear: If x is positive, then $x^{2m} \cdot x$ is also positive; if x is negative, then $x^{2m} \cdot x$ *is* also positive. And since this expression is nothing but x^n, we have thus shown that the graph of this function is negative for negative x and positive for positive x. Well, and if $x = 0$, then x^n is *so* zero!

Overall, you can see that the graphs of functions x^n for odd n all look qualitatively the same; the simplest and therefore rather boring representative is $p_1(x) = x$, and just as this function is invertible on all of $\mathbb{R}$ (remember that?), so is every power function for odd n.

What this inverse function is, you also already know at the latest since reading the first chapter, although functions were not yet mentioned there: To reverse the effect of x^n I need a function, let us call it for the moment $w_n(x)$, which has the property

$$w_n(x^n) = x$$

because that is the very definition of an inverse function. But according to the rules of power arithmetic, the function

$$w_n(x) = x^{\frac{1}{n}}$$

does this because it holds for all $x \in \mathbb{R}$ and all odd $n \in \mathbb{N}$:

$$w_n(x^n) = (x^n)^{\frac{1}{n}} = x^{n \cdot \frac{1}{n}} = x$$

and that is exactly what I was looking for.

n-th Root Function
The function $w_n(x) = x^{1/n}$, which can also be written as $w_n(x) = \sqrt[n]{x}$, is called the ***n*-th root function.**

The *n-th* root function, being the inverse of the strictly monotonically increasing function x^n, is also strictly monotonically increasing. Its graph is obtained by mirroring that of x^n at the first bisector.

I would like to point out a small problem in the practical handling of roots of negative numbers: according to the above considerations, it is perfectly okay to use an expression like

$$\sqrt[3]{-8}$$

because 3 is an odd number, but the word has not yet gotten around to some programmers, and therefore some software packages report an error when entering $\sqrt[3]{-8}$ because of the negative radicand and refuse to continue. (The good programs, on the other hand, calculate the correct result as -2.)

If something like this happens to you, then you have to reach into the bag of tricks: Since $-8 = (-1) \cdot 8$ and $-1 = (-1)^3$, you can successively transform the expression $\sqrt[3]{-8}$, which is unloved by the computer, strictly according to the rules of the square root or power calculation, as follows:

$$\sqrt[3]{-8} = \sqrt[3]{(-1) \cdot 8} = \sqrt[3]{(-1)^3 \cdot 8} = \sqrt[3]{(-1)^3} \cdot \sqrt[3]{8} = (-1) \cdot \sqrt[3]{8} = (-1) \cdot 2 = -2,$$

because the third root of the *positive* number 8 is 2, even the dumbest computer program can do that.

Did you go too fast? Take another look at it step by step, it is a very simple transformation, just several in a row; I am afraid this will happen to you more often in the further course of your mathematical life.

What I did here for $n = 3$ and $x = -8$ of course works for every odd number n and every negative number x *in the* same way: The *nth* root $\sqrt[n]{x}$ can always be calculated by

$$\sqrt[n]{x} = -\sqrt[n]{|x|}.$$

Here |x| is the absolute value of the negative number x, i.e. its "positive part".

2.4 Polynomials and Rational Functions

Now, if you think that I cannot shock you with anything after studying the previous chapters and sections, I may have a surprise for you after all: I will now, without much preamble, simply define the term "polynomial"; without pizza, without verbiage, without anything; a real boon, isn't it?

Polynomials

Let n be a natural number or zero and let $a_0, a_1, \ldots, a_n$ be real numbers. A function $p(x)$, which can be expressed in the form

$$p(x) = a_n x^n + a_{n-1} x^{n-1} + \cdots + a_1 x + a_0 \tag{2.2}$$

is called a **polynomial** of degree at most n.

The numbers $a_0, a_1, \ldots, a_n$ are called the **coefficients** of the polynomial.

It may well be that the first coefficient a_n or, if worse comes to worst, several of the first coefficients are equal to zero. For example

$$s(x) = 0x^5 + 0x^4 + x^3 - x + 1$$

is a polynomial of degree 5 according to the above definition. Of course, no one writes these leading zeros all the time, but one simply writes

$$s(x) = x^3 - x + 1,$$

and so our polynomial of the fifth degree has become one of the third degree. In order not to have to consider any such pathological cases separately, the word "at most" was inserted into the definition: $s(x)$ thus belongs by definition to the set of polynomials of at most degree five, and this is true; the fact that this function gives away two high powers is its own fault. If, on the other hand, one wants to emphasize that a polynomial really has a certain degree, say n, one says that it is *of the exact degree n*.

Examples of polynomials are:

$$p(x) = 3x^4 - 2x^2 + 5x + 2,$$
$$q(x) = -17x^{100} + 1,$$
$$r(x) = x^3 + x^2 + x + 1,$$

which – in this order – are of the exact degree 4, 100 and 3 respectively.

On the other hand

$$f(x) = 2x^3 - 4x^{\frac{4}{3}} + 1$$

is *not a* polynomial, because only natural numbers may appear in the exponent of a polynomial; and 4/3 can disguise itself as much as you like, it still appears as a fraction, and also

$$f(x) = 2x^2 - x^{-1}$$

is *not a* polynomial, because negative exponents are not allowed either.

Speaking of "camouflage": A polynomial does not necessarily have to be written down in the pure form (2.2), but may well be camouflaged to begin with; the important thing is that it is *possible to* put it into the form (2.2).

Example 2.6 The function

$$f(x) = (x^2 + 2x)^2$$

does not have the form of a polynomial desired in the definition; but by multiplying it out, or by applying the first binomial formula, it turns out that

$$f(x) = (x^2 + 2x)^2 = (x^2)^2 + 2 \cdot 2x \cdot x^2 + (2x)^2 = x^4 + 4x^3 + 4x^2$$

is a polynomial after all, and there is really nothing wrong with it.▪

Exercise 2.6 Which of the following functions are polynomials?

(a) $f(x) = x^3 + 3^x$
(b) $g(x) = \sqrt{x^2 + 2x + 1}$
(c) $h(x) = x^2 - x^{-2}$

At this point, since we have been talking so nicely about polynomials for quite a while now (well, admittedly, in a book like this a conversation is more like a monologue by the author, but you have to admit that at least I am trying to include you in the deliberations), I have to say that I have been annoyed by this word for 25 years. This is because I once learned both Greek and Latin in my youth; not too much of it is left – my old language teachers may forgive me – but at least it is still enough to be able to say that the word "polynomial" is actually an impossible conglomeration of these two languages: *polys*(πολυσ) is Greek and means "much" or "many", while the word component "nom" comes from the Latin word "nomen", which means "expression" or "sentence member". So actually we should rather say "multinom" (because then both word components come from Latin), but I am afraid we are all unlikely to change that in this lifetime. So let us leave it at the term polynomial, and let us be happy together when once again a lecturer uses this term without knowing that it is actually wrong. By the way, my dictionary of foreign words explains the term polynomial as follows: "A mathematical expression consisting of more than two elements connected by plus and minus signs". If you find that better than the mathematical definition given above, then let us leave it at that and you please study classical philology. But please tell your colleagues there that this definition is not correct, because a polynomial may well consist of only two or even only one member.

The rest of us come back to mathematics: Polynomials are in many ways the simplest functions imaginable, and you will encounter them again at many points in your studies. "Simple" because they are only composed of power functions, and these in turn, as you will see later, can be derived, integrated or otherwise mathematically doctored quite easily. You can fill whole books with statements about polynomials (and some do), but in this section, in which we only want to get a taste of polynomials, I will content myself with introducing polynomials with small degrees and, in particular, saying something about their zeros.

The smallest value allowed for n is certainly $n = 0$. Polynomials of degree zero – the simplest polynomials, in other words – are not really polynomials at all, because they are constant functions, i.e. not dependent on x; a polynomial of degree zero is of the form

$$p(x) = a,$$

thus has the same value for all $x \in \mathbb{R}$, which I have simply called a here. Such a function is called a **constant function**. The diagram of such a function is hard to beat for simplicity: It is simply a horizontal line, that is, parallel to the *x-axis*; at least if the value a is not itself zero. But if it is, i.e. if $f(x) = 0$, the graph of f is identical to the *x-axis*; not very thrilling either.

There is not much more to say about polynomials of degree zero, which is why I will now move on to those of degree one. A polynomial of degree one is of the form

$$p(x) = ax + b.$$

Such a function is called a **linear function**. Again, similar to the constant function above, I have simply called the coefficients a and b, since it seems excessive to me to work with indices for two.

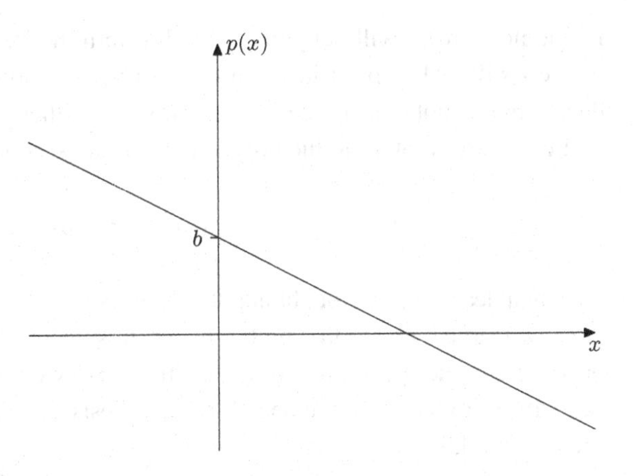

Fig. 2.6 Linear function with *y-axis intercept b*

The reason for the term "linear" can be seen immediately, for example, if one looks at the diagram of such a function in Fig. 2.6: It is a straight line whose slope is given by a and which intersects the *y-axis* at the point b, because for $x = 0$ the value $p(0) = b$ results.

I now want to determine the zeros of a linear function; to do this, it might not be a bad idea to first define the term zero. This is hugely important to all of mathematics, though not as important as pizza or potato chips, and you will certainly encounter it again and again throughout your studies. Now then:

Zero
A real number $\overline{x}$ is called the **zero of a** function f: $\mathbb{R} \rightarrow \mathbb{R}$ if

$$f(\overline{x}) = 0$$

applies.

For example, $\overline{x} = -1$ is a zero of the function

$$f(x) = (x + 1)^3,$$

because $f(-1) = (-1 + 1)^3 = 0 = 0.$[3]

There is one function that consists only of zeros, namely the constant function $f(x) = 0$; it is often called the *zero function.* All other constant functions have no zeros, which completely explains the zero behaviour of this class of functions, i.e. the polynomials of degree zero.

Polynomials of the *exact* degree one, i.e. linear functions, always have a zero, because a non-constant, i.e. non-horizontal straight line always intersects the *x-axis* somewhere. If you feel like it, you can also calculate this zero, but since I do not have one right now (feel like it, not zero), I will put you off for this until the chapter on equations and inequalities, where I will make up for it. For now, it suffices to note that a polynomial of at most first degree that is not exactly the zero function has either none (if constant) or a zero.

Let us now venture to the polynomials of the second degree, which have the form

$$p(x) = ax^2 + bx + c \tag{2.3}$$

have or at least can be brought into this form. Such a function is also called a **parabola**. We already learned about the prototype of this class of functions, the normal parabola $p(x) = x^2$, in the section on power functions. How does a general parabola like the one in (2.3) differ from this special one? Well, my thesis is: "Not at all! If you've seen one, you've seen them all."

To substantiate this, I first attach a prefactor a to the normal parabola, i.e. I consider functions of the form

$$f(x) = ax^2.$$

As long as a is positive, this function does not differ qualitatively from the normal parabola at all, it just gets a bit wider (for $a < 1$) or slimmer (for $a > 1$); by the way, it would be nice if there was something like that for humans too, because thanks to potato chips and pizza I personally could use such a slimming prefactor from time to time quite well. But back to parabolas. The best way to see that what I have just said is true is to look at an example: If you look at the function $f(x) = 1/4\ x^2$ (Fig. 2.7 (black)), for example, you will see that it does not have the value 1 for $x = \pm 1$, as the normal parabola does, but the value 1/4; it only reaches the function value 1 for $x = \pm 2$. And so it goes on: Overall, this function comes up much more slowly than the normal parabola, which gives it a broader picture overall. The situation is different, for example, with $f(x) = 4x^2$ in Fig. 2.7 (dashed): This function has already grown to 4 at $x = \pm 1$, and at $x = \pm 2$ it has already reached 16. Therefore it has a slimmer appearance than the normal parabola.

So much for positive prefactors; a *negative prefactor* mirrors the graph of the parabola with the corresponding positive prefactor just at the *x-axis*, it folds the graph down, so to speak. A picture is probably worth a thousand words here, so please consider Fig. 2.8.

But with this you have in principle already seen all parabolas of the general type, because the coefficients b and c in the representation (2.3) only cause shifts of the graph defined by the term ax^2; as an example you can see in Fig. 2.9 the graph of the function $p(x) = 2\,x^2 - 3\,x - 2$.

The basic shape of a parabola can therefore be recognized by the coefficient of x: If it is positive, the parabola is said to be "open upwards", if it is negative, it is said to be "open downwards", and if it is zero, the parabola has degenerated to a straight line, because then it has the form $b\,x + c$.

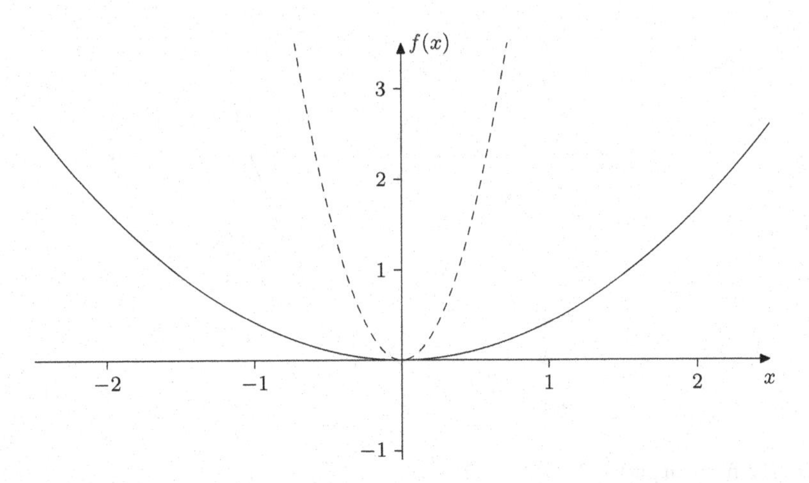

Fig. 2.7 The parabolas 1/4 x^2 (*black*) and 4 x^2 (*dashed*)

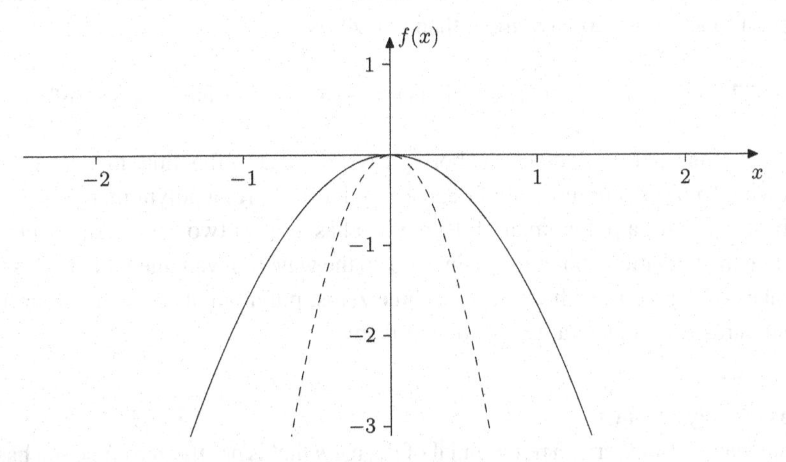

Fig. 2.8 The parabolas – ¼ x^2 (*black*) and – 4 x^2 (*dashed*)

What about the zeros of a second degree polynomial, a parabola? Certainly there are some without zeros, for example the function $p(x) = x^2 + 1$; since x^2 never becomes negative, there is no function value that would be smaller than 1, in particular none that becomes zero.

Furthermore, there are parabolas that have exactly one zero, and we do not even need to think of the degenerate case of the straight line, the good old normal parabola $p(x) = x^2$ will do: This has one zero in $x = 0$, and that is it, all other function values are positive.

If I subtract 1 from the normal parabola, I end up with the parabola $p(x) = x^2–1$. This now has two zeros, namely 1 and -1, as you can easily test by inserting these values.

But no parabola in the world can have more than two zeros, because if you look at the graphs of the examples shown above, you will see that you can fold, stretch or move them

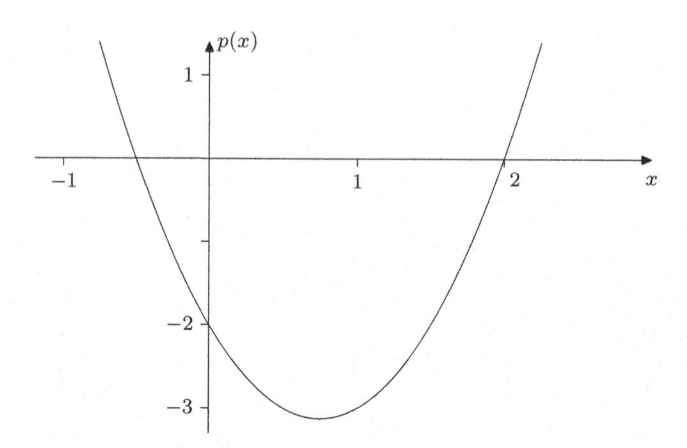

Fig. 2.9 The function $p(x) = 2\,x^2 - 3\,x - 2$

as long as you want, the graph will never intersect the *x-axis* more than twice, and that means for a function not to have more than two zeros.

Exercise 2.7 Why must the polynomial $p(x) = -28\,x^2 + 117$ have two zeros?

Notice anything? If we exclude the boring case of the zero function, then we have seen the following so far: A polynomial of degree zero has no zero, a polynomial of degree one has at most one zero, a polynomial of degree two has at most two zeros. And believe it or not, even in mathematics, sometimes things go on the way you want them to! That is to say: polynomials of degree three have at most three zeros, polynomials of degree four have at most four zeros,. . ., in general the following is true:

Zeros of Polynomials

For any natural number n, a polynomial of degree n that is not the zero function has at most n zeros.

I do not want to prove that here, you have to learn something in the course of your studies.

You may have wondered why I have not said anything about the definition range of polynomials all this time; well, that is because polynomials accept any real number as an input value, so you do not have to pay much attention when defining the definition range.

The situation is different for the class of functions that I will now introduce to you at least briefly, the rational functions:

Rational Function

Let p and q be two polynomials and let D be a subset of the real numbers that does not contain a zero of q.

One calls a function of the form

$$r : D \to \mathbb{R}, \quad r(x) = \frac{p(x)}{q(x)}$$

a **rational function.**

Not only the name, but also the procedure reminds of the construction of the rational numbers in the first chapter: These are defined as quotients of two integers, where the number in the denominator must not be zero, while the rational functions are defined as quotients of two polynomials, where the polynomial in the denominator must not become zero, which is why I banished its zeros from the definition range of the function.

Let us look at examples first: The functions

$$r_1(x) = \frac{3x^3 - 4x^2 + 2}{x^2 + 1}, \quad r_2(x) = \frac{17x^{100} - 2x^{33} + 2x}{x - 1}$$

and

$$r_3(x) = \frac{1}{x^3 - 2x^2 - x + 2}$$

all three have the correct form, thus represent rational functions, because in the numerator as well as in the denominator there are only polynomials; however, we still have to think about their domain of definition, because without the specification of this the definition of a function is never complete.

Before that, however, I will show you two examples of function rules that are *not* rational functions: The function

$$g(x) = \frac{3x^3 - 4x - 2}{x^2 - 3x^{1/2} + 4}$$

is not a rational function, because the exponent 1/2 has crept into the denominator and the function in the denominator is therefore not a polynomial. But also a candidate like

$$h(x) = \frac{3^x - x^3}{x^2 + x - 1}$$

fails, because the expression 3^x causes the numerator not to be a polynomial.

Back to our positive examples from above; to complete the function definition, we still need to determine the domain of definition. With rational functions, one usually proceeds in such a way that one first defines the function prescription and then gets clear about the definition range, in other words: examines the zeros of the denominator; if it is all right with you, I will start with that already.

The denominator of the first rational function is $x^2 + 1$, and you already know that this never becomes zero, so the function $x^2 + 1$ has no zero: The expression x^2 never becomes negative, we have already seen that several times, and if I add 1 to it, the whole thing becomes positive, so in particular not zero. The domain of definition of the function $r_1(x)$ is therefore completely $\mathbb{R}$.

The denominator of the function $r_2(x)$ is $(x - 1)$, so it is a linear function; now I have successfully avoided giving a formula for the zeros of linear functions already in this chapter, but that this simple function has the number $x = 1$ as zero, I think you can see with the naked eye. As a polynomial of the first degree, it cannot have any other zeros, so the largest possible domain of the function $r_2(x)$ is equal to the set of real numbers without the one; this is written down in the short form $\mathbb{R} \setminus \{1\}$.

The third function I actually only wrote down to give you an idea that recognizing the zeros of the denominator and thus the definition gaps of the rational function is not always a pure pleasure; well, as far as mathematics is ever a pleasure for you at all, but I would not get into *that* debate now!

At this point I would like to confront you with three statements, the first two of which you can easily verify, and the third you will believe me anyway:

Statement 1 The denominator of the function $r_3(x)$, i.e. the function $x^3 - 2x^2 - x + 2$, has the zeros -1,1, and 2; you can check this immediately by inserting, which I would like to ask you to do, since you should never believe anything I say without checking it.

Statement 2 This function has no further zeros; but you already know that, because the denominator is a polynomial of degree three and such a polynomial cannot have more than three zeros according to the above.

Statement 3 If I had not come up with this function myself and constructed the denominator using a method I would like to leave in the "dirty tricks" box for textbook writers at the moment, I would not have come up with these three zeros so easily; I am sure you will take my word for it, and I am only saying it here to build you back up if you are now desperately thinking, "I never would have thought of that!" As I said, there are already two of us with this fate.

Overall, we have now figured out with joint powers that the largest possible domain of definition of the function $r_3(x)$ is equal to the set $\mathbb{R} \setminus \{-1, 1, 2\}$.

Exercise 2.8 Check which of the following are rational functions and, if so, give the largest possible domain of definition.

$$f(x) = \frac{2x^2 + x^x}{3x^2 - x + 1}$$

$$g(x) = -\frac{x^3 + 5x - 3}{(x-1)^4} \qquad \blacksquare$$

As you may remember, or as you may reconsider quite quickly, integers – if you really want it complicated – can also be thought of as rational numbers with a denominator of one. It is no different with polynomials and rational functions: You can think of any polynomial as a rational function with a denominator of one, for example

$$3x^4 - 3x^3 + 5x^2 - 2 = \frac{3x^4 - 3x^3 + 5x^2 - 2}{1}.$$

Pretty silly, you are right, but in that sense polynomials are just the simplest rational functions. So if we want to leave them to the left, then we have to allow at least one x in the denominator. So the simplest "correct" rational function is then

$$h(x) = \frac{1}{x}.$$

This function, or more precisely its graph, is also called a **hyperbola** or **normal hyperbola**. It is obviously defined for all real numbers except zero and maps every real number to its reciprocal.

At least once I should probably show you a graph of a rational function, and that is what I am doing now with the hyperbola.

You can see that this looks quite different from the polynomials, and this is due to the fact that for rational functions there are zeros of the denominator somewhere on the real axis, which have to be banished from the domain of definition. If the rather improbable case does not occur that the zero of the denominator is also one of the numerator, then this place is called a pole of the function.

Polarity of a Rational Function

A real number $\bar{x}$, which is zero of the denominator but not of the numerator of a given rational function $r(x)$, is called a **pole** or **pole point of** $r(x)$. A pole is therefore *not* part of the domain of definition of the rational function.

Zero points of the denominator of a rational function should be avoided like the devil avoids holy water, because even if you get close to them, the function becomes very unpleasant. The diagram of the hyperbola with the only pole $\bar{x} = 0$ shows what happens in such a case: The function values become incredibly large near the pole location and can no longer be plotted. Admittedly, "incredibly large" is not a strictly mathematical term, but you will forgive me if I withhold the exact definition of "growing beyond all limits".

It is more important to be clear about whether the function tends to the positive area (i.e. upwards) or to the negative area (i.e. downwards). In the hyperbola shown (Fig. 2.10), the function graph tends upward when approaching the pole from the right, and downward when doing so from the left. To understand this, I stalk the zero numerically from the right and insert the *x-values* 1, 1/10, 1/100 and 1/1000 one after the other; the corresponding function values are 1, 10, 100 and 1000, and one believes quite quickly that the function disappears here into the positive area, i.e. upwards. If, on the other hand, you approach zero from the left, for example by inserting the *x-values* −1, −1/10, −1/100 and −1/1000 one after the other, you get the function values −1, −10, −100 and − 1000, so the function becomes *very* negative quite quickly (which is of about the same linguistic quality as "quite pregnant", but you know what I mean).

In the case of the power functions x^n we had seen that their diagrams in principle break down into two varieties, namely those for even n and those for odd n. It is no different with the functions $1/x^n$, whose first representative, the normal hyperbola, we have just examined.

If you look at the diagram of the function $1/x^2$ in Fig. 2.11, you will see a fundamental difference to the hyperbola, because this new diagram only runs in the positive area, so it grows from the left as well as from the right into the unbelievably large *positive area*, especially when approaching the pole zero. But this is not surprising, because we have known for a long time that the denominator x^2 is always positive, and the inverse does not change this.

But with this you have in principle already seen all functions of the type $1/x^n$, because qualitatively they all look like $1/x$ when n is odd, and like $1/x^2$, when n *is* even. This is no different from power functions: If you have seen one, you have seen them all!

In the next section, I deal with exponential functions, which are functions of greatest interest to both nuclear physicists and capital investors. – With this I have probably only directly addressed 10% of the readership of this book, but hopefully I have at least made the remaining 90% curious.

Fig. 2.10 The hyperbola 1/x

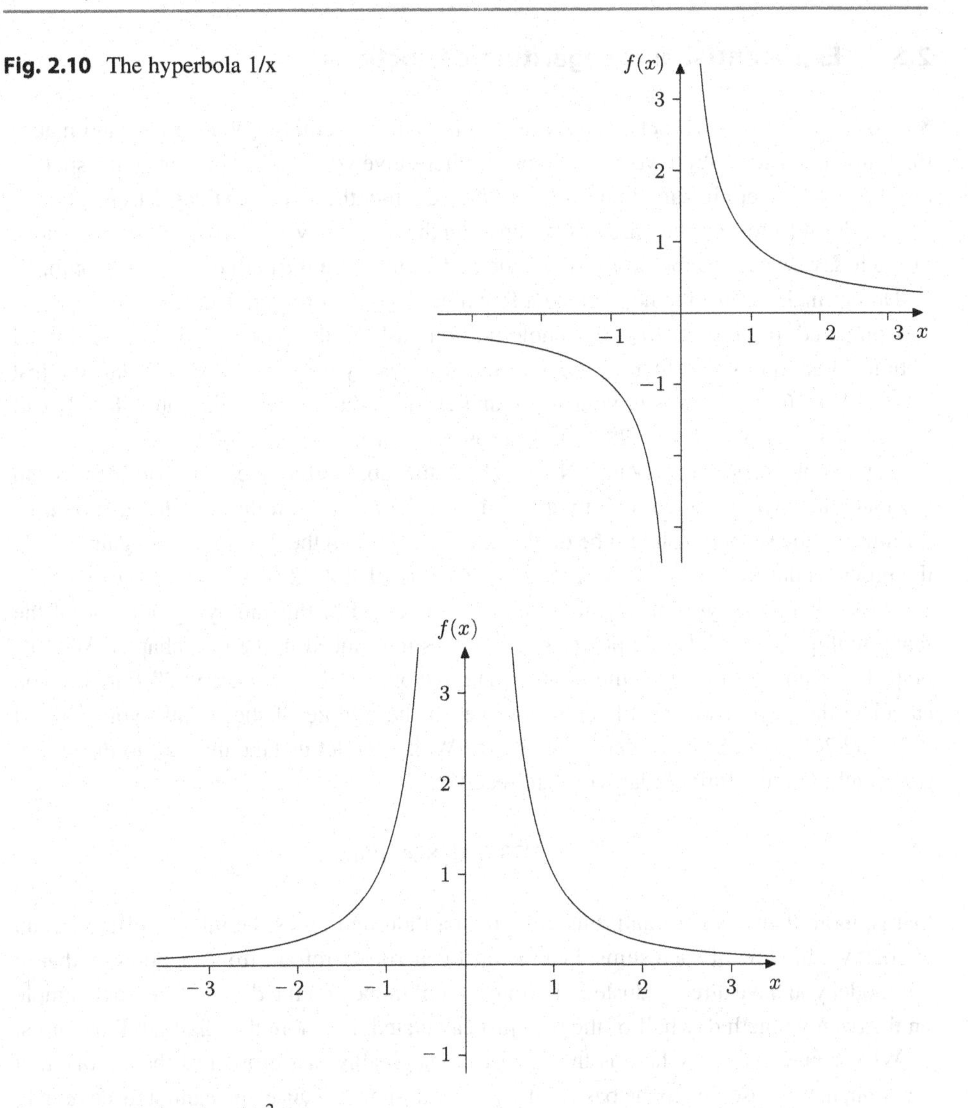

Fig. 2.11 The function $1/x^2$

But first it is my duty (well, there is a little bit of freestyle, too) to confuse you a little bit: I have to mention, that the classes of functions, which are called polynomials or rational functions by me here, are sometimes called differently, especially in the somewhat older literature. There, the polynomials are called “wholly rational functions”, while the rational functions are called “fractional rational functions”.

I do not want to go into this in detail, and I would not use these terms any further, but I wanted to mention them to you at least once, so that if, for example, you ever come across the name fractional rational functions in your studies, you would not look at this book in annoyance and think, “Well, he could *have* explained that!” He just did.

2.5 Exponential and Logarithmic Functions

Suppose you have a rich benefactor who wants to do something good for you and makes the following offer: "Well, you can choose: Either I give you a million euros on the spot, or else I pay you a certain sum on each day of the next month; on the first day 2 Euros, on the second day 4 Euros, on the third day 8 Euros, on the fourth day 16 Euros, and so on, that is, on each day twice the amount of the previous day, until the thirtieth day of the month."

Do not make a mistake now! Read a few more lines before you decide!

Compared to the offer to get a whole million cash on the paw, of course, the second option looks puny at first glance: After 4 days, you will have a total of just $2 + 4 + 8 + 16 = 30$ Euros in your hand, and even if you wait a week, you will only call $2 + 4 + 8 + 16 + 32 + 64 + 128 = 254$ Euros your own.

But do not worry, it gets better! Now, to back this up, I will first calmly consider how to calculate the payout amount on any given day of the month. On the first day it should be 2 Euros, on the second it should be double, i.e. $2 \cdot 2 = 4$, on the day after that again double the previous day, i.e. $2 \cdot 2 \cdot 2 = 8$, then the day after that $2 \cdot 2 \cdot 2 \cdot 2 = 16$, and so on.

Does this remind you of anything? Exactly to get rid of this annoying notation of the many multiplication dots, the power notation was introduced in the first chapter. With its help, I can now formulate quite simply: The payout on the *nth* day is 2^n Euros.If you calculate the payout on the fifteenth day, i.e. in the middle of the month, you will get $2^{15} = 32{,}768$ Euro. Still not very impressive? Well, then let us take the leap to the end of the month: On the thirtieth day you will receive

$$2^{30} = 1{,}073{,}741{,}824 \text{ Euro},$$

that is, more than a billion and thus more than a thousand times the million offered as an alternative. In view of such sums, I can almost generously refrain from mentioning that in our model you have already dusted off quite a lot of money on the days before, for example on the twenty nineth day half of the sum just calculated, i.e. more than half a billion euros.

What comes into play here is the almost unimaginably fast growth of the exponential function, in this case that to the base 2. If you now hesitate because you cannot discover any function far and wide, it is probably because I have written the variable as n above, i.e. 2^n, and one always identifies n with a natural number. In the notation

$$f : \mathbb{R} \to \mathbb{R}, \quad f(x) = 2^x$$

the whole thing looks more like a function. In fact, you can use any real number for x and get a reasonable function value.

In the case where x is a rational, in particular an integer, you learned in the first chapter how to calculate the function value 2^x. For example, $f\left(\frac{1}{2}\right) = 2^{1/2} = \sqrt{2}$ and $f\left(\frac{4}{3}\right) = 2^{4/3} = \sqrt[3]{2^4} = \sqrt[3]{16}$.

But what if x is a proper real number, that is, not a rational number? What on earth, for example, is $f(\sqrt{2}) = 2^{\sqrt{2}}$ supposed to be? Well, mathematicians over the centuries have solved this problem too, and they have solved it the same way Anna and Joe would: You take a few (strictly speaking, infinitely many, but let us not go there) rational numbers r_1, r_2, r_3,... here, which are getting closer and closer to the number $\sqrt{2}$, and calculate the corresponding values $2^{r_1}, 2^{r_2}, 2^{r_3} \ldots$ Once you feel it is not changing much, you stop and take the last calculated value as an approximation for $2^{\sqrt{2}}$ (Before you lose faith in the exactness of mathematics altogether: Of course, one actually does this more exactly, one uses a so-called sequence of rational numbers convergent to $\sqrt{2}$ $\{r_i\}$ and determines the limit of the resulting sequence $\{2^{r_i}\}$, but for home use, the version described above does just fine).

Now do not be afraid (you should never be in math, by the way) that you would actually have to do this yourself: if you ever have the problem of calculating $2^{\sqrt{2}}$, rely on the people who programmed your calculator and use that to do the calculation. I do it exactly like this and get

$$2^{\sqrt{2}} = 2.66514414\ldots$$

What I have done here all the time with the number 2 as a basis can of course also be done with any other positive real number. The rich benefactor quoted at the beginning as an example could also say that he triples the payout sum every day, which would lead to the function 3^x, or he could decide to halve the sum day by day, which would be modeled by the function $(1/2)^x$. In the latter case, however, if I were you, I would assume the million mentioned as an alternative, and if you ever calculate a few values of this function, I think you will agree.

But now it is high time for an exact definition:

Exponential Function to General Base

Let a be a positive real number. The function

$$\exp_a : \mathbb{R} \to \mathbb{R}, \quad \exp_a(x) = a^x$$

is called an **exponential function** or, more precisely, an **exponential function with base** a.

And for something else it is high time:

Exercise 2.9 For $a = 1/2, 1, 2, 4$, calculate the function values of $\exp_a(x)$ at the points $x = -2, 0, 1, 2, 10$. ▪

If you have worked on this task, you will have noticed a significant difference, for example, in the behaviour of $\exp_{1/2}(x)$ and $\exp_2(x)$: One falls, the other grows, and the function $\exp_1(x) = 1^x$ again convinces by boredom, because it is constant equal to 1, since just 1^x for all real numbers x is equal to 1.

In fact, the number 1 as a base is a boundary between the two fundamentally different types of exponential functions:

Behavior of the Exponential Function to Base *a*
The function $\exp_a(x) = a^x$ is strictly monotonically increasing on all of $\mathbb{R}$ if $a > 1$, and strictly monotonically decreasing on all of $\mathbb{R}$ if $a < 1$. In any case, the function values are always positive numbers, i.e., the image set of the exponential function is the set of positive real numbers $\mathbb{R}^+$.

This is not surprising, because if I exponentiate a number that is greater than 1, that is, multiply it a few times by itself, then it naturally becomes larger and larger. If, on the other hand, I exponentiate a number that is positive but smaller than 1, then yes, I make it smaller, and more so the higher this power is; if you have not already done so anyway, just give it a try with $a = 1/2$.

From the calculation rules for the simple exponentiation, which you can read again in the first chapter, one derives directly the following regularities:

Calculation Rules for the Exponential Function
For all real numbers x and y and for each positive base a *the* following calculation rules apply:

1. $\exp_a(x + y) = \exp_a(x) \cdot \exp_a(y)$
2. $\exp_a(-x) = \exp_{1/a}(x)$.

In case the second rule blew you away at first, let me briefly rebuild it: This is just a compact (and therefore not immediately obvious at first glance) notation for the equation that holds according to the power laws

$$a^{-x} = \frac{1}{a^x} = \left(\frac{1}{a}\right)^x.$$

Mathematics can be that simple.

In the introduction to this chapter, I promised to show you that the exponential function is of interest to both nuclear physicists and capital investors, and I will now keep this promise. Since there are probably more capital investors than nuclear physicists among the readers of this book – at the latest after successful graduation – I will start with an example for the former:

Example 2.7 Suppose you invest a certain capital *K*, it does not have to be the billion mentioned above, at an annual interest rate of *p* percent. After 1 year your capital is then

$$K + \frac{p}{100} \cdot K = K \cdot \left(1 + \frac{p}{100}\right)$$

increased; for example, a capital of 100 Euros at an interest rate of 3% increases to $100 \cdot (1 + 3/100) = 100 \cdot 1.03 = 103$ Euros.

If you – now again in the general case – leave the interest in the account, it will also earn interest in the next year and at the end of the second year you will hit

$$K \cdot \left(1 + \frac{p}{100}\right) \cdot \left(1 + \frac{p}{100}\right) = K \cdot \left(1 + \frac{p}{100}\right)^2$$

to the account. And so it goes on: At the end of the *nth* year, where *n is* to be any natural number, the initial capital *K has* grown to

$$K \cdot \left(1 + \frac{p}{100}\right)^n$$

And now you can see what this has to do with our topic: The factor by which the initial capital is multiplied is just the function value of the exponential function with base $a = (1 + p/100)$ at the point *n*. And just as with any real function, you do not have to limit yourself to the natural numbers as the defining set, but may substitute any positive real number in place of *n*. This is of practical importance, because if you or your bank want to know the value of your capital on any day in the middle of the year (for example, because you want to withdraw your money), you have to use non-integer values; for example, the net present value after 2 years and 5 months, i.e. after 29/12 years, is equal to

$$K \cdot \left(1 + \frac{p}{100}\right)^{\frac{29}{12}}.$$

In principle, this also works for non-rational exponents; why do not you ask your bank advisor what the value of your capital is after $\sqrt{2} \approx 1.4142\ldots$ years, but please take your camera with you and send me a picture of his facial expression.

Exercise 2.10 Mrs. Broke wants to invest 8000,– Euro optimally. Bank A offers her to pay interest on the money at 4.5% per annum. Bank B offers her to fix the money for 4 years; at the end of the fourth year she will receive 9431.07 Euro. Which bank should Mrs. Broke choose if she wants to have maximum assets at the end of the fourth year?▪.

Example 2.8 In the meantime, I turn to the second current target group, the nuclear physicists. Here, in radioactive decay, there is the phenomenon of the **half-life**: after an exactly equal period of time, the half-life, just half of the radioactive material still present before has decayed. So if at the beginning of the decay process there is a mass *M of* material, after one half-life period half of it, i.e. *M/2, is* left, and after another half of it again, i.e. *M/4.* You can see how this works: After *n* half-lives there is still

$$M \cdot \left(\frac{1}{2}\right)^n$$

material is present. So now, too, it is an exponential function, in this case the one in base 1/2, which makes the decisive contribution to this mass function. Again, one can insert non-integer values and thus calculate, for example, how much material is still present after 2.3451 half-life spans, which, however, does not generally baffle nuclear physicists at all, unlike bank advisors.

Exercise 2.11 Radioactive carbon^{14} C, used to date fossil finds, has a half-life of about 5776 years. Calculate how much of 1 g of^{14} C will still be there after 10,000 years.

To conclude this little excursion into the world of exponential functions, I would like to familiarize you once again with a special term in mathematics: You will certainly come across the term "exponential function" at some point in a mathematics book without any further addition, i.e. without specifying the base. In this case, one means a very special exponential function, namely the one to the base *e*.

Are you okay? No, it cannot be, I guess I have to say what this *e* is supposed to mean. Within mathematics, it is just as fixed a designation as, for example, the notation π for the number of a circle. *e* denotes the **Eulerian number**, named after Leonhard Euler, one of the giants of mathematics.

By a conservative estimate on my part, there are about 117 different definitions or derivations of this number, and they all lead to the same value, of course. The problem is that not a single one of these definitions is so short and crisp that I could expect you to read it here; so let us leave it at naming it and giving the numerical value of *e*; it is

$$e = 2.718281828459045\ldots,$$

where the sequence of these decimal numbers never terminates, since *e* is not a rational number. So when someone talks or writes about *the* exponential function, they mean the function $f(x) = e^x$ with *e* as just defined.

This is how far I had written the text when the publisher, who by definition is always right, objected and asked me to motivate the definition of the number *e* at least a little bit. Since I am the last person to object to motivating those interested in mathematics, I am happy to oblige. Now then:

A little further up we went into the depths of compound interest together. Suppose you found a bank that gives you the fabulous annual interest rate of 100%. Being justifiably suspicious of this offer, you invest only 1 Euro. At the end of the year you then receive

$$K_1 = \left(1 + \frac{100}{100}\right)^1 = (1+1)^1 = 2$$

paid out.

The next step in improving your financial situation is for the bank to pay interest on your investment at the end of each month, i.e. 12 times a year, but of course only at 1⁄12 of the annual interest rate. At the end of the year, you will receive in this case

$$K_{12} = \left(1 + \frac{1}{12}\right)^{12} = 2.61303$$

paid out.

Why would a bank that pays interest at the end of each month not also pay interest at the end of each week, i.e. 52 times a year? If it did, at the end of the year you would receive

$$K_{52} = \left(1 + \frac{1}{52}\right)^{52} = 2.69259$$

paid out.

You know the game now: I now move to daily interest at 1⁄365 of the annual rate: You would then receive at the end of the year

$$K_{365} = \left(1 + \frac{1}{365}\right)^{365} = 2.71456$$

paid out.

And if we have already found a bank crazy enough to pay interest on your capital daily, let us also find one that pays interest hourly, that is, one that allows $24 \cdot 365 = 8760$ interest dates at 1⁄8760 of the annual rate each; here you will get at the end of the year

$$K_{8760} = \left(1 + \frac{1}{8760}\right)^{8760} = 2.71813$$

paid out.

I now ask you to look again at the calculated numerical values $K_2, \ldots, K_{8760}$ in turn. Surely you will notice that these values approach more and more the Eulerian number e defined above; in fact, it can be proved that they come arbitrarily close to the Eulerian number for ever decreasing interest intervals, and one can also define this number as the so-called limit value of the sequence K_n for n towards infinity – one then also speaks of *continuous interest*.

I wrote down more in passing above that the exponential function is a strictly monotonic function for any positive base except 1. But that (not the fact that I wrote it down in passing, but the fact that it is there at all) means that there exists an inverse function of the exponential function that has the same monotonic behavior as it.

This inverse function I want to introduce to you now at the end of this chapter, it is the **logarithm function**. Most people shudder at the mere mention of this name, but I will try to convey that this is something quite normal (as far as mathematics can be something quite normal, but I would not get into *that* discussion now). Perhaps the best way I can start is by explaining this somewhat strange word: It is composed of two Greek parts of speech, namely *logos*(λογος) and *arithmos*(αριθμος). "Arithmos" is the Greek word for "number" and is also in "arithmetic", for example; "logos" is a strongly philosophically inclined term and has very many meanings, for example "word", "teaching", but also "meaning". So "logarithm" could be translated in numerous different ways, but since I am a mathematician and not a classical philologist, I will not even try that, but go directly to the mathematical-precise definition of logarithm:

Logarithm to Base *a*

Let a and x be positive real numbers and $a \neq 0$. That real number y which satisfies the equation

$$a^y = x$$

is called the **logarithm** or **logarithm function of** x **to the base** a and is denoted by $\log_a(x)$.

So, for example, $\log_2(8) = 3$, because $2^3 = 8$, and $\log_5(1/5) = -1$, because $5^{-1} = 1/5$. Another characteristic example is $\log_4(2) = 1/2$, because $4^{1/2} = 2$. Finally, I want to point out that for any base a, $\log_a(1) = 0$, because for all positive real numbers a, $a^0 = 1$.

I do not dare tackle more difficult things, I would like to leave that to you:

Exercise 2.12 Without a calculator, calculate the following logarithms:

$$\log_{10}(0.001), \quad \log_7\left(\sqrt[4]{7^3}\right). \quad ■$$

According to the above definition, $a^{\log_a(x)} = x$ applies, which is why the logarithm function can rightly be regarded as the inverse of the exponential function. This in turn implies the following statement due to the monotonicity properties of the exponential function:

Monotonicity Behavior of the Logarithm Function to the Base *a*

The domain of definition of the logarithm function $\log_a(x)$ is the set of positive real numbers, denoted $\mathbb{R}^+$. The function $\log_a(x)$ is strictly monotonically increasing on all $\mathbb{R}^+$ if $a > 1$, and strictly monotonically decreasing on all $\mathbb{R}^+$ if $a < 1$.

Not only the monotonicity behavior, but also the calculation rules for the exponential function have direct effects on the logarithm function, because they lead directly to the following rules:

Calculation Rules for the Logarithm Function

For all real numbers x, y and p and for each positive base a *the* following calculation rules apply:

1. $\log_a(x \cdot y) = \log_a(x) + \log_a(y)$
2. $\log_a(x/y) = \log_a(x) - \log_a(y)$
3. $\log_a(x^p) = p \cdot \log_a(x)$

"Pretty dry and unworldly the whole thing," you are thinking right now, right? That is probably because of me (well, who else?), because so far I have only been throwing around monotony and calculation rules; but that is going to change now, here come the practical examples:

Example 2.9 Do you remember the capital investors and nuclear physicists? As we have seen above, they can make very good use of the exponential function; but the logarithm is also of great use to both professions. For the capital investors, I will demonstrate it to you once now; I will leave the nuclear physicists to you then!

Once again, you have a capital K at your disposal, which you invest in a bank you trust at the interest rate of 4% per year. Question: When did the capital double?

To answer this question, let us briefly recall together the formula for compound interest that I gave above in connection with the exponential function: at the interest rate of 4%, the capital K has grown after p years to $K \cdot (1.04)^p$, and since we are asking when it doubled, we must therefore calculate when (i.e. for which p)

$$(1.04)^p = 2$$

holds. But this is a clear case for the logarithm, because by definition

$$p = \log_{1.04}(2) \approx 17.673.$$

So you have to wait about 17 years and 8 months for your capital to double.

Of course, I determined the numerical value 17.673 with my calculator. Before you now desperately search your calculator for the key for the logarithm to the base 1.04: Mine does not have such a key either, and very few of them do. Usually only one or two prominent logarithms are implemented, for example those to the base 2 or 10, and these also have special names, which I will reveal to you in a moment.

But first we have to solve the problem of calculating a logarithm in, say, base 1.04, when our calculator can only handle the one in base 2 or 10. Fortunately, this is quite easy, because you can easily convert logarithms between two bases (this does not denote relatives here, but is the plural of base) using the following formula:

Conversion of Logarithms to Different Bases

For any positive numbers a and b different from 1 and positive values x the conversion formula applies

$$\log_a(x) = \frac{\log_b(x)}{\log_b(a)}.$$

To apply this formula to our example above, I now set $a = 1.04$ and b to a value for which the calculator has a logarithm ready, say $b = 10$. Note that on the right-hand side of the conversion formula I only need to calculate logarithms of tens, which I assume my calculator can do; it gives.

$\log_{1.04}(2) = \frac{\log_{10}(2)}{\log_{10}(1.04)} = \frac{0.301029995}{0.017033339} \approx 17.673,$

i.e. the value given above. If you want to recalculate this, you may have to look again in the operating manual of your calculator, because the key for the logarithm of ten is probably not labeled with "$\log_{10}$", but with "log" or "lg". I will tell you why in a moment... – no, not after the advertising, but after the next exercise:

Exercise 2.13 A radioactive substance has a half-life of 500 years. After how many years is still

(a) one quarter
(b) fifth part

of the original material available?

Also at the end of this section I want to arm you with a few special designations, so that you do not stumble over them unnecessarily in the later course of your mathematics career, because for the very frequently occurring bases 2, 10 and e (the Eulerian number) special designations have been introduced, which you will probably also find on your calculator in this way: The logarithm to base 2 is called **logarithmus dualis** and is denoted by $ld(x)$, i.e.

$$ld(x) = \log_2(x),$$

the logarithm to the base 10 is called the **decadic logarithm** and is denoted by $\lg(x)$, i.e.

$$\lg(x) = \log_{10}(x).$$

For the probably most common logarithm, the one to the base e, two names have been found, one calls it the **logarithmus naturalis** or the **natural logarithm** and denotes it by $\ln(x)$, i.e.

$$\ln(x) = \log_e(x).$$

Once again, I could get upset that they fabricated a mix of Greek (logarithm) and Latin (naturalis and dualis) here, but what the heck, the chapter has been successfully brought to a close, and that is where the sheer joy should prevail. It is the same with me, how are you?

Equations and Inequalities

In addition to the concept of function, which you struggled with in the last chapter, that of equation is almost synonymous with "mathematics" for many people. For example, in politics and administration one speaks of "equations with several unknowns", of "equations that do not add up" and so on. I will refrain from making the obligatory side blow at politicians and their ability to think mathematically, i.e. logically, because we have to get on.

But what *is* an equation? Well, for example, $2 + 2 = 4$ is an equation, a fairly straightforward one at that. But I am not talking about equations like that here, but about equations that contain (at least) one variable; an example of such a thing is

$$\frac{2x^2}{1+x^3} - 7(x-1)^2 = \frac{\sqrt{3+x}}{2x^5} \tag{3.1}$$

In very general terms, the concept of an equation can be defined as follows:

Equation
An equation consists of two mathematical expressions (terms) connected by an equal sign.

If at least one of the expressions contains a variable, look for **solutions** to **the equation.** These are numbers that, when substituted for the variables, cause the two terms connected by the equal sign to have the same value. The set of all solutions, which can also be empty, is called the **solution set of** the equation.

So wherever there is a variable – often called x in the following – you insert the same number and calculate the values of the expressions to the right and to the left of the equals

G. Walz et al., *Foundations of Mathematics: A Preparatory Course*,
https://doi.org/10.1007/978-3-662-67809-1_3

sign; if both times the same value results, the inserted number is a solution of the equation. In the above example (3.1), by the way, $x = 1$ is a solution, because this gives the value 1 on the left as well as on the right side of the equal sign.

In the following, I will introduce you to some solution procedures for equations, i.e. methods for calculating their solution(s); do not worry, we would not deal with such monsters as (3.1), that just served to wake you up.

Almost all solution methods for equations are based on the fact that it is allowed to transform equations according to certain rules without changing their solution set. One then applies these transformations – possibly in several steps – in such a way that one can read off a solution directly at the end.

I will show you this with two very simple examples: The equation

$$x - 2 = 4$$

is to be solved. To do this, I add 2 to both sides of the equation; this gives $x - 2 + 2 = x$ on the left and $4 + 2 = 6$ on the right, i.e.

$$x = 6.$$

And there it is. If you do not trust me, then insert the solution thus determined into the initial equation and see that your mistrust was unfounded for once.

As a second example I take the also very simple equation

$$3x = 12$$

before. Here I multiply both sides of the equation by 1/3 and get the solution

$$x = 4.$$

With these two examples, I have in principle already presented to you all the permitted transformations (which one usually applies several times, of course, and thus can solve much more complex equations than those just exemplified); I will now write this down again in formal terms.

Transformation of Equations

The following transformations of an equation do not change its solution set and are called **allowed transformations:**

- Addition or subtraction of the same number on both sides of the equation
- Multiplication of both sides by the same non-zero number

You may be wondering about the fact that adding zero is allowed, but multiplying by zero is not. Well, if you add zero on both sides, nothing at all happens to the equation, and this undoubtedly does not change your solution set, so it is allowed. The fact that it does not get us any further in our search for the solution is another matter.

What about multiplication by zero? Consider the equation

$$\sin\left(x^{2/7} + 2^{x^2}\right) = 528x^{-3/5} - 3^{1/x}. \tag{3.2}$$

I do not have the slightest idea what the solution set of this equation looks like, but if I multiply both sides (forbidden) by zero, I get the equation

$$0 = 0.$$

Like almost everything in life, this has an advantage and a disadvantage: the advantage is that the equation thus transformed is undoubtedly true, the disadvantage is that it does not help one step in the search for the solution set of (3.2). And since this would be the same for any other equation, you can see that it is a good idea not to allow multiplication by zero as an allowed transformation.

I will now show you methods for solving some very common equations, and sensibly start with the simplest, the linear ones.

3.1 Linear Equations

Let us assume that you have just taken a bath and filled the tub with 120 L of water; now you let the water run off, with 8 L of water per minute going down the drain (old house, calcified pipes ...). If x denotes the number of minutes after opening the drain, the bathtub drain function calculates

$$b(x) = 120 - 8x$$

the remaining contents after x minutes. When is the tub empty? Well, “being empty” means mathematically formulated that 0 L are in the tub (mathematicians can express everything in a complicated way!), so I have to solve the equation

$$120 - 8x = 0$$

solve. Such an equation is called *linear*, because there is no power of x or anything else dangerous to see here. To solve it, I first add $8x$ on both sides, which gives $120 = 8x$, and then multiply both sides by $1/8$, which leads to the solution $x = 15$.

Putting these two steps together, we gain the insight that we can solve this equation by calculating

$$x = -\frac{120}{-8} = 15$$

can solve. I am now writing this down as a general principle:

Solution of Linear Equation
An equation of the form

$$ax + b = 0 \tag{3.3}$$

with $a \neq 0$ is called a **linear equation**, more precisely a **linear equation in normal form**. Such an equation has exactly one solution, this is

$$x = -\frac{b}{a}. \tag{3.4}$$

The fact that a must not be zero has two consequences here, which also complement each other wonderfully: First, the linear equation for $a = 0$ would be either trivial (if $b = 0$) or simply wrong (if $b \neq 0$). For another, you would not be allowed to divide by a if it were zero. So you see: everything is in perfect order. Except for the fact that I have not yet calculated that $-b/a$ solves the equation. Let us do that just before we eat: If you put the given value into the left side of the equation, you get

$$a \cdot \left(-\frac{b}{a}\right) + b = -b + b = 0,$$

i.e. the value of the right side.

So if you have a linear equation in normal form, you do not need to do much, but can give the solution directly. However, by far not every linear equation is directly available in normal form; as an introduction to this, take a look at Exercise 3.1.

Exercise 3.1 It is been a long time since we talked about pizza here. So then: You want to calculate exactly what it will cost you if you invite a few friends to your house for pizza. To do this, you asked that a pizza costs 6 Euros, so if you buy x pizzas, it will cost you $6x$ Euros. In addition, the invitation will incur fixed costs, independent of the number of guests, such as heating the oven and lighting the rooms, of 12 Euros. This gives you a total of what is known as the pizza party cost function $P(x) = 12 + 6\,x$. With a budget of 60 Euros, how many guests can you invite if everyone eats a pizza?

In the following lines I want to show you some examples of equations, which are linear equations, but partly so far away from the normal form above, that one does not recognize them as such at first sight. At the same time, of course, you will learn how to bring these equations into normal form and thus solve them.

Let us start with a very simple case and consider the expression

$$2x - 3(5 - x) = 3(2x - 5) + 9. \tag{3.5}$$

Linear it is for sure, but not in normal form. To get this, I first have to multiply out the brackets according to the rules I learned in the first chapter and get

$$2x - 15 + 3x = 6x - 15 + 9.$$

Now I summarize on each side the terms afflicted with x and likewise the terms not afflicted with x, resulting in

$$5x - 15 = 6x - 6$$

Note: So far I have treated each side of the equation separately, so in particular I have not had to use any of the rules for transforming equations. That is coming now, because if I want to get normal form, I have to force a zero on the right-hand side; to do that, I add the negative of the right-hand side *to both sides* and get

$$-x - 9 = 0.$$

So the solution to Eq. (3.5) is $x = -9$, which I calculated using (3.4).

As a next increase, let us consider the equation

$$(x - 1)^2 + (x + 1)(x - 3) = 2(x - 2)^2. \tag{3.6}$$

Does not look very linear, do you think? You are absolutely right, but it *only looks that way*. If you follow the motto "first multiply out and summarize, then think", you get

$$x^2 - 2x + 1 + x^2 + x - 3x - 3 = 2\left(x^2 - 4x + 4\right),$$

so

$$2x^2 - 4x - 2 = 2x^2 - 8x + 8.$$

Now I add again the negative of the right side on both sides and get

$$4x - 10 = 0,$$

the x^2 term has therefore disappeared and the solution of the equation is calculated by (3.4) as

$$x = \frac{10}{4} = \frac{5}{2}.$$

If this is now substituted back into (3.6) for checking purposes, the result is 1/2 on both sides, i.e. an equation.

But now, before you fall into false euphoria and believe that every equation is a linear one after multiplying it out, let me show you:

$$x^2 - 4x + 1 = -2.$$

Here you can sort as long as you like, the x^2 term does not vanish and the equation is not linear; you will learn what to do in this case (if you do not already know) a few lines below.

"Couldn't get any worse," you mean? Wrong! Also the equation

$$\frac{2(x+1)}{x-2} = \frac{2(2-x)}{1-x} \tag{3.7}$$

is in fact a linear one. To show this, I multiply by the common denominator (remember?) $(x - 2)(1 - x)$ and get

$$2(x+1)(1-x) = (x-2)2(2-x).$$

Multiplying out and summing up as practiced above transforms this equation into

$$2 = 8x - 8,$$

thus a linear equation of the purest water, the solution of which can be calculated with the aid of (3.4); it is $x = 5/4$.

But now comes another important point: The multiplication of both sides of an equation with the same number is only allowed if this number is not zero. But who tells us that the term $(x - 2)(1 - x)$, with which we had multiplied through in the course of the transformations, is not equal to zero? First of all, you really do not know, you have to do the just shown transformations according to the motto "Close your eyes and go through!"

and find the solution. *Now* you put this solution (which you do not yet know if it really is one, which is why some people call it a **bogus solution**) into the initial equation and test whether it solves it, and in particular whether the denominator does not become zero. If yes, it is a solution, if no, the equation is unsolvable. By the way, later on, in the so-called root equations, we will often have to deal with such spurious solutions.

Now, to finally conclude the present example, I put $x = 5/4$ into the initial Eq. (3.7): Both sides then get the value -6, and in particular the denominators are far from being zero.

In order to show you an example of a fraction equation where this goes wrong, I have come up with

$$\frac{x^2 - 1}{x - 1} = x \tag{3.8}$$

in the first place. The multiplication with the denominator $x - 1$ gives here

$$x^2 - 1 = x^2 - x,$$

and if I now subtract x^2 on both sides, I can directly read the apparent solution $x = 1$. But if I put this into (3.8), the left side is not defined, because the denominator would be zero; so $x = 1$ is really only an apparent solution and the equation itself has no solution.

That is all I want to say about linear equations from my side, but invite you to practice a little:

Exercise 3.2 Calculate the solutions of the following equations:

(a) $2(x - 3) + 4 = 3(1 - x)$
(b) $3(x + 1)(1 - x) = 1 - x - 3\,x^2$
(c) $\frac{x-1}{2(x+3)} = \frac{x}{2x-1}$⍰

3.2 Quadratic Equations

If the variable in an equation is no longer linear but has an exponent of 2, i.e. "squared", the equation itself is also called quadratic:

Quadratic Equation
An equation of the form

$$ax^2 + bx + c = 0$$

is called a **quadratic equation**.
An equation of the form

$$x^2 + px + q = 0$$

is called a quadratic **equation in normal form.**

The fact that I have labeled the coefficients differently in the two equations should not bother you, it is just convention. The important thing about normal form is that the coefficient of x^2 is equal to 1. In the following, I will only deal with quadratic equations in normal form; if you now think: "Well, he's making it easy for himself!", you are right, but I may do so, not only because I am the author, but because this actually also covers the general case: If you are dealing with a quadratic equation of the form $a\,x^2 + b\,x + c = 0$, you can, if a is not zero, first divide the whole by a and get

$$x^2 + \frac{b}{a}x + \frac{c}{a} = 0,$$

so an equation in normal form. But if $a = 0$, you do not need to do anything, because then the equation is simply $b\,x + c = 0$, so it is a linear one.

So let us focus on the solutions of quadratic equations in normal form.

For once I do not want to give a big preface, but give you directly the solution formulas for quadratic equations, which you probably already know anyway.

Solution of Quadratic Equations
To solve the quadratic equation

$$x^2 + px + q = 0 \tag{3.9}$$

one calculates the two numbers

(continued)

$$x_1 = -\frac{p}{2} + \sqrt{\frac{p^2}{4} - q}$$

and

$$x_2 = -\frac{p}{2} - \sqrt{\frac{p^2}{4} - q}.$$

If the expression under the root, the *radicand*, is negative, then both expressions are undefined, and Eq. (3.9) has no solution.

If the radicand is zero, the value of the root is also zero; in this case, the two numbers x_1 and x_2 coincide and represent the only solution to Eq. (3.9).

If the radicand is positive, the value of the root is also a positive number; in this case, x_1 and x_2 are two distinct real numbers and represent the two solutions of Eq. (3.9).

I should add that the two formulas for the calculation of x_1 and x_2, which differ only in the sign of the root, are usually written in the short form

$$x_{1/2} = -\frac{p}{2} \pm \sqrt{\frac{p^2}{4} - q}$$

summarizes. It is also called the *p–q formula*(s) because you pretty much always refer to the coefficients of this equation by those letters, and I will bet that is what you will find in every textbook you will pick up in the course of your studies.

I hope you believe me when I say that I could also derive these formulas or prove their correctness in a serious case; but since I do not see any compelling necessity for this here, I will just leave it at that and prefer to illustrate the whole thing with a few examples.

I start with the quadratic equation

$$x^2 + 3x + 2 = 0,$$

which conveniently already is in normal form. The application of the mentioned formulas (in the compact version) yields

$$x_{1/2} = -\frac{3}{2} \pm \sqrt{\frac{9}{4} - 2} = -\frac{3}{2} \pm \sqrt{\frac{1}{4}} = -\frac{3}{2} \pm \frac{1}{2},$$

so

$$x_1 = -\frac{3}{2} + \frac{1}{2} = -1 \quad \text{and} \quad x_2 = -\frac{3}{2} - \frac{1}{2} = -2.$$

So this equation has two different solutions.

As a second example I examine the equation

$$-2x^2 - 3x + 2 = 0.$$

To get this into normal form, I divide it by −2, the coefficient of x^2, and get

$$x^2 + \frac{3}{2}x - 1 = 0.$$

Now everything goes its way, I apply the solution formula, and this yields

$$x_{1/2} = -\frac{3}{4} \pm \sqrt{\frac{9}{16} - (-1)} = -\frac{3}{4} \pm \sqrt{\frac{9}{16} + \frac{16}{16}} = -\frac{3}{4} \pm \sqrt{\frac{25}{16}} = -\frac{3}{4} \pm \frac{5}{4},$$

hence

$$x_1 = -\frac{3}{4} + \frac{5}{4} = \frac{1}{2} \quad \text{and} \quad x_2 = -\frac{3}{4} - \frac{5}{4} = -2.$$

So this equation also has two different solutions.

My third example in this context is

$$x^2 - 4x + 4 = 0.$$

Applying the solution formula to this case yields

$$x_{1/2} = 2 \pm \sqrt{4 - 4} = 2.$$

So the radicand is zero here, and consequently the two solutions x_1 and x_2 are identical, so there is only one solution to the equation.

Up to this point everything went well, and perhaps you think that there are actually no unsolvable equations or that these look so pathological that I do not want to present you with one. Far from it, here is one and it actually looks quite harmless:

$$x^2 + 2x + 3 = 0.$$

But if I apply the solution formula to this, I get

$$x_{1/2} = -1 \pm \sqrt{1-3} = -1 \pm \sqrt{-2}.$$

However, the square root of the negative number −2 is not defined in the real and thus no solution of the quadratic equation studied here exists.

With that, we have seen at least one example of each of the three possible cases, and it is time to practice it some more; you guess what is coming? You are right:

Exercise 3.3 Determine the solutions of the following quadratic equations:

$$x^2 + \sqrt{8}x + 2 = 0$$

$$x^2 + \left(3 - \sqrt{3}\right)x - \sqrt{27} = 0$$

Just like linear equations, quadratic equations rarely occur in normal form right away, but must first be converted to it in order to be solved. Again, I will show you this with some examples; I think I can start with a more complex one right away:

$$x(x+2)(x-1) + 2x^3 - x = 3\left(x^2 - 1\right)(x+2) + 1 \tag{3.10}$$

Multiplying out the left and right sides makes this:

$$x^3 + x^2 - 2x + 2x^3 - x = 3x^3 - 3x + 6x^2 - 6 + 1$$

and summarizing results in

$$-5x^2 = -5.$$

So it must be

$$x^2 = 1$$

and either by applying the solution formulas or by just looking at them, you can see that the solutions hereof are

$$x_1 = -1 \quad \text{and} \quad x_2 = 1$$

are. To test this, you can now substitute these two numbers into the initial Eq. (3.10); you will see that the same value is obtained on each side.

In the case of linear equations, I had shown you that they sometimes even disguise themselves as fractional equations, and this is no different with quadratic equations. As an example for this I examine the fraction equation

$$\frac{x}{x+1} = \frac{2-x}{2x}. \tag{3.11}$$

Multiplying by the main denominator $(x + 1)\ 2\ x$ *gives* the form

$$2x^2 = (x+1)(2-x).$$

Multiplying out the right-hand side and summing up the terms associated with x^2, the terms afflicted with x, and the constant terms, I think I can get you down to one step by now. I get the quadratic equation

$$3x^2 - x - 2 = 0,$$

in normal form

$$x^2 - \frac{1}{3}x - \frac{2}{3} = 0.$$

Of course I apply the solution formula again and get

$$x_{1/2} = \frac{1}{6} \pm \sqrt{\frac{1}{36} + \frac{2}{3}} = \frac{1}{6} \pm \sqrt{\frac{25}{36}},$$

hence

$$x_1 = \frac{1}{6} + \frac{5}{6} = 1 \quad \text{and} \quad x_2 = \frac{1}{6} - \frac{5}{6} = -\frac{2}{3}.$$

Equation (3.11) therefore possibly has two different solutions, whereby the "possibly" refers to the fact that they could also be bogus solutions. But this is not the case, as you can see by substituting x_1 and x_2.

Before I leave you alone with the next exercise – perhaps I should rather say leave you undisturbed – one last example on hidden quadratic equations: Let us determine the solutions of

$$\frac{1}{x} + \frac{1}{x-1} = \frac{1}{x+1}. \tag{3.12}$$

You already know what comes next: to get rid of the awkward fractions, I multiply by the main denominator $x(x - 1)(x + 1)$ and get the transformed equation

$$(x-1)(x+1) + x(x+1) = x(x-1).$$

Again, I multiply out and summarize what is on

$$x^2 + 2x - 1 = 0$$

leads. Conveniently, this equation is already in normal form, so I can immediately apply the solution formulas and get the solutions

$$x_1 = -1 + \sqrt{2} \quad \text{and} \quad x_2 = -1 - \sqrt{2}$$

For these two numbers certainly none of the denominators in (3.12) becomes zero, so they are real solutions of this equation and not bogus solutions. Perhaps you are surprised that the Eq. (3.12), which initially looks so smooth and integer, has such "crooked" solutions, but this can happen at any time with quadratic equations because of the root formation in the solution formula and unfortunately there is no legal claim to integerity of the solutions of exercise and examination tasks.

Exercise 3.4 Determine the solutions of the following equations:

(a) $(x - 1)^2 + x(x + 2) = 2(x - 1)(x + 2)$
(b) $\frac{x-1}{x+2} + \frac{x}{x-1} = \frac{2x^2+1}{(x+2)(x-1)}$

Expressions of the form $a\,x^2 + b\,x + c$, which I examine here as the "left side" of quadratic equations, you had already got to know in the chapter about functions, there they were called polynomials of the second degree or also parabolas; and there you had also met the concept of the zero of a function; at least if you read this book in the given order of the chapters; if not, you should take a short look at the functions chapter to understand the following lines.

If you have, then you may have noticed that solving a quadratic equation is nothing more than determining the zeros of the second-degree polynomial, that is, the function that represents the left-hand side of the equation. And the fact that a quadratic equation can have no, one or two solutions also coincides splendidly with the statement that a second-degree polynomial can have no, one or two zeros.

So this would bridge the gap between solving equations and dealing with functions; and what for? Well, for example to better understand what follows, namely the so-called **factoring of polynomials**. Again, I start with an example and examine the polynomial

$$p(x) = x^2 + 2x - 3$$

to zeros. So that means that I have to solve the equation

$$x^2 + 2x - 3 = 0$$

which, however, is no problem with the help of the *p–q formulas I* just practiced sufficiently: There are two different zeros, namely $x_1 = 1$ and $x_2 = -3$. Now I will do something quite audacious at first and define a new function, let us call it $q(x)$, by subtracting these two numbers from the variable x, respectively, and then multiplying these two factors; you have not quite understood that yet, have you? Do not worry, it is just my fuzzy expressions, here a formula is actually worth a thousand words. So I define the function

$$q(x) = (x - x_1)(x - x_2),$$

hence

$$q(x) = (x - 1)(x + 3).$$

Obviously $q(x)$ has the two zeros $x_1 = 1$ and $x_2 = -3$, because if I insert these two numbers, one of the two factors becomes zero and so does the whole function. Since I calculated x_1 and x_2 as zeros of p, this means that p and q have the *same* zeros. And here it comes: With this, they are already completely identical.

In the present example, you can recalculate this by multiplying out $(x - 1)(x + 3)$: You should then get back the term of p formulated at the beginning. Just do this once, and then read the following text box, which formulates in general what has just been explained by way of example.

Decomposition of Second Degree Polynomials into Linear Factors
If a second degree polynomial $p(x) = a\,x^2 + b\,x + c$ has two real zeros x_1 and x_2, it has the representation

$$p(x) = a(x - x_1)(x - x_2).$$

(continued)

This is also true if these two zeros are identical, so the radicand in the *p-q formula* is zero.

This is called the **decomposition of second degree polynomials into linear factors**.

This not only calls for, it literally screams for at least one more example. For this I examine the polynomial

$$p(x) = 2x^2 + 32x - 34$$

to zeros, so solve the equation

$$2x^2 + 32x - 34 = 0.$$

According to my recommendation at the beginning to put every equation into normal form, I first divide this equation by the coefficient of x^2, i.e. 2. This results in

$$x^2 + 16x - 17 = 0,$$

and the solution formulas, which I had to strain now already several times, deliver here the zeros

$$x_1 = -8 + \sqrt{64 + 17} = -8 + 9 = 1 \quad \text{and} \quad x_2 = -8 - \sqrt{64 + 17} = -8 - 9 = -17.$$

Thus

$$x^2 + 16x - 17 = (x - 1)(x + 17).$$

Since I had divided by 2 to apply the zero formula, I now have to multiply by 2 again to get the initial polynomial back. So it is valid

$$p(x) = 2x^2 + 32x - 34 = 2(x - 1)(x + 17).$$

Once again, I challenge you to check the correctness of this equation by multiplying out the right-hand side.

This decomposition into linear factors does not only serve the nicer representation of the polynomial (which would be a matter of taste anyway), but has for example quite practical use in the simplified representation of rational functions. I will show you this best with an example: The function

$$r(x) = \frac{x^2 - 6x + 5}{x^2 + 4x - 45}$$

does not seem like it could be truncated in any way; after all, there is the old saying "only the stupid truncate differences and sums", so just keep your hands off the x^2 term!

It is much better to take the path practiced above and decompose both numerator and denominator into linear factors. I think I can now slowly dispense with the detailed presentation of the solution formulas and give the decomposition directly: The numerator has zeros 1 and 5, so is

$$x^2 - 6x + 5 = (x - 1)(x - 5),$$

and the denominator has the zeros 5 and -9, therefore

$$x^2 + 4x - 45 = (x - 5)(x + 9).$$

Altogether we have found out that the function $r(x)$ can also be written as

$$r(x) = \frac{(x - 1)(x - 5)}{(x - 5)(x + 9)},$$

and this literally cries out to truncate the factor $(x - 5)$: you get

$$r(x) = \frac{x - 1}{x + 9}.$$

Exercise 3.5 Truncate the following rational functions as much as possible:

(a) $r_1(x) = \frac{x^3 + 7x^2 - 60x}{3x^2 - 27x + 60}$

(b) $r_2(x) = \frac{x^2 - 4}{2x - 4}$?

3.3 Higher Order Polynomial Equations

In the course so far, I had always tacitly assumed that one already knew whether the equation presented was a linear or a quadratic equation, and one of the two cases occurred in any case, however puzzling the equation may have looked at first; just think of the equation in (3.12), which turned out to be a quadratic one.

But in real life – as far as it has to do with mathematics – this is unfortunately very seldom the case: Here you often come across equations which, if you are lucky, only contain powers of x, but whether it is a linear, a quadratic or an equation of higher degree, unfortunately no one tells you, and you often cannot easily tell.

Nothing helps: First of all, you have to multiply out any bracket expressions, make any denominators disappear by expanding them, and then sort the equation by *x powers to* convert it to normal form.

$$a_n x^n + a_{n-1} x^{n-1} + \cdots + a_1 x + a_0 = 0$$

Since the left-hand side has the form of a polynomial, such equations are also called **polynomial equations**. If you are very lucky, all higher powers of x disappear, so that at most x^2 remains, and you can apply one of the solution methods practiced above. If not, then one can only hope that one of the special cases to be presented in a moment is present, because polynomial equations whose degree is greater than two are in general practically impossible to solve. For degree three and degree four there are still solution formulas, but they are so complicated that I actually do not know any living person who has ever used them. If you are nevertheless interested in what, say, those for degree three look like, you can find them in the literature or on the Internet under the name *Cardanian formulas*; they are named after and presumably found by Geronimo Cardano, who also invented the Cardan joint named after him.

If the degree of the equation is five or greater, no solution formulas exist, not because mathematicians are too clumsy to find them, but because the Norwegian mathematician Niels Henrik Abel proved that no such formula *can* exist. Please do not ask me how he proved it, bad enough that he succeeded.

But I ha've digressed too much now, I wanted to first show you how to put an equation into polynomial form using two examples. To warm up I take the equation

$$(x^2 + x - 2)(x^2 - 2) - 2 = x^3(x + 1) - 3(x^2 - 1)$$

before. In this form, one can certainly not recognize any solutions, but no one has claimed that. After multiplying out, things look more friendly, we get

$$x^4 + x^3 - 4x^2 - 2x + 2 = x^4 + x^3 - 3x^2 + 3,$$

and if we now sort by *x-potencies,* we see that both the x^4- and the x^3-term lift away and a simple quadratic equation remains, namely

$$x^2 + 2x + 1 = 0,$$

which has the only solution $x = -1$. Consequently, the initial equation also has only this one solution.

As a second example I look at the equation

$$\frac{x^2+2}{x+5}+\frac{1}{x^3+x+1}=\frac{x}{x+1}$$

an. Here one will have to multiply by the main denominator, i.e. the product of the three denominators; this yields

$$(x^2+2)(x^3+x+1)(x+1)+(x+5)(x+1)=x(x+5)(x^3+x+1),$$

and if you multiply this out and sort by *x-powers*, you get

$$x^6-2x^4+3x^3-2x^2+5x+7=0.$$

You do not see how it should go on here? To be honest, I do not either. Nobody said that every equation turns into a quadratic or a linear one if you transform it long enough, and here we have the case that an equation of the sixth degree comes out, which is inaccessible to simple solution methods. That is life, at least sometimes.

Exercise 3.6 Determine all solutions of the equation

$$\frac{-2(1+x)}{x^2+1}+x^2=x^2-1,$$

by first turning it into as simple a polynomial equation as possible.

I had already indicated above that in special situations you can also find the solution of polynomial equations of higher than second degree, although the solution formulas for equations of third and fourth degree are extremely complicated and there are no general formulas for equations of higher degree. At the end of this section, I will now present you with examples of some such situations, in order to sharpen your eye if you should ever encounter something like this in an exercise or even an exam problem.

The first case has actually less to do with mathematical considerations, but with the fact that also the authors of exercise or exam problems know on the one hand about the non-existence of higher solution formulas and on the other hand are mostly friendly people. This has the effect that one can usually see a solution with the naked eye, i.e. *guess it,* when equations of the third (or even higher) degree are presented.

For example, for the equation

$$x^3-x^2+x-1=0$$

actually the solution $x=1$ comes to mind, because 1–1 + 1–1 = 0.

For the second example

$$x^3 - 2x^2 + x - 2 = 0$$

one finds relatively quickly that $x = 2$ is a solution. Just as with the decomposition into linear factors shown in Sect. 3.2, one can also split off a linear factor in this case, namely $(x - 2)$, and one obtains

$$x^3 - 2x^2 + x - 2 = (x - 2)\left(x^2 + 1\right).$$

Since $x^2 + 1$ cannot become zero, the equation has

$$x^3 - 2x^2 + x - 2 = 0$$

except 2 no solution.

I now give you a well-meant piece of advice motivated by many years of experience: If you have to solve an equation of third or higher degree in an exercise or exam, first test the values $x = \pm 1$ and $x = \pm 2$ as well as, if that has not yet led to the goal, the integer divisors of the absolute member (i.e. the coefficient not associated with x) of the equation.

Most of the time, the solution you are looking for can be found below. Mind you: This is in no way a mathematical theorem or anything like that, but falls into the category of exam writer psychology; but that can be quite useful sometimes.

The next special case, which I want to present to you, is now again of a strictly mathematically provable nature. As a first example I show you the equation

$$23x^7 + 77x^4 - 17x^2 + 3x = 0.$$

Already by the not quite everyday coefficients one sees that it will not depend on these at all. Rather it is decisive with this equation that it has no absolute member a_0 and/or that this is zero. This means that every occurring term is multiplied by x and thus $x = 0$ is a solution of the equation.

I do not think I need to back that up with any more examples, rather I want to state it as a general rule:

Polynomial Equations Without Absolute Element
If a polynomial equation does not contain an absolute element, then $x = 0$ is always a solution of this equation.

However, an equation without an absolute element has another advantage closely related to the one just mentioned: You can factor out x and thereby reduce the degree of the remaining equation. If this is then two or one, one can apply the solution formulas presented above.

A first simple example: The equation

$$x^3 - 9x = 0$$

has $x = 0$ as solution and therefore one can also factor out x; the result is

$$x(x^2 - 9) = 0.$$

However, the remaining factor $x^2–9$ is now a quadratic polynomial, and the corresponding equation is

$$x^2 - 9 = 0$$

has the two solutions $x_1 = -3$ and $x_2 = 3$, which can be recognized using either the *p–q formula* or the third binomial formula.

A second example is the fourth degree equation

$$x^4 + 4x^3 + 3x^2 = 0.$$

Here you can even factor out x^2 (strictly speaking, according to the current state of affairs, I would have to factor out x first and then be "amazed" to find that I can immediately factor out another x, but that would be silly) and you get

$$x^2(x^2 + 4x + 3) = 0.$$

The remaining quadratic equation has, as you can quickly calculate, the two solutions $x_1 = -1$ and $x_2 = -3$, with which we would have found all solutions in this example as well. I think more examples are unnecessary here.

Finally, there is a special situation in which an equation of higher, in this case fourth, degree is amenable to a fairly simple solution possibility; I am talking about so-called biquadratic equations, which are equations of fourth degree in which the odd powers of x are missing:

Biquadratic Equation
An equation of the form

$$ax^4 + bx^2 + c = 0 \tag{3.13}$$

is called a **biquadratic equation**.

To get closer to the solution of such an equation, I first perform the *substitution*

$$u = x^2$$

So I forget, so to speak, for the moment that the expression x^2 arises by squaring a certain variable, and first of all I simply take it as a fixed quantity, and to document this I rename it, in this case to u. If I now replace x^2 by u *in* the biquadratic equation, I get the new equation

$$au^2 + bu + c = 0, \tag{3.14}$$

since $x^4 = (x^2)^2 = u^2$.

But this new equation is a very simple quadratic equation, for the solution of which we have seen a handy formula above. If we have now calculated these solutions – assuming existence – and if they are non-negative numbers, we can again take the root from them and obtain the desired solutions of the biquadratic equation (3.13). To be more precise, both the positive and the negative root of u represent a solution of (3.13), because the following applies

$$\left(\pm\sqrt{u}\right)^2 = \left(\sqrt{u}\right)^2 = u = x^2.$$

Since there are at most two different solutions u_1 and u_2 of the Eq. (3.14), and from each of these again up to two different roots are to be drawn, in this way one obtains up to four solutions of the biquadratic equation, and since this is, after all, an equation of the fourth degree, this is just as well.

I know it is high time for an example, but I want to write this down again first in the form of a memorable rule; if you do not think this is necessary, feel free to skip the box below and go straight to the examples.

Solution of Biquadratic Equations
To find the solutions of the biquadratic equation

$$ax^4 + bx^2 + c = 0$$

one performs the substitution $u = x^2$ and calculates the solutions u_1 and u_2 of the quadratic equation

$$au^2 + bu + c = 0$$

in the usual way. If u_1 is not negative, one calculates

$$x_{11} = \sqrt{u_1} \text{ and } x_{12} = -\sqrt{u_1},$$

and if u_2 is not negative, then one also calculates

$$x_{21} = \sqrt{u_2} \text{ and } x_{22} = -\sqrt{u_2}.$$

The numbers x_{11}, x_{12}, x_{21} and x_{22} calculated in this way are then solutions of the biquadratic equation.

All right, all right, examples are coming. First, let us consider the equation

$$x^4 - 6x^2 + 8 = 0.$$

Again it happened, I got into the typical mathematician slang: Considering alone is not enough, of course; I actually want to solve the equation. To do this, I substitute $u = x^2$ and solve the resulting equation

$$u^2 - 6u + 8 = 0$$

using the solution formulas from Sect. 3.2. This gives

$$u_1 = 2 \quad \text{and} \quad u_2 = 4.$$

Both are positive numbers, so I can blithely draw roots and get the following four solutions:

$$x_{11} = \sqrt{2}, \quad x_{12} = -\sqrt{2}, x_{21} = 2, \quad x_{22} = -2.$$

Feel free to distrust me again and substitute these four values into the original biquadratic equation.

As a second example I take the equation

$$-2x^4 + 16x^2 - 32 = 0$$

First I have to bring this into normal form by dividing by −2; this yields the equation

$$x^4 - 8x^2 + 16 = 0,$$

which is now due to the already familiar substitution in

$$u^2 - 8u + 16 = 0$$

If one applies the solution formulae to this, one obtains

$$u_1 = u_2 = 4.$$

So this equation has only *one* solution, but that does not really matter, after all, this one is positive, so you can take the square root of it. The final result is the two solutions

$$x_{11} = 2 \quad \text{and} \quad x_{12} = -2$$

of the initial equation, which you may of course optionally denote by x_{21} and x_{22}.

As a third and – let this be said for your comfort and encouragement – penultimate example, I give the equation

$$x^4 - 8x^2 - 9 = 0.$$

Substitution and solution of the resulting quadratic equation leads to

$$u_1 = -1 \quad \text{and} \quad u_2 = 9.$$

Now caution is required, because u_1 is negative, and in everything that was said above, it was always stated as a precondition that the number in question must not be negative. But I am not allowed to draw roots from negative numbers and therefore no x-solutions of the initial equation can be obtained from this $u._1$.

The situation is different with u_2, which is positive, and thus we have the result that the initial equation has only two solutions, viz.

$$x_{21} = 3 \quad \text{and} \quad x_{22} = -3.$$

In order to confirm you in your presumably existing prejudice that in mathematics also everything can go wrong, I examine at the end the equation

$$x^4 + 2x^2 + 4 = 0.$$

Substitution here leads to the equation

$$u^2 + 2u + 4 = 0,$$

and the solution formula provides

$$u_{1/2} = -1 \pm \sqrt{1 - 4}.$$

So the substituted equation does not have a single solution from which one could possibly draw roots and thus the initial equation does not have a real solution.

Exercise 3.7 Determine all the solutions of the following biquadratic equations:

(a) $x^4 + x^2 + 1 = 0$
(b) $2\,x^4 + 32\,x^2 – 450 = 0$
(c) $x^4 – 13\,x^2 + 36 = 0$

3.4 Root and Exponential Equations

In mathematics, as in every scientific discipline, there is a plethora of initially incomprehensible and sometimes even misleading terms; unfortunately, this is undeniable. Sometimes, however, mathematical life also writes beautiful stories and a way of designation is easily understandable and self-explanatory; so it is in the case of root equations, which I will now bring closer to you:

Root Equation
An equation where the variable is under a root is called a **root equation**.

A first example is the equation

$$\sqrt{x-1}=3.$$

To solve this, i.e. to find an x for which both sides of the equation have the same value, I have to get rid of the root somehow. To do this, it makes sense to square both sides of the equation, i.e. to raise them to the power of 2; this results in

$$x-1=9,$$

so a linear equation whose solution $x=10$ you can immediately recognize. If you substitute this into the left side of the root equation, you get

$$\sqrt{10-1}=\sqrt{9}=3,$$

thus the value of the right-hand side, and thus $x=10$ is confirmed as the solution of the root equation.

In the case of root equations, this substitution into the initial equation is by no means just a test to rule out the possibility that you have miscalculated, but an indispensable *part of the solution process*. You will see why this is the case in a moment:

To solve the equation

$$\sqrt{x-1}=-1$$

I again square both sides and obtain the linear equation

$$x-1=1$$

with the solution $x=2$. But if I now insert this into the initial equation, I get on the left side

$$\sqrt{2-1}=\sqrt{1}=1,$$

which unfortunately does not correspond to the right side -1.

Messy business, the whole thing: solving the transformed equation led us to a number that, unfortunately, *does not* solve the initial equation.

This often happens with root equations; as in the case of fraction equations, we speak of **bogus solutions**, and in order to exclude such solutions, one must always insert the "solutions" obtained by the transformation process into the initial equation for checking purposes.

Do not worry, numerous examples are already waiting, but you know by now that I like to summarize the most important things first in a little mnemonic box:

Solution of Root Equations
To solve a root equation, you square or exponentiate it – possibly several times – to transform it into a polynomial equation.

Then, if possible, solve the resulting polynomial equation and then substitute the solutions of the polynomial equation into the initial equation (root equation) to exclude bogus solutions.

What can happen with root equations, I will show you now, as usual, with some examples.

First, I solve the root equation

$$\sqrt{17 - x} = \sqrt{2x + 14}. \tag{3.15}$$

The recommended squaring of both sides leads to

$$17 - x = 2x + 14.$$

Now this is undoubtedly a polynomial equation, and a fairly easy one to solve, because it is linear. Sorting it by x-terms and constants, we get

$$-3x = -3,$$

so $x = 1$.

However, this is first of all only a candidate for the solution of the initial equation. In order to verify that this is indeed a solution and not just a bogus solution, I have to insert $x = 1$ into (3.15); this results in the value $\sqrt{16} = 4$ on both sides, and thus I have actually found a solution with $x = 1$.

You probably guessed it, here comes an example where not everything goes smoothly; I will try my hand at the equation

$$\sqrt{2x - 3} = \sqrt{x^2 - 2x}. \tag{3.16}$$

Again, squaring both sides leads directly to a polynomial equation, in this case a quadratic one:

$$2x - 3 = x^2 - 2x,$$

so

$$x^2 - 4x + 3 = 0.$$

To solve this, I once again use the *p–q formula* and get

$$x_{1/2} = 2 \pm \sqrt{4-3} = 2 \pm 1,$$

hence

$$x_1 = 3 \quad \text{and} \quad x_2 = 1.$$

If we now insert $x_1 = 3$ into the initial equation, we get the value $\sqrt{3}$ on both sides, so everything is fine and $x_1 = 3$ is a solution of the equation.

But if we insert $x_2 = 1$, we get a negative radicand on the left side, namely -1, and since the root of a negative number is not defined, neither is the left side of this equation, and thus $x_2 = 1$ is only a spurious solution.

The third example is the root equation

$$\sqrt{x - \sqrt{x-2}} = \sqrt{2}. \tag{3.17}$$

Again, I start in the now familiar way by squaring both sides and get

$$x - \sqrt{x-2} = 2.$$

So here it is no longer done with simple squaring, because there is still a root expression standing around in the way, which I have to get rid of. But if I square the equation as it is here, I have to apply the binomial formula on the left side and there is still a root expression standing around even after squaring (try it out!). It is better to first *isolate* the term $\sqrt{x-2}$ on one side of the equation; I do this by adding $\sqrt{x-2}$ on both sides, which gives me

$$x = 2 + \sqrt{x-2}$$

and then subtract 2. The result is

$$x - 2 = \sqrt{x-2}.$$

Now nothing stands in the way of successful squaring, and this leads to the quadratic equation

$$x^2 - 4x + 4 = x - 2$$

or in normal form

$$x^2 - 5x + 6 = 0.$$

I really do not want to bore you with the *p–q-formulas* now, I just apply them silently and secretly and get as solutions of the quadratic equation the numbers

$$x_1 = 2 \quad \text{and} \quad x_2 = 3.$$

But I am *not done with* that yet, because I still have to check that they are not bogus solutions by substituting in (3.17).

If I insert $x_1 = 2$ on the left-hand side, the inner root becomes zero, and the outer root remains, applied to $x_1 = 2$, so the value of the left-hand side is equal to $\sqrt{2}$, which is firstly defined and secondly consistent with the right-hand side. Substituting the second candidate, $x_2 = 3$, into the left-hand side, I now perform somewhat formulaically for once: it becomes

$$\sqrt{3 - \sqrt{3 - 2}} = \sqrt{3 - 1} = \sqrt{2}.$$

So also x_2 is a real solution of this root equation, this one has no bogus solutions.

As a final example in the context of root equations, let us look at the following:

$$\sqrt{x + 1} = \sqrt{x} + 3 \tag{3.18}$$

Again, I recommend squaring first to get rid of at least one of the troublesome roots; this results in

$$x + 1 = x + 6\sqrt{x} + 9.$$

(If you have a different result on the right side at this point, look again under "binomial formulas"!) Here, of course, it suggests itself to subtract x on both sides, whereby this term simply disappears and the residual equation

$$-6\sqrt{x} = 8$$

remains. Squaring again gives

$$36x = 64,$$

so $x = 16/9$. Again, I must not forget to check this candidate by substituting it into the initial equation; by the way, it would be better to check candidates more closely from time to time in politics as well, but that is just by the way.

Putting 16/9 in the left-hand side of (3.18), I get

$$\sqrt{\frac{16}{9} + 1} = \sqrt{\frac{25}{9}} = \frac{5}{3},$$

still a smooth value. But on the right side results

$$\sqrt{\frac{16}{9}} + 3 = \frac{4}{3} + 3 = \frac{13}{3},$$

and this is not identical with 5/3, even if algebraic rules are interpreted in a generous way. So I have only determined an apparent solution, the initial Eq. (3.18) has no solution at all.

Perhaps you have been thinking all along that this is all well and good, but that these are purely inner-mathematical brainteasers that have no relation to real life. But that is wrong, as I will now briefly show:

Suppose you are standing on a high cliff or a lighthouse looking out to sea. At what distance can you still see an object on the surface of the water (assuming better eyes than mine), i.e. at what distance does the surface of the water disappear behind the horizon?

Contrary to everything I set out to do when writing a textbook, I will slap you with the formula that solves this one first and say something explanatory about it afterwards. It is

$$l = \sqrt{12,740,000 \cdot h + h^2}.$$

Here l is the distance asked and h *is* the height at which you (or more precisely, your eyes) are above sea level. The strange coefficient 12,740,000 is approximately the diameter of the earth, also given here in meters.

To prove or derive this formula, we would have to use our colleague Pythagoras and his theorem; you probably already know it, but since it has not yet appeared in this book, I do not want to strain it here and ask you to accept the formula as it is. With its help you can calculate, for example, that from a 20-m-high cliff you will fall about

$$\sqrt{12,740,000 \cdot 20 + 20^2} = 15,962.47$$

meters, or almost 16 km away.

Very pretty, but at least as interesting is the reverse question: How high do you have to climb, for example on a lookout tower, to be able to see a desired distance l, say $l = 20$ km, far away?

To answer this question, you need to solve the equation

$$20,000 = \sqrt{12,740,000 \cdot h + h^2}$$

to h, that is, solve a root equation!

I will do this for you, after all I am the author here. To do this, of course, I square both sides of the equation again to get rid of the unloved root, and get the quadratic equation

$$400,000,000 = 12,740,000 \cdot h + h^2,$$

whose solutions I calculate using the *p–q formula*; they are

$$h_1 = 31.40 \quad \text{and} \quad h_2 = -12,740,031.40.$$

Since I do not suppose you want to dig through the earth, the only solution is h_1; so you have to climb about 31.40 m to see 20 km far.

Exercise 3.8 Determine all the solutions of the following root equations:

(a) $\sqrt{x+4} = \sqrt{x^2+x}$
(b) $\sqrt{x} + \sqrt{5-x} = \sqrt{2x+7}$
(c) $\sqrt{x - \sqrt{x+2}} = 2$

I now come to the second part of this chapter, the exponential equations; here the naming is as nice and easy as with the root equations, it is more or less self-explanatory.

Exponential Equation
An equation where the variable occurs in the exponent is called an **exponential equation.**

The simplest case of an exponential equation is the equation of the form

$$a^x = b,$$

where a and b are supposed to be positive real numbers and a is also not equal to 1.

Such an equation can be solved directly by applying the logarithm: it is

$$x = \log_a(b).$$

If you are not quite so used to dealing with the logarithm, then take another look in the second chapter. At the latest afterwards you will agree with me that the solution of the exponential equation

$$3^x = 9$$

$x = 2$, because $\log_3(9) = 2$, and also that the exponential equation

$$10^x = 117$$

is solved by $x = \log_{10}(117) = 2.06818586\ldots$

As an aside, I would just like to note that there is no need to bother with the logarithm if the right-hand side of the equation, i.e. b, is itself a power of a. For example, one reads at the exponential equation

$$17^{x+1} = 17^5$$

immediately deduces that $x + 1 = 5$, so x must be equal to 4.

I do not think that I should make further examples to this simple case, rather we can, as you are already used to, successively increase the degree of difficulty of the examples. Should you ever come across an exponential equation in the course of your studies or in any other contact with mathematics, you will certainly find a suitable example among the following ones that you can use as a guide.

The first, still tiny difficulty occurs if the left side a^x is still afflicted with a prefactor, so the equation is for example

$$2 \cdot 5^x = 50$$

But you can see immediately that in such a case you only have to divide by the prefactor, which in this case leads to the equation $5^x = 25$, whose solution $x = 2$ again immediately jumps out at you.

Furthermore, it can happen that the variable x occurs in the exponent of several summands. If you are lucky, and I want to assume that for now (luckily I can control luck here), then it is the same base (i.e. the same a). An example for such an equation is

$$2 \cdot 3^x + 3^{x+1} = 135. \tag{3.19}$$

In this case, it pays to have paid attention to power arithmetic; you then know that $3^{x+1} = 3 \cdot 3^x$, and so I can also write the equation as

$$2 \cdot 3^x + 3 \cdot 3^x = 135$$

or

$$5 \cdot 3^x = 135.$$

But now we are in the fairway, because we have already studied this form of an exponential equation above. You now divide both sides by 5, which comes to

$$3^x = 27$$

and reads off the solution $x = 3$ from this. Put this into the initial Eq. (3.19) for a test, you will see that $x = 3$ actually makes this a true statement.

The general principle, which you might already have recognized from this example, is the following: You look for the smallest of all the addends that contain x in the exponent (here that is 3^x), and put the others (here that is just 3^{x+1}) into the same form by decomposing them using the power rules.

Another example is the exponential equation

$$2^{x+1} - 3 \cdot 2^x + 5 \cdot 2^{x-1} = 48.$$

Here the smallest of the summands afflicted with x is 2^{x-1}, so I decompose the other two as follows:

$$4 \cdot 2^{x-1} - 3 \cdot 2 \cdot 2^{x-1} + 5 \cdot 2^{x-1} = 48.$$

Combining the left side results in

$$3 \cdot 2^{x-1} = 48,$$

or

$$2^{x-1} = 16.$$

It follows that $x - 1 = 4$ and finally that $x = 5$.

Unfortunately, to pick up the above, in life you are not always lucky, and it may happen that x occurs in the exponent of *different* bases. In this case, nothing helps, you have to logarithmize. Let us look at a typical example, namely the exponential equation

$$2^{x-1} = 3^{x+1}.$$

I now apply the logarithm on both sides and use – this is the crucial trick – the calculation rule $\log(a^b) = b \cdot \log(a)$. It follows

$$(x-1) \cdot \log(2) = (x+1) \cdot \log(3).$$

Suddenly, the unloved logarithm has become a simple coefficient of x. Now I multiply both sides out and sort by x-terms and constants; this yields

$$x(\log(2) - \log(3)) = (\log(2) + \log(3)).$$

So it is

$$x = \frac{\log(2) + \log(3)}{\log(2) - \log(3)} = \frac{\log(6)}{\log\left(\frac{2}{3}\right)} = -4{,}419,$$

where I again made use of a logarithm rule, in this case $\log(a) + \log(b) = \log(a\,b)$.

You may have noticed that I have not even said *which* logarithm I am using. The reason is not so much my advancing senility, but rather the fact that it does not *matter at all* which logarithm you use, all that matters is that it is the same on both sides of the equation. For purely practical reasons, I recommend that you always use a logarithm that is also implemented on the calculator, and usually this is the one for the base 10. This is also how I calculated the above solution; you can check the result using a different logarithm.

I have already shown you the decisive trick for solving exponential equations with different bases by means of this example, and there is actually not much more to say here; except perhaps for the fact that you cannot use this trick directly if there is still a sum on at least one side of the equation. In such a case, you always have to achieve by transformations, mostly factoring out, that only products and powers remain.

An example is the equation

$$3^x + 3^{x+1} = 5^{x-1}.$$

Here I must first transform the left side by factoring out as follows:

$$3^x + 3^{x+1} = 3^x + 3 \cdot 3^x = 4 \cdot 3^x.$$

This again inserted into the equation gives

$$4 \cdot 3^x = 5^{x-1},$$

logarithmizes this into

$$\log(4) + x \cdot \log(3) = (x - 1) \cdot \log(5),$$

and after some tidying up, this equation is

$$x \cdot (\log(3) - \log(5)) = -\log(4) - \log(5),$$

hence

$$x = \frac{-\log(4) - \log(5)}{\log(3) - \log(5)} = 5.865.$$

Just as with the root equations, it would be wrong to assume that the exponential equations have no practical relevance; the reference can even be very close to everyday life, at least when it comes to money. In the context of logarithms, this has already been presented once; I will briefly take it up again here in the language of equations.

In the second chapter it was shown that a capital K invested at the rate of interest of p per cent per annum will, after a period of x years, has grown to

$$K_x = K \cdot \left(1 + \frac{p}{100}\right)^x \tag{3.20}$$

I have deliberately not designated the exponent as n here, but as x, *in order* to make it clear that this does not necessarily have to be an integer, because banks also use this formula if the capital is withdrawn on any day of the year.

Very practically relevant, you will admit, is the following question in this context: How long must a capital K *be* invested until it has grown to a given size K_x? In this situation, then, K, K_x and, of course, p are given, and x is sought. Thus the Eq. (3.20) becomes an exponential equation of purest water.

Probably once again a good example is the tool of choice: Suppose you had 12,000 Euros at your disposal and your bank offered you to pay interest on this sum at 3% per year. How long do you have to leave the money to get a payout sum of 15,000 Euros?

The exponential Eq. (3.20) in this case is:

$$15,000 = 12,000 \cdot 1.03^x.$$

Dividing by 12,000 and then truncating turns this equation into

$$\frac{5}{4} = 1.03^x$$

and subsequent logarithmizing yields

$$x = \log_{1.03}\left(\frac{5}{4}\right) \approx 7.55. \tag{3.21}$$

So you need to fix the money for about seven and a half years. That is exactly how the banks calculate it. By the way, if you look in vain on your calculator for the key "logarithm to the base 1.03" to recalculate this, please remember the conversion formula

$$\log_a(x) = \frac{\log_b(x)}{\log_b(a)}.$$

You will already have some logarithm on your calculator, probably the one to base 10, and with its help you can calculate the number x in (3.21) as follows:

$$x = \log_{1.03}\left(\frac{5}{4}\right) = \frac{\log_{10}\left(\frac{5}{4}\right)}{\log_{10}(1.03)}.$$

With this, I leave – hopefully still together with you – the realm of equations and turn at least briefly to inequalities. Before that, however, I would like to encourage you to think for yourself a little:

Exercise 3.9 Determine all the solutions of the following exponential equations:

(a) $7^x = 17^x$
(b) $2^{x+1} + 16 - 12 \cdot 2^{x-1} = 0$
(c) $2^{x+1} + 2^{x+2} = 2 \cdot 3^{x-1}$ ⍰

Exercise 3.10 How long do you have to invest a capital of 1 Euro at 4% interest to get paid 1 million Euros?

3.5 Inequalities

Inequalities are so called because they are *incomparably* more difficult to deal with than equations.

No, do not keep scrolling, I was just kidding! In fact, inequalities have this name because they consist of two terms that are not connected by an equal sign like equations, but by an inequality sign. So an **inequality** has the form

$$a < b,$$

spoken "*a* is smaller than *b*", which should mean that *a is* a smaller number than *b*, or

$$a > b,$$

spoken "*a* is greater than *b*" in the reverse case. Since by interchanging the two sides of an inequality one can always transform a "less than" inequality into a "greater than" inequality and vice versa, it is sufficient to deal with one of the two cases; in the following I will confine myself to the former case.

Besides the two sharp inequality signs just introduced, there are also the signs $\leq$ and $\geq$, pronounced "less than or equal to" and "greater than or equal to" respectively, which you surely already know, and which also allow the two terms connected by them to be equal. Since the treatment of such inequalities is identical to that of the sharp inequalities, I will limit myself to the case of the real inequality in this respect as well.

As in the case of equations, one usually looks at inequalities that contain a variable and tries to *solve* them:

Inequality
An inequality consists of two mathematical expressions (terms) connected by an inequality sign.

If at least one of the expressions contains a variable, one looks for **solutions to the inequality**. These are numbers which, when substituted for the variables, cause the values of the two terms connected by the inequality sign to be related according to it. The set of all solutions, which can also be empty, is called the **solution set of** the inequality and symbolized by $\mathbb{L}$.

Unlike equations, an inequality usually has an infinite number of numbers as a solution, which is why from now on I will make more use of writing it as a solution set.

A very first example should at the same time point out a little trickiness to you: The inequality

$$x + 1 < 2$$

is solved – no solution method is needed yet – by all numbers x that are smaller than 1, so the solution set of this inequality is

$$\mathbb{L} = \{x \in \mathbb{R} | x < 1\}.$$

Now the pitfall I just wrote about is that this does not at all mean that the elements of $\mathbb{L}$ have to be *smaller in* amount than 1, they just have to lie somewhere to the left of one on the number line. In other words: Not only do 1/2, 0, −3/4, for example, lie in $\mathbb{L}$, but also −2 and −1,000,000. And if you are thinking right now, "What's the point, it's obvious!", please believe me that this is very often done wrong in the heat of the moment, as obvious as it may be.

But now to the solution procedures for inequalities; here, too, there is a great deal of agreement with the procedure for equations, for here, too, one subjects the inequality to be solved to certain permitted transformations:

Transformation of Inequalities

The following transformations of an inequality do not change its solution set and are called **allowed transformations:**

- Addition or subtraction of the same number on both sides of the inequality
- Multiplication of both sides with the same *positive* number

Have you noticed the difference to equations? I have printed it in italics: In the case of inequalities, it is no longer sufficient to require that the number by which the multiplication is carried out is different from zero, it must also not be negative!

But what if you are forced to multiply by a negative number? Do not worry, here comes a remedy:

Multiplication of an Inequality by a Negative Number

Multiplying both sides of an inequality by the same negative number does not change its solution set, so it is allowed if you simultaneously invert the inequality sign connecting the two sides, that is, change "<" to ">" and vice versa.

This is by no means as crazy as it may sound at first: for example, the set of all numbers x for which $x > 1$ holds is identical to the set of all numbers for which $-x < -1$ holds; feel free to try it out with a few values.

But now finally to the presentation of the solution methods, in which I (and you) will limit myself to linear inequalities, since one can already explain everything necessary on these: To determine the solution set of the inequality

$$3x + 4 < 1 \tag{3.22}$$

I first subtract 4 on both sides, which yields

$$3x < -3,$$

and then multiply by the positive number 1/3. The solution set of the inequality (3.22) is therefore

$$\mathbb{L} = \{x \in \mathbb{R} | x < -1\}.$$

Wasn't so bad, was it? I use this positive basic mood and lead you to the somewhat more complicated inequality

$$x^2 - 2x + 3 < (x - 1)(x + 2) \tag{3.23}$$

At the beginning I had promised to limit myself to linear inequalities, and now something like this! But you probably already see that here, of course, one also proceeds as with equations and first simplifies both sides of the inequality; in this case, this results in

$$x^2 - 2x + 3 < x^2 + x - 2.$$

If I now subtract x^2 on both sides, the linear inequality remains

$$-2x + 3 < x - 2$$

which I still bring in the more manageable form

$$-3x < -5$$

Now I still have to divide by -3 or multiply by $-1/3$ and care must be taken. Since this is a negative number, I have to invert the inequality sign, as I said above, so the inequality (3.23) takes the form

$$x > \frac{5}{3}$$

their solution set is therefore

$$\mathbb{L} = \left\{x \in \mathbb{R} | x > \frac{5}{3}\right\}.$$

So in principle, everything seems to run just as smoothly as with the equations – with this little bobble in the multiplication with negative numbers. Well, that is also true, but this bobble can have quite complicated consequences, which I will now illustrate in a last instructive example, before I let you out of the world of equations and inequalities again, not without having suggested a few exercises to you.

The example mentioned is the inequality

$$\frac{x-1}{2x+1} < 1. \tag{3.24}$$

To get this into standard linear inequality form, I want to multiply by the denominator of the left-hand side. Now the question is, "Is this denominator positive or negative?" Because, after all, what happens to the inequality sign depends on that. The answer to that question is, "I haven't the faintest idea!"

Accordingly, there is no alternative but to investigate both possibilities separately and to combine the results at the end.

Case 1 $2\,x + 1 > 0$, so $x > -1/2$: In this case I can multiply through the inequality (3.24) without changing anything at the inequality sign and get

$$x - 1 \; < \; 2x + 1,$$

so the condition $-\mathrm{x} < 2$ or

$$x > -2.$$

Now, for good measure, we need to concentrate hard to understand the following: So in this first case we are studying, x must satisfy two conditions: It must be greater than $-1/2$ due to the case distinction, and it must be greater than -2 as a result of the transformation. So, in general, these are two conditions that will contribute to the formulation of the solution set, but in this example – and most of the time this will be the case – the first condition is sharper than the second (since any number greater than $-1/2$ is automatically greater than -2), so in reality there is only one condition left here, viz.

$$x \; > \; -\frac{1}{2}. \tag{3.25}$$

Case 2 $2\,x + 1 < 0$, so $x < -1/2$: Again I multiply the inequality (3.24) by, but now I have to invert the inequality sign and get

$$x - 1 > 2x + 1,$$

so $x < -2$. As in case 1, we have two conditions on x, namely $x < -1/2$ and $x < -2$, one of which – here it is the second named – automatically entails the other. So in case 2 we have the condition

$$x < -2. \tag{3.26}$$

There are no more cases, since $2\,x + 1$ must surely not be zero, so I can obtain from (3.25) and (3.26) the solution set of the inequality (3.24); it is

$$\mathbb{L} = \left\{ x \in \mathbb{R} \middle| x < -2 \text{ or } x > -\frac{1}{2} \right\}.$$

Now, you:

Exercise 3.11 Determine the solution sets of the following inequalities:

(a) $2\,x + 1 < 3\,x - 5$
(b) $2(x + 1)(x - 2) < 2\,x^2 + 5$
(c) $\frac{1-2x}{x+1} < -1$

Geometry

4

Since ancient times, people have been intensively occupied with triangles, quadrilaterals and more general figures such as circles and ellipses. The reason for this is certainly *not* that people had much more free time at that time and the invention of television was far from being thought of. Rather, such figures appeared in very many situations of daily life, and the mastery of these objects made it possible to economize with limited resources in order to make a living, which was generally meager at the time, with a halfway reasonable amount of work. In particular, the ancient Greeks developed knowledge about the length and angle relationships of such objects, called **plane figures, in order** to be able to estimate, for example, the length of a canal to be built, distances between distant places, the height of a building to be constructed, or even the area of a plot of land to be cultivated. In the meantime, mankind has developed decisively – also, and especially, because of this knowledge. Today, we need it if, for example, we want to lay a carpet efficiently, know how big the pieces of a pizza are that we share with our friends, or get to the bottom of the phenomenon of hamburgers smoldering together on the grill. All kidding aside – in this chapter, I am going to explain the basics of calculations on the **triangle** in the context of **trigonometric functions**. Here I will remind you of a few key statements that I am sure you remember from your school days. Then I will show you how to treat more general, straight-line bounded figures called **polygons** – the main trick here is to break them down into triangles and apply the things I know about triangles. Finally, I will chat with you a bit about **circles** and **ellipses**.

G. Walz et al., *Foundations of Mathematics: A Preparatory Course*,
https://doi.org/10.1007/978-3-662-67809-1_4

4.1 Triangles and Trigonometric Functions

In this section, I intend to remind you basic things about particularly simple geometric figures. These figures are **triangles**. I consider three points A, B and C *in* the plane, which do not lie on a common straight line. Then I can form a triangle with the **vertices** A, B and C by intersecting the three straight lines by two of these points each. The lines between each two vertices of triangle of lengths a, b, and are then called **sides of** triangle, while the angles α, β, and γ inside lltriangle at the vertices **are** called **interior angles of** triangle. Since two straight lines intersecting at a point enclose at most two distinct angles, these angles α, β and γ also occur outside the triangle. However, **exterior angles of** triangle are called the corresponding angles α', β', and γ', each of which occurs twice outside lltriangle. Look at Fig. 4.1 – you can see that the side opposite the vertex A has length a, and I denote the angles there by α and α'.

In Fig. 4.2 I show you different types of triangles. The classification made there distinguishes between **acute-angled**, **right-angled**, **obtuse-angled**, **isosceles** and **equilateral** triangles. In the acute triangle all three interior angles are smaller than 90° (degree measure), while for right triangles one interior angle has 90° (this is marked by a dot) and for obtuse triangles one interior angle exists with more than 90°. For isosceles triangles two sides are equal in length and equilateral triangles are a special case of isosceles triangles where even all three sides are equal in length.

Depending on the nature of a triangle at hand, as I will show you below, I can make statements about the relationship of various sizes of the triangle. For example, two of the three interior angles of an isosceles triangle agree, and for equilateral triangles, all three interior angles are even equal. I will be able to show you later quite easily why this is so. I will now start by telling you a few basic relationships that are valid for *any* triangle.

To clarify a first such relationship, look again at Fig. 4.1: In order to get from vertex A to vertex B *by* the shortest possible route, I must travel a distance of length c. Any other choice represents a detour, because the shortest connection between two points is determined by the distance between them. For example, it is a detour if I go from A to C first, and then from C to B, because then I have covered the greater distance $a + b$ *in* total. I record this observation as a **triangle inequality**:

Triangle Inequality
The sum of the length of two sides a and b *of* a triangle is always greater than the length of the remaining side c: $a + b > c$.

Next, I will look at Fig. 4.3. Here I show you a triangle together with a (dashed) straight line that is parallel to one side of the triangle and passes through the opposite corner point C. The interior angle α at point A appears here a *second time* as an **alternating angle** – this is, in fact, also the angle between this straight line and the side of the triangle connecting A and C. Similarly, the interior angle β at B. Figure 4.3 now shows you that the addition of

Fig. 4.1 The points, side lengths and interior and exterior angles of a triangle

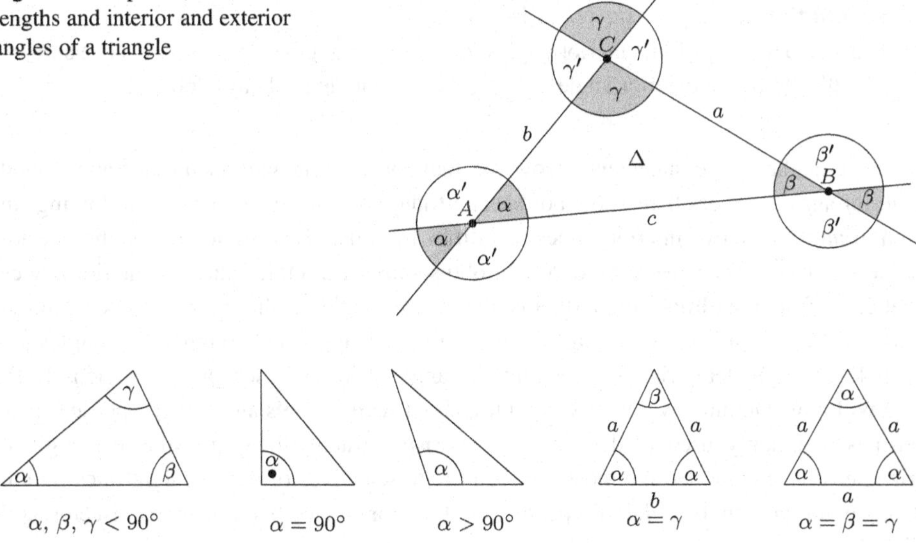

Fig. 4.2 Classes of triangles: acute-angled, right-angled, obtuse-angled, isosceles and equilateral (from left to right)

Fig. 4.3 The sum of the interior angles in the triangle is 180°

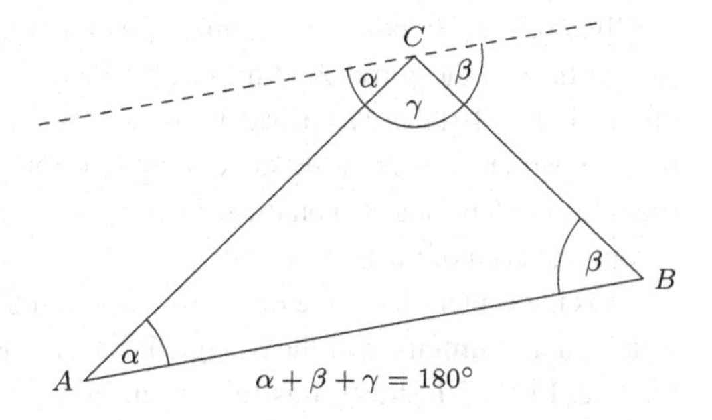

the remaining interior angle γ to the sum of these two duplicates of α and β is always equal to 180°. Notice, moreover, that the exterior angles are in relation with the interior angles:

$$\alpha' = 180^\circ - \alpha, \quad \beta' = 180^\circ - \beta \quad \text{and} \quad \gamma' = 180^\circ - \gamma,$$

so I get in addition

$$\alpha' + \beta' + \gamma' = 3 \cdot 180^\circ - \overbrace{(\alpha + \beta + \gamma)}^{180^\circ} = 360^\circ.$$

You can remember that.

Sum of the Angles on the Triangle

The sum $\alpha + \beta + \gamma$ of the interior angles α, β, γ of a triangle is always 180°. The sum $\alpha' + \beta' + \gamma'$ of the exterior angles α', β', γ' of a triangle is always 360°.

In particular, these statements show me that for any triangle, all interior angles and exterior angles are less than 180°. For special triangles, I also recognize the following: In right triangles, the two interior angles that differ from the **right angle** (that is, the interior angle that has 90°) always sum to 90°. In obtuse-angled triangles, the two interior angles that differ from the **obtuse angle** (that is, the interior angle that has more than 90°) sum to less than 90°. If an isosceles triangle is also a right triangle, the two remaining angles are both 45°. In equilateral triangles, each interior angle is 60° and each exterior angle is 120°.

I will now continue by telling you a little about **transversals** and their properties. This term is commonly used to describe characteristic straight lines that intersect a given triangle. I have compiled the most important representatives of such straight lines on the triangle for you in Fig. 4.4: Perpendicular bisector angle, bisector side, bisector and altitude.

A straight line is called the **median perpendicular** of a triangle if it passes through the center of one side of the triangle and is **perpendicular** (also: **orthogonal**) to that side, that is, the angle between the straight line and that side is 90°.

Obviously every point on a central perpendicular has the property that the distance to the two vertices of the corresponding side of the triangle is always the same – I have indicated this in Fig. 4.4 (left) by dashed lines with lengths d_1 and d_2. The intersection S of *two* median perpendiculars of the same triangle is thus *equidistant from* all three vertices of the triangle, and S *is* thus also on the remaining, third median perpendicular. Since the distance r from S to each of the three vertices of a given triangle triangle is thus equal, I can now see that S is the center of a circle of radius r through the three vertices of triangle. This circle is called the **circumcircle of** the triangle because it is the smallest circle I can put around the triangle. I have illustrated this for you in Fig. 4.5 (left). I will show you later in Example 4.15 how to explicitly state and calculate the circumcircle from three given vertices. There you will find more information about circles.

Intersection of the Central Perpendicular and the Circumcircle

The three medians of a triangle intersect at the center of the perimeter of the triangle.

Bisectors are transversals that pass through a vertex of the triangle and bisect the interior angle there. Figure 4.5 (right) motivates that the three angle bisectors of a given triangle also intersect at a point. This is the midpoint of the **incircle of** the triangle. This circle is called that way because it is the largest circle I can put *inside* the triangle.

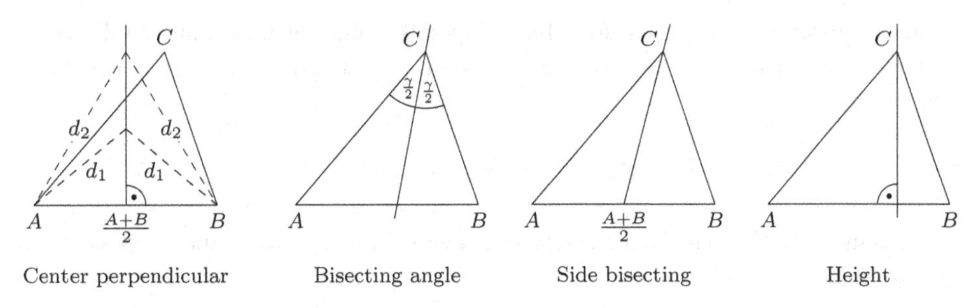

Fig. 4.4 Transversals – from left to right: Perpendiculars pass through the center of one side of the triangle and are perpendicular to this side. Angle b bisectors pass through a vertex of the triangle and bisect the interior angle there. Side bisectors pass through the midpoint of one side of the triangle and the opposite vertex. Elevations pass through a vertex of the triangle and are perpendicular to the side opposite this point

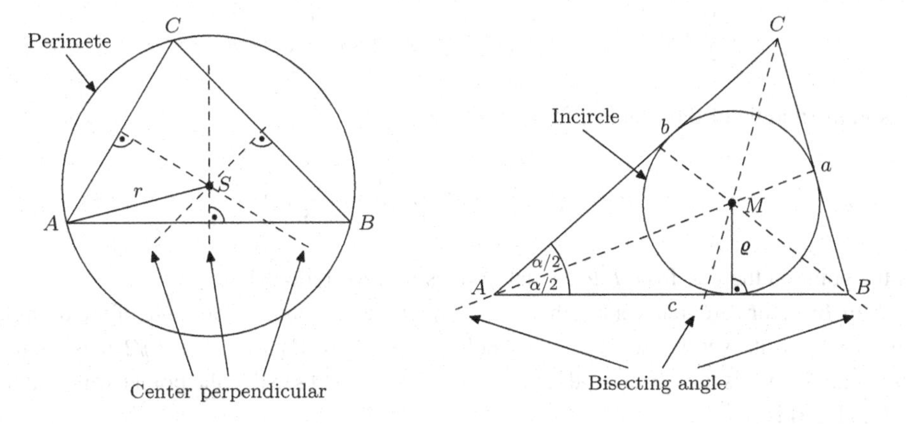

Fig. 4.5 The median perpendiculars of a triangle intersect at the center of the circumcircle (*left*), while the angle bisectors of a triangle intersect at the center of the incircle (*right*)

Intersection of the Angle Bisector and the Incircle

The three angle bisectors of a triangle intersect at the center of the incircle of the triangle.

I will tell you at this point that the radius ρ (pronounced "ro") of the **incircle** (also: **incircle radius**) of any triangle with side lengths *a*, *b*, and *c is given by*

$$\rho = \sqrt{\frac{(s-a)(s-b)(s-c)}{s}},$$

where $s = (a + b + c)/2$.

In the following I will "calculate" a bit with points in the plane. I do this as follows:

If $P = (p_1|p_2)$ and $Q = (q_1|q_2)$ are two points, I can add them by summing their components

$$P + Q = (p_1 + q_1|p_2 + q_2)$$

and I can multiply P by a real number λ (called **scalar**) by doing this for the components as well like this

$$\lambda P = (\lambda p_1|\lambda p_2).$$

These operations form the basis of **analytic geometry**. Here it is enough to recognize that the point

$$\frac{1}{2}(P + Q) = \frac{1}{2}P + \frac{1}{2}Q = \left(\frac{1}{2}(p_1 + q_1)|\frac{1}{2}(p_2 + q_2)\right)$$

lies in the middle of the distance from P to Q and

$$\frac{1}{3}P + \frac{2}{3}Q = \left(\frac{1}{3}p_1 + \frac{2}{3}q_1|\frac{1}{3}p_2 + \frac{2}{3}q_2\right)$$

is the point on the line from P to Q that divides it in the ratio 2:1.

Side bisectors are transversals that pass through the midpoint of one side of the triangle and the opposite vertex (see Fig. 4.4 (middle, right)). If $M_{AB} = (A + B)/2$ it is such a midpoint, then C is the corresponding opposite vertex, and I see that the point (weighted by 1/3 and 2/3) is

$$P = \frac{1}{3}C + \frac{2}{3}M_{AB}$$

lies on this bisector and divides the distance from C to M_{AB} in the ratio 2:1. But because of $M_{AB} = (A + B)/2$ I can also write P as follows:

$$P = \frac{1}{3}C + \frac{2}{3}(A + B)/2 = \frac{1}{3}C + \frac{1}{3}(A + B) = \frac{1}{3}(A + B + C).$$

I thus get the same point by looking at the midpoints of the remaining sides and the corresponding opposite vertices and weighting them in the same way along the two remaining side bisectors.

Intersection of the Bisectors
The three bisectors of a triangle intersect at the **centroid** $P = (A + B + C)/3$ of the triangle. This divides each distance of a vertex of triangle to the center of its opposite side in the ratio 2:1.

Finally, **altitudes** are transversals through the vertices, each perpendicular to the side of the triangle opposite that point (see Fig. 4.4 (right)). For your information, I will tell you that these also intersect at a point.

If I know the **length of a height** h, that is, the distance from the vertex belonging to a height to the intersection point on the opposite side of the triangle, and the length of this side c, I can calculate the area of this triangle. In fact, Fig. 4.6 (left) shows you that this is just half of the area of a rectangle with side lengths h and c.

Area of the Triangle
The area of a triangle F is half the product of the length of a height h with the length c *of* the side on which this height is perpendicular: $F = hc/2$.

This simple formula is valid for every triangle – especially for *obtuse* triangles. With this fact I can now immediately see that the area of a **parallelogram is** given by hc, where c is the length of a base side and h is the length of the height to this base side. This is shown in Fig. 4.6 (middle), because the grey area of the parallelogram there is twice the area of the triangle with vertices A, B, and C. In the next section I will give you some more information about parallelograms and more general quadrilaterals.

Another prominent formula for calculating the area of a triangle has the advantage that you do not have to determine a height, but can directly use the lengths of the three sides of triangle. To see how to arrive at this, I look at Fig. 4.5 (right). I can calculate the area of the partial triangle with vertices A, band M simply as $\rho c/2$ using the above formula, because the length of the incircle radius ρ is just the length of the height to the side of this triangle with vertices A and B. Similarly, for the area of the triangle with vertices B, C, and M, and of the triangle with vertices C, A, and M, respectively, I get $\rho a/2$ and $\rho b/2$. Since the area of the triangle with vertices A, B, and C *is* obviously just the sum of the areas of these three partial triangles, I get the following result:

Heron's Formula
The area of a triangle F with side lengths a, bund c is half the product of the incircle radius ρ with the sum of the side lengths: $F = \rho(a + b + c)/2$.

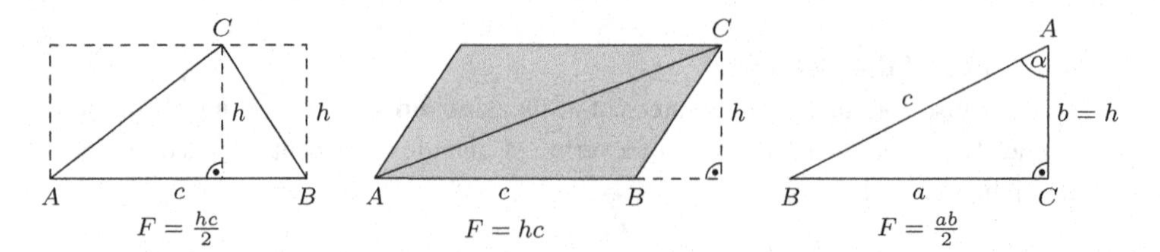

Fig. 4.6 The area of each triangle with vertices A, B and C is half the area of a rectangle with side lengths h and c (*left*). The formula also holds true if the altitude is outside triangle, and so the area of a parallelogram is $h\,c$ (*middle*). For right triangles, the application of this formula is simplified because you do not have to calculate the length of the altitude: The triangle's area is then half the product of the lengths of the triangle's cathets (*right*)

Above I mentioned that the inscribed radius of a triangle is given by the formula$\rho = \sqrt{(s-a)(s-b)(s-c)/s}$, where $s = (a + b + c)/2$. If I replace ρ in Heron's formula by this expression, I get for the area F of triangle

$$\begin{aligned} F &= \rho s = \sqrt{s(s-a)(s-b)(s-c)} \\ &= \frac{1}{4}\sqrt{(a+b+c)(b+c-a)(a+c-b)(a+b-c)}. \end{aligned}$$

So you only need the three side lengths a, b and c of triangle to calculate F – by specifying (reasonable) side lengths of a triangle, its area is determined.

Sometimes it is now so that I know for a triangle only the three corner points $A = (a_1|a_2)$, $B = (b_1|b_2)$ and $C = (c_1|c_2)$ and I would have to determine the area F with the above procedures either the length of a side and the length of the height standing perpendicularly on this side or however all three side lengths. I will now tell you a formula for calculating the area F of triangle, which saves me this possibly somewhat tedious detour:

$$F = \frac{1}{2}(a_1(b_2 - c_2) + b_1(c_2 - a_2) + c_1(a_2 - b_2)).$$

The calculation of the area of a triangle is especially easy when the triangle is *right-angled.* In this case two of the three sides lie on heights of the triangle and I do not have to calculate the height needed to calculate the area. It is then sufficient to determine only half of the product of the lengths a and b of these two sides (see Fig. 4.6 (right)). These two sides intersect at the vertex with the right interior angle and are called **cathetes**, while the remaining side of the right triangle (with length c in Fig. 4.6 (right)) is called the is called the **hypotenuse**.

Exercise 4.1 Let $A = (0|0)$ and $B = (4|0)$. Determine the area of the triangle by the points A, B and C if (a) $C = (2|2)$, (b) $C = (0|2)$, (c) $C = (6|2)$ and (d) C *is* the point at distance $\sqrt{5}$ from A and at distance $\sqrt{13}$ from B.

In my opinion, it is now a good idea for me to spend a little time with you on other statements about areas of *right* triangles. One of the absolute classics of such statements is the **theorem of Pythagoras** (after Pythagoras of Samos, around 580–496 BC), which I would like to remind you of first. To derive it, consider Fig. 4.7 (right). The area of the (whole) square there with side length a + b is according to the first **binomial formula**

$$(a+b)^2 = a^2 + 2ab + b^2,$$

while c^2 is the area of the (tilted) square with side length c inside it. Each of the four grey triangles is right-angled and according to the above statement their area is $ab/2$. If I add the sum of the areas of the four grey triangles to c^2, I get the alternative formula $c^2 + 2ab$ for the area of the (whole) square. But since this area has a unique value, I get

$$a^2 + 2ab + b^2 = c^2 + 2ab.$$

Now I can subtract the term $2ab$ on both sides of the last equation and arrive at the statement I have illustrated in Fig. 4.7 (left):

Pythagorean Theorem
For right triangles, the sum of the squares of the lengths of the cathetes a, b is always equal to the square of the length of the hypotenuse c: $a^2 + b^2 = c^2$.

Exercise 4.2 In a right triangle, the hypotenuse has length 5 and the area of the square over one of the two cathets is 16. What is the length of the remaining cathetus?

The Pythagorean theorem shows in particular that the length c of the hypotenuse is in each case greater than the lengths of the two cathets

$$c > \max\{a, b\},$$

because $c^2 = a^2 + b^2 > \max\{a^2, b^2\}$ (here I use max as an abbreviation for the maximum of two numbers). This remark will be of importance further below in a small discussion in the context of the sine theorem.

Exercise 4.3 In a right triangle, let the sum of the lengths of the cathets be 2 and the hypotenuse be twice as long as either of the cathets. Determine the lengths of the cathets and the length of the hypotenuse of triangle.

Example 4.1 In equilateral triangles with given side length a, the length of each height h is equal. Since in this case the median perpendiculars coincide with the heights, by the Pythagorean theorem $h^2 + (a/2)^2 = a^2$. This gives $h^2 = a^2 - a^2/4 = 3\,a^2/4$ and hence

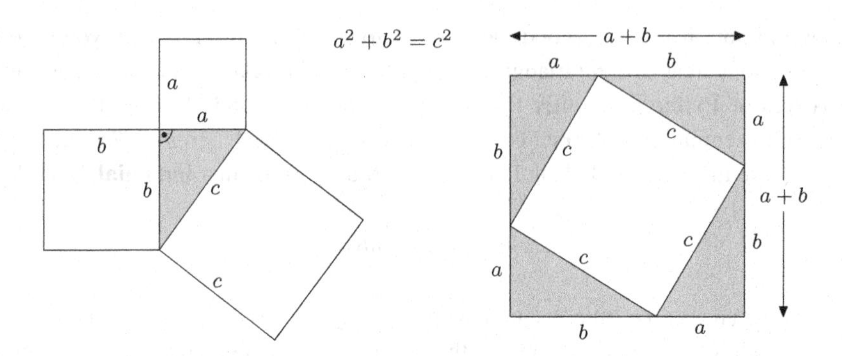

Fig. 4.7 Illustration of the Pythagorean theorem (*left*) and one of the approximately 100 known derivations of this statement (*right*)

$h = \sqrt{3}\,a/2$. With this, I can now conveniently calculate the area F *of* such a triangle: The formula $F = ha/2 = \sqrt{3}\,a^2/4$ for the **area of the equilateral triangle** applies.

Exercise 4.4 Show that the **area** F **of isosceles triangle is** given by $F = b\sqrt{4a^2 - b^2}/4$, where a **is** the length of the two legs and b **is** the remaining side length.

An often occurring application of the Pythagorean theorem is the determination of the **distance** d **between two points** – let us call them $(x_1|y_1)$ and $(x_2|y_2)$ – in the plane. This is given by the length of the distance between these two points, which I can take to be the hypotenuse of a right triangle with cathetus lengths $x_1 - x_2$ and $y_1 - y_2$. Thus, it results

$$d = \sqrt{(x_1 - x_2)^2 + (y_1 - y_2)^2}.$$

Another classic that I have to talk about is **Euclid's theorem**. For this you can look at Fig. 4.8. If I consider the grey triangle, the cathetus with length a forms one side of this triangle and the length of the height $\tilde{h}$ (lying outside triangle) of triangle above this cathetus has the same length: $\tilde{h} = a$. So it holds, from what I told you above about calculating the area of triangles, that $a^2/2$ is the area of triangle. But now the parallelogram with vertices B, C, D, and E has twice the area of triangle – that is, a^2. Moreover, the area of this parallelogram coincides with the area of the rectangle with side lengths c (hypotenuse length of triangle) and p (length of the hypotenuse segment). Thus, the following statement is obtained:

Euclid's Theorem

For right triangles, the square of the length of a cathetus a is equal to the product of the length of the hypotenuse c with the length of the **hypotenuse segment** p that has a point in common with that cathetus: $a^2 = cp$.

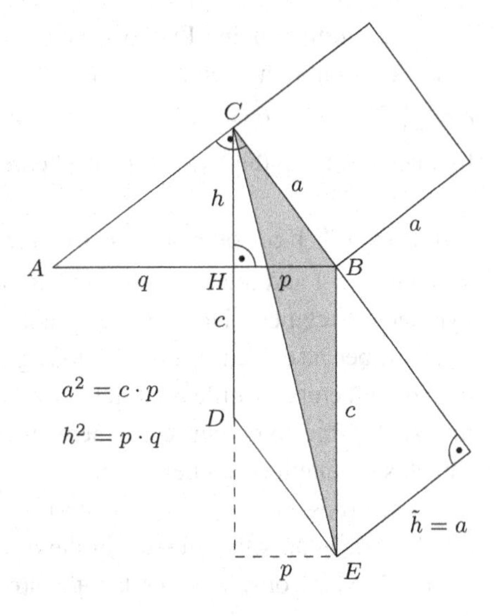

Fig. 4.8 Designations for deriving Euclid's theorem and the height theorem

Another interesting relationship of areas on the right triangle is expressed by the **height theorem**. Looking again at Fig. 4.8 and noting that the triangle with vertices H, B, and C is right-angled, I first note that by Pythagorean theorem $h^2 + p^2 = a^2$ holds. According to Euclid's theorem, I can now replace a^2 by cp here and I obtain by writing p^2 on the right side in this equation,

$$h^2 = cp - p^2.$$

Since the length of the hypotenuse c *can* be written as the sum of the lengths of the hypotenuse sections p and q, $c = p + q$, it finally follows that

$$h^2 = (p + q)p - p^2 = p^2 + qp - p^2 = qp.$$

Height Theorem

For right triangles, the square of the height h on the hypotenuse is equal to the product of the two hypotenuse segments p and q: $h^2 = p\,q$.

With the three theorems above, the lengths of the remaining sizes at the right triangle can be determined in a simple way from certain given sizes.

Example 4.2 For example, if I know that for a right triangle the hypotenuse has length $c = 6$ and the two hypotenuse segments have length $p = 2$ and $q = 4$, I use the height theorem to calculate the length h *of* the height on the hypotenuse from $h^2 = 2 \cdot 4 = 8$ as

$h = 2\sqrt{2}$. Further, using Euclid's theorem, I then determine the length of the cathetus that has a common point with the hypotenuse segment of length p as $a = \sqrt{p \cdot c} = \sqrt{2 \cdot 6} = \sqrt{12} = 2\sqrt{3}$. Finally, I can calculate the length b *of* the remaining hypotenuse using the Pythagorean Theorem: $b = \sqrt{c^2 - a^2} = \sqrt{36 - 12} = \sqrt{24} = 2\sqrt{6}$.▢

Example 4.3 I consider a right triangle for which I know the length of a hypotenuse segment $p = 1$ and the length $b = \sqrt{2}$ of the cathetus that has *no* point in common with this hypotenuse segment. Now the determination of the remaining lengths becomes a bit more difficult, because each of the formulas given by the last three theorems still contains two unknown factors. A little trick helps me here: first I remember that $c = p + q$ holds, which means that the hypotenuse segment of length q that has a point in common with the cathetus of length b can be written as $q = c - p = c - 1$. Euclid's theorem now gives me $b^2 = qc = (c - 1)c = c^2 - c$ and because of $b^2 = 2$ I arrive at the **quadratic equation** $c^2 - c - 2 = 0$. I can solve this using the $\boldsymbol{p - q}$ **formula** and I get the solutions $c_1 = 2$ and $c_2 = -1$. Since only positive lengths are allowed, then $c = c_1 = 2$. Now, I could easily determine the remaining quantities q, h and a. However, I will leave that to you:

Exercise 4.5 Continue Example 4.3 by determining the missing quantities q, h, and a there.

I have one more standard statement about right triangles. For this I consider the three median perpendiculars of a *right* triangle. Above I had explained to you that for each triangle these intersect at the center S of the circumcircle. The peculiarity of right triangles lltriangle is that this point S is now the midpoint of the hypotenuse of triangle (see Fig. 4.9 (left)). So the length of the radius of the circumcircle of such triangles is half the length of the hypotenuse: $c/2$. Thus, if I only decide to fix the two vertices A and B *of* the hypotenuse of a right triangle, the remaining vertex C always lies on a circle of radius $c/2$ around the center $S = (A + B)/2$ (see Fig. 4.9 (right)).

Thales' Theorem

For right triangles with common hypotenuse (of length c formed by the vertices A and B) the remaining vertex C the common circumcircle of these triangles. This circumcircle has the center $S = (A + B)/2$ and the radius $c/2$.

I will now explain a statement related to Thales' theorem using Fig. 4.10. This states: The **peripheral angles** over a fixed circular arc are always equal. I will come back to this later when I talk about the **sine theorem**, which relates the interior angles to the side lengths of the general triangle.

Speaking of the general triangle. Have you noticed it too? That is right, in the statements I have made so far, I have pretty much not dealt with the (interior) angles on the triangle at all. Until now, I have often assumed right triangles and the remaining two angles $\neq 90°$ in these triangles have interested me less – it was sufficient for me to determine the essential

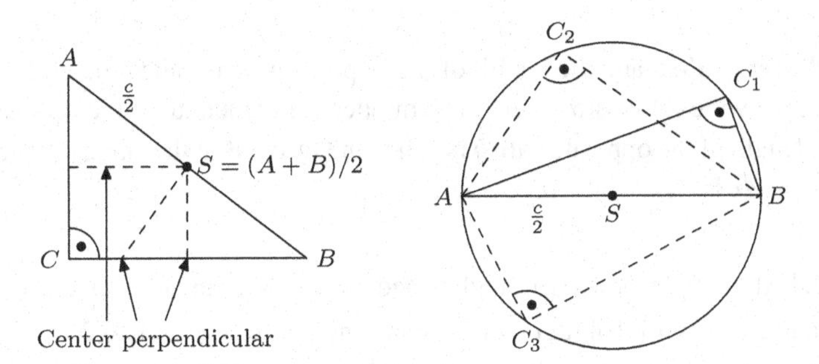

Fig. 4.9 For right triangles, the center of the circumcircle (intersection of the median perpendiculars) is the midpoint $S = (A + B)/2$ of their hypotenuse (*left*). Thales' theorem (*right*) states that the remaining vertex C always lies on the common circumcircle of such triangles

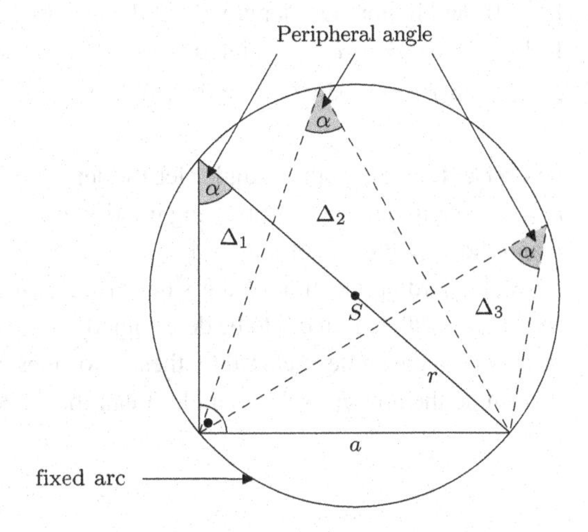

Fig. 4.10 Peripheral angles over a fixed arc are always the same. The triangles triangle_1, triangle_2 and triangle_3 have the common side with length a over the fixed arc – therefore the angle opposite to this side is always the same α

remaining quantities at the triangle as for example the lengths of the cathetes a and b or the length of the hypotenuse $c = \sqrt{a^2 + b^2}$, from given quantities. However, I do not want to be satisfied with this now, because firstly, it might also be useful for right triangles to calculate the remaining two angles $\neq 90°$, and secondly, of course, I would like to be able to perform similar calculations for general triangles as for right triangles. It will turn out that I can do this in a convenient way if I include the angles at the triangle. For this, however, I need **trigonometric functions**. Interestingly, I arrive at these by first again considering only *right* triangles as in Fig. 4.6 (right).

Sine, Cosine and Tangent

The quotient of the length of the **opposite side** a and the length of the hypotenuse c is called the **sine of** the angle α at the vertex A: $\sin(\alpha) = a/c$. The quotient of the length

(continued)

of the **adjacent side** b and the length of the hypotenuse c is called the **cosine of the** angle α at the vertex A: cos(α) = b/c. The quotient of the length of the **opposite side** a and the length of the opposite **side** b is called the **tangent of** the angle αat the vertex A: tan(α) = a/b.

Example 4.4 If triangle is a right-angled and isosceles triangle with cathetus lengths $a = b = 1$ and hypotenuse length $c = \sqrt{2}$, then $\sin(\alpha) = \cos(\alpha) = 1/\sqrt{2} = \sqrt{2}/2$ and tan (α) = 1 and α = 45° apply. I get this angle by entering $\sqrt{2}/2$ on the calculator and then pressing (in degrees) INV and COS, for example, in this order.⊡

Example 4.5 Considering a right triangle with remaining interior angles α = 30° and β = 60° and hypotenuse length $c = 2$, I can calculate the two remaining cathetus lengths as follows: $a = c \cdot \sin(\alpha) = 2 \cdot \sin(30°) = 2 \cdot 1/2 = 1$ and $b = c \cdot \sin(\beta) = 2 \cdot \sin(60^\circ) = 2 \cdot \sqrt{3}/2 = \sqrt{3}$.⊡

Exercise 4.6 In a right triangle, let the lengths of the cathetes be known: $a = 3\sqrt{3}$ and $b = 3$. Determine the two remaining interior angles α and β as well as the hypotenuse length c and the area F of this triangle.⊡

A standard application of the sine for general triangles is when the length of a height h on a side with length c *is* to be determined. If one knows the length of another side a *of* the triangle and the angle enclosed by these two sides, then $h = a \sin(\alpha)$ holds. If I then want to determine the area F of the triangle, I can then calculate this according to

$$F = \frac{hc}{2} = \frac{ac\sin(\alpha)}{2}$$

Exercise 4.7 For a triangle, know the lengths of two sides $a = 4$ and $c = 5$ and the angle enclosed by these sides β = 30°. Determine the area F of this triangle.⊡

From the definitions of the trigonometric functions it is immediately obvious that

$$\frac{\sin(\alpha)}{\cos(\alpha)} = \frac{\frac{a}{c}}{\frac{b}{c}} = \frac{a}{b} = \tan(\alpha)$$

holds true: The tangent of an angle is therefore just the quotient of the sine and the cosine of this angle. Furthermore, from the Pythagorean theorem I get $(a/c)^2 + (b/c)^2 = 1$, and thus the next important relation:

Sum of the Squares of the Sine and Cosine

For any angle α holds:

$$\sin^2(\alpha) + \cos^2(\alpha) = 1.$$

You will sometimes encounter this equality in the form $\sin(\alpha) = \sqrt{1 - \cos^2(\alpha)}$ or $\cos(\alpha) = \sqrt{1 - \sin^2(\alpha)}$. Do not worry – there is not suddenly something negative under the root here, because I also see from the last statement that the values of the sine and cosine are always between −1 and 1:

$$-1 \leq \sin(\alpha) \leq 1 (short: |\sin(\alpha)| \leq 1) \text{ and } -1 \leq \cos(\alpha) \leq 1 (short: |\cos(\alpha)| \leq 1).$$

For if for an angle α $|\sin(\alpha)| > 1$ is satisfied, then of course $\sin^2(\alpha) > 1$, and because of $\cos^2(\alpha) \geq 0$, the sum of $\sin^2(\alpha)$ and $\cos^2(\alpha)$ could not be equal to 1. But this cannot be, because otherwise something would be wrong with the Pythagorean theorem – and this would shake the world of mathematicians like an earthquake.

I already said it: With the help of the trigonometric functions I plan to find relations for the quantities at the general triangle. However, at the moment there is a minor obstacle in my way to do this immediately in a successful way. Namely, above I told you that the two remaining angles in the right triangle can only take values (real) between 0° and 90°. But if I plan to use, for example, the sine for general triangles, including, for example, obtuse-angled triangles, it would be useful for this purpose to know how the sine can be extended to arbitrary angles. To find out, I look at a **unit circle** as in Fig. 4.11. For angles α between 0° and 90°, as the figure shows, the sine is just the length of the *cathetus* opposite the angle α (hence: *opposite cathetus*), because here the length of the hypotenuse is 1. Similarly, for these angles, the cosine is just the length of the cathetus *adjacent to* the angle α (hence: *adjacent cathetus*).

If I now wish to determine the value of the sine or cosine, respectively, for any other angle (I have labeled an example of such angles α′ in Fig. 4.11), I can do so by considering the point $(b|a)$ on the unit circle corresponding to the angle and setting $\sin(\alpha') = a$, and $\cos(\alpha') = b$. For such more general angles α′, I then fix the tangent by relation $tan(\alpha') = sin(\alpha')/cos(\alpha') = a$. The value of the tangent is the slope of the straight line through the two points (0|0) and $(b|a)$. I have to be a little careful here, though: since the cosine of 90° and 270° has a value of 0, I must not use this formula of tangent for these particular angles, because you know that I must *never never never* divide by 0. Moreover, it is worth mentioning here that the above relations $\sin^2(\alpha') + \cos^2(\alpha') = 1$ and $|\sin(\alpha')| \leq 1$ as well as $|\cos(\alpha')| \leq 1$ are still unrestrictedly fulfilled for such more general angles α′.

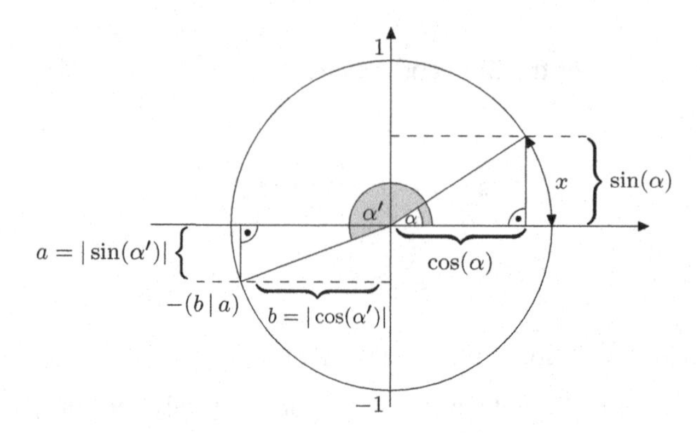

Fig. 4.11 Illustration of the sine and cosine on the unit circle

Another point may come to the mind of the attentive reader when comparing with the contents of Chap. 2. I am talking about trigonometric *functions* here and used – contrary to common usage – **angles** instead of **real numbers** as **arguments.** This is of course allowed – but it would be nice, if I could understand the trigonometric functions as functions of a **real variable** – say x – as I am used to do with other functions. I succeed in doing this by using the **radian**, for whose subsequent explanation it is permitted to take another cursory glance at Fig. 4.11. Furthermore I point to the next section, where the radian will appear again.

Radians

The radian x *of* an angle α is the signed length of the arc associated with α on the unit circle, where the sign of x *is* positive exactly when the arc is traced counterclockwise. The conversion formula between degrees and radians is $x = \pi\, \alpha/180°$.

Example 4.6 If I want to know the value of the radian of $\alpha = 360°$, I calculate in this way the circumference of the unit circle $x = \pi 360°/180° = 2\pi$ (see also the next section – there I give hints as to why this is so and what the number $\pi \approx 3.14$ exactly is). If I want to know the radian of $\alpha = 45°$, I find $x = \pi\, 45°/180° = \pi/4$ – this is one-eighth of the circumference of the unit circle. If $\alpha = -45°$, I expect the sign of x to be negative as well, because to subtract this angle I must proceed clockwise. This is also the case, because I *calculate* $x = \pi\,(-45°)/180° = -\pi/4$. *Conversely*, if someone gives me the radian x, I determine the corresponding angle α according to $\alpha = 180°\, x/\pi$. So for $x = \pi/3$, for example, I get $\alpha = 180°\pi/(3\pi) = 60°$. This is – I think – quite clear.▯

Trigonometric Functions
With the help of the radian I can now understand the trigonometric functions as functions of the real variable x in the usual sense by

$$\sin(x) = \sin(\alpha),\ \cos(x) = \cos(\alpha) \text{ and } \tan(x) = \tan(\alpha),$$

where x is the radian associated with the angle α.

The graphs of the **sine** and **cosine functions can** be found in Fig. 4.12. In Table 4.1 I have listed some values of the trigonometric functions for characteristic angles together with their radians.

You can see from the graphs in Fig. 4.12 that the sine and cosine functions are each 2π-**periodic**, i.e.

$$\sin(x) = \sin(x + 2\pi) \text{ and } \cos(x) = \cos(x + 2\pi)$$

for all x – little surprising, because after passing (for example, counterclockwise) the complete edge of the unit circle (which, as you will learn in the next section, has length 2π), the values of the sine and cosine repeat. Thus, these are functions that have infinitely many **zeros.** For example, if I consider the sine, I first see that $\sin(0) = 0$ and $\sin(\pi) = 0$ hold. However, the sine is 2π-periodic and so $\sin(2\pi) = 0$ and $\sin(-\pi) = 0$, for example, also hold. If I want to describe *all the* zeros of the sine function, I can do so as follows:

$$\sin(x) = 0 \text{ exactly when } x = k\pi, \text{ where } k \text{ is any integer.}$$

For example, $\sin(311\pi) = 0$ and $\sin(-1022\pi) = 0$. Similarly, I can describe the zeros of the cosine:

$$\cos(x) = 0 \text{ exactly when } x = \frac{\pi}{2} + k\pi, \text{ where } k \text{ is any integer.}$$

Exercise 4.8 Describe all x where (a) the sine function has value 1 (that is, $\sin(x) = 1$ holds), (b) the cosine function has value -1 (that is, $\cos(x) = -1$ holds).⍰

Exercise 4.9 Describe all real numbers x on which the tangent function $\tan(x) = \sin(x)/\cos(x)$ is well defined.⍰.

Furthermore, it is worth mentioning that the **graph of** the cosine function is **axisymmetric to** the *y-axis,*

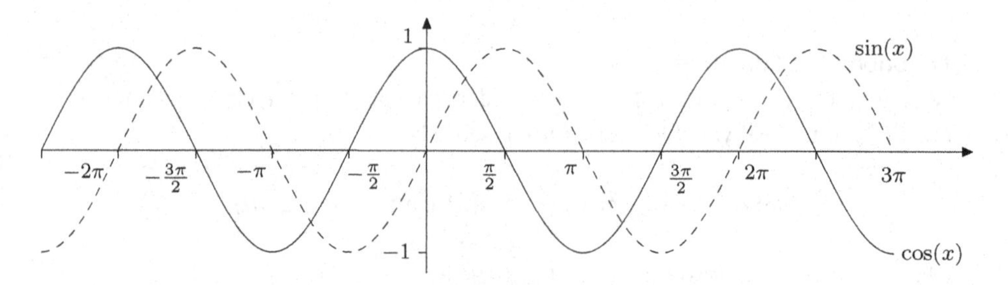

Fig. 4.12 Sine and cosine as a function of the (real) radian x

Table 4.1 Values of the trigonometric functions at characteristic points x

A	−90°	−60°	−45°	−30°	0°	30°	45°	60°	90°
X	$-\pi/2$	$-\pi/3$	$-\pi/4$	$-\pi/6$	0	$\pi/6$	$\pi/4$	$\pi/3$	$\pi/2$
$\sin(x)$	-1	$-\sqrt{3}/2$	$-\sqrt{2}/2$	$-1/2$	0	$1/2$	$\sqrt{2}/2$	$\sqrt{3}/2$	1
$\cos(x)$	0	$1/2$	$\sqrt{2}/2$	$\sqrt{3}/2$	1	$\sqrt{3}/2$	$\sqrt{2}/2$	$1/2$	0
$\tan(x)$	–	$-\sqrt{3}$	-1	$-\sqrt{3}/3$	0	$\sqrt{3}/3$	1	$\sqrt{3}$	–

$$\cos(x) = \cos(-x),$$

for all x, while the graph of the sine function is **point-symmetric** with respect to the **origin** (0|0):

$$\sin(x) = -\sin(-x)$$

for all x. Thus, the cosine function is an **even function** and the sine function is an **odd function.** Moreover, I can observe that by shifting the graph of the cosine by $\pi/2$ along the *x-axis,* the graph of the sine function is formed. In terms of formulas, I can write this down as follows:

$$\cos\left(x - \frac{\pi}{2}\right) = \sin(x) \text{ respectively } \cos(x) = \sin\left(x + \frac{\pi}{2}\right)$$

for all x. The trigonometric functions satisfy quite a number of other relations which an ordinary person cannot quickly recognize at once. For example, the proof of the following relations requires somewhat more elaborate arguments, but I will spare you those now.

Addition Theorems of Sine and Cosine
For all x and y, the following relationships hold:

$$\sin(x+y) = \sin(x)\cos(y) + \cos(x)\sin(y) \text{ (1.Additiontheorem)}$$
$$\cos(x+y) = \cos(x)\cos(y) - \sin(x)\sin(y) \text{ (2.Additiontheorem)}$$

Example 4.7 I set $x = -\pi/3$ and $y = \pi/3$ and use Table 4.1. Obviously, $\sin(x + y) = \sin(-\pi/3 + \pi/3) = \sin(0) = 0$ and $\cos(x + y) = \cos(-\pi/3 + \pi/3) = \cos(0) = 1$. On the other hand, for the more complicated term on the right-hand side of the first addition theorem, I obtain

$$\sin(x)\cos(y) + \cos(x)\sin(y) = \overbrace{\sin\left(-\frac{\pi}{3}\right)}^{-\frac{\sqrt{3}}{2}}\overbrace{\cos\left(\frac{\pi}{3}\right)}^{\frac{1}{2}} + \overbrace{\cos\left(-\frac{\pi}{3}\right)}^{\frac{1}{2}}\overbrace{\sin\left(\frac{\pi}{3}\right)}^{\frac{\sqrt{3}}{2}} = 0$$

respectively of the second addition theorem

$$\cos(x)\cos(y) - \sin(x)\sin(y) = \overbrace{\cos\left(-\frac{\pi}{3}\right)}^{\frac{1}{2}}\overbrace{\cos\left(\frac{\pi}{3}\right)}^{\frac{1}{2}} - \overbrace{\sin\left(-\frac{\pi}{3}\right)}^{-\frac{\sqrt{3}}{2}}\overbrace{\sin\left(\frac{\pi}{3}\right)}^{\frac{\sqrt{3}}{2}} = \frac{1}{4} - \left(-\frac{3}{4}\right) = 1. \blacksquare$$

I can use the addition theorems to derive many other relationships of the sine function with the cosine function. For example, if I put the term $-y$ for y into the first addition theorem, I get $\sin(x - y) = \sin(x)\cos(-y) + \cos(x)\sin(-y)$, and because the cosine is an even function and the sine is an odd function, it follows that

$$\sin(x-y) = \sin(x)\cos(y) - \cos(x)\sin(y)$$

for all x and y. In the same way I get with the second addition theorem

$$\cos(x-y) = \cos(x)\cos(y) + \sin(x)\sin(y)$$

for all x and y. It is interesting that if I now set $y = x$ in the last equation, $\cos(x - x) = \cos(x)\cos(x) + \sin(x)\sin(x)$, i.e. $1 = \cos^2(x) + \sin^2(x)$, follows – a relation which I have already shown you above in a different way for merely certain angles α. If I do something similar in the addition theorems – i.e. if I set $y = x$ there – I see that the relations

$$\sin(2x) = 2\sin(x)\cos(x) \text{ and } \cos(2x) = \cos^2(x) - \sin^2(x)$$

respectively

$$\frac{\sin(2x)}{2} = \sin(x)\cos(x) \text{ and } \sin^2(x) = \cos^2(x) - \cos(2x)$$

are fulfilled. If I replace in the last equation $\sin^2(x)$ by $1 - \cos^2(x)$, I get $1 - \cos^2(x) = \cos^2(x) - \cos(2x)$ and from this I get

$$\cos(2x) = 2\cos^2(x) - 1.$$

If I want, I can use these equations to express the tangent (at the well-defined locations x, see Exercise 4.9) differently:

$$\tan(x) = \tan\left(2\frac{x}{2}\right) = \frac{\sin\left(2\frac{x}{2}\right)}{\cos\left(2\frac{x}{2}\right)} = \frac{2\sin\left(\frac{x}{2}\right)\cos\left(\frac{x}{2}\right)}{2\cos^2\left(\frac{x}{2}\right) - 1}.$$

Finally, possibly the formulas

$$\sin(x) + \sin(y) = 2\sin\left(\frac{x+y}{2}\right)\cos\left(\frac{x-y}{2}\right)$$

and

$$\cos(x) + \cos(y) = 2\cos\left(\frac{x+y}{2}\right)\cos\left(\frac{x-y}{2}\right)$$

cross your future (study) path. Do not worry – these equations can also be derived quite easily from the addition theorems. I will show you using the first formula – you then get to derive the second formula. In both derivations, you do the same little trick. Namely, I write

$$x = \frac{x+y}{2} + \frac{x-y}{2}$$

and then apply the first addition theorem

$$\sin(x) = \sin\left(\frac{x+y}{2} + \frac{x-y}{2}\right) = \sin\left(\frac{x+y}{2}\right)\cos\left(\frac{x-y}{2}\right) + \cos\left(\frac{x+y}{2}\right)\sin\left(\frac{x-y}{2}\right).$$

Also, I use $y = (x + y)/2 + (y - x)/2$ to get, in the same way.

$$\sin(y) = \sin\left(\frac{x+y}{2} + \frac{y-x}{2}\right) = \sin\left(\frac{x+y}{2}\right)\cos\left(\frac{y-x}{2}\right) + \cos\left(\frac{x+y}{2}\right)\sin\left(\frac{y-x}{2}\right).$$

Noting that because of the symmetry properties of the graph of the sine and cosine functions, $\sin((y - x)/2) = -\sin((x - y)/2)$ and $\cos((y - x)/2) = \cos((x - y)/2)$ hold, I see that adding $\sin(x)$ and $\sin(y)$ in these representations actually produces the first of the above formulas.

Exercise 4.10 Prove the second of the above formulas.

Of course, it is a lot of fun chatting with you about trigonometric functions, and we could – if it were only up to me – happily continue our conversation for a bit longer. For example, we could talk about the interesting relationship

$$e^{\sqrt{-1}\alpha} = \cos(\alpha) + \sqrt{-1}\sin(\alpha)$$

with **Euler's number e**, and I could show you elegantly, for example, how to derive the addition theorems. However, **complex numbers** (roughly speaking, numbers for which taking roots from negative numbers, i.e. also from -1, is allowed) are not yet our topic in this chapter (cf. Chap. 9), and we should not lose sight of our common goals. Above I promised you some time ago, that I will use the trigonometric functions to put the quantities at the general triangle into clear relations, especially including the (inner) angles at the triangle. In this sense I will continue now.

So, in the following I consider an arbitrary triangle with vertices A, B and C and side lengths a, b and c as in Fig. 4.13, and my aim is now first to establish a relationship between these quantities and the interior angle α at A. To do this, I consider the height to the side joining A and B. This height has length h, intersects this side with length c at a point D, and I obtain two right-angled (partial) triangles with vertices A, D and C, respectively B, D and C. If I use the labels from Fig. 4.13, the **Pythagorean theorem** states that the equations $p^2 + h^2 = b^2$ and $q^2 + h^2 = a^2$ must be satisfied. Since $q = c - p$ holds, I use the **binomial formula to** calculate $q^2 = (c - p)^2 = c^2 - 2\,p\,c + p^2$ and I get $c^2 - 2\,p\,c + p^2 + h^2 = a^2$ by substituting into the second eq. I can further simplify this by using the first equation ($p^2 + h^2 = b^2$):

$$b^2 + c^2 - 2pc = a^2.$$

This equation relates the lengths of the sides of the general triangle – unfortunately, however, the length of the section p appears here: A quantity I have not assumed to

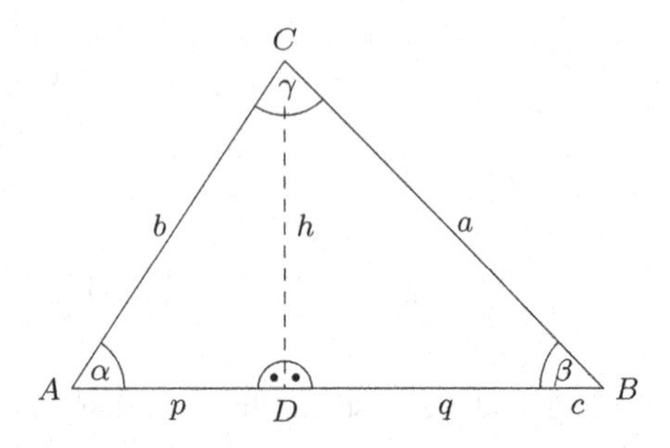

Fig. 4.13 Designations for deriving the cosine theorem

know here. In contrast, I assume here that I know the angle α, or cos(α). Therefore I replace the unknown quantity p by

$$p = b\cos(\alpha).$$

I am allowed to do this because the distance from A to D is the adjacent in the right triangle with the additional vertex C *and* the cosine of α is known to be given by the formula cos (α) = p/b. Altogether I get now

$$b^2 + c^2 - 2bc\cos(\alpha) = a^2.$$

This is the equation. I was looking for, because the three lengths a, b and c *of* the sides of the general triangle are here related to an angle α – more precisely, the angle of the triangle enclosed by the sides with length b and c.

Cosine Theorem

For any triangle, the square of the length a *of* a side is the difference of the sum of the squares of the lengths b, c of the remaining sides with the term $2bc$ cos(α), where α is the angle enclosed by the sides with lengths b and c: $a^2 = b^2 + c^2 - 2bc\cos(\alpha)$.

Just as with the angle α, I can of course argue for the angles β and γ in Fig. 4.13, and thus I see that also the formulas

$$b^2 = a^2 + c^2 - 2ac\cos(\beta) \text{ and } c^2 = a^2 + b^2 - 2ab\cos(\gamma)$$

are valid. I should mention that the cosine theorem is a *generalization of* the **Pythagorean theorem for** general triangles. If the special case of a right triangle with vertices A, B and C is present – for example, if γ = 90°, then cos(γ) = cos(90°) = 0, and the term $2ab$ *cos*(γ) will also be 0. So in this case, the cosine theorem gives $c^2 = a^2 + b^2$, which is nothing more than the statement of Pythagoras' theorem for right triangles.

Example 4.8 A classic example of the application of the cosine theorem is when given the length of two sides of a triangle, say $b = 2$ and $c = 3$, and the angle enclosed by those sides, say $\alpha = 30°$, and you would like to work out the length of the remaining side a. The formula in this case gives

$$a^2 = 2^2 + 3^2 - 2 \cdot 2 \cdot 3 \cdot \cos(30^\circ) = 4 + 9 - 12 \cdot \sqrt{3}/2 = 13 - 6 \cdot \sqrt{3}$$

and thus $a = \sqrt{13 - 6 \cdot \sqrt{3}} \approx 1.6148$. Conversely, I can now wonder what the two remaining angles β and γ are in this triangle. To do this, I convert one of the two remaining formulas above $-\cos(\beta) = (a^2 + c^2 - b^2)/(2\,ac)$ – and insert the lengths of the sides of the triangle:

$$\cos(\beta) = \frac{13 - 6 \cdot \sqrt{3} + 9 - 4}{2\left(\sqrt{13 - 6 \cdot \sqrt{3}}\right)3} = \frac{3 - \sqrt{3}}{\sqrt{13 - 6\sqrt{3}}} \approx 0.7852.$$

Pressing INV and COS on the calculator gives $\beta \approx 38.26°$. I could now calculate the remaining angle γ in the same way. However, I make this easier for myself by remembering that the sum of the interior angles in the triangle is always 180°. So here the remaining angle γ is approximately $180° - 30° - 38.26° = 111.74°$. So in this example, it is an obtuse triangle.

The cosine theorem thus teaches us that by prescribing the lengths of two sides a and b of a triangle and the angle α enclosed by them, the remaining side length and the two remaining angles of this triangle are fixed. In other words, this means that by these three quantities a, b and α the shape of the triangle e is already completely fixed. I can create other triangles of this shape and size by moving triangle in the plane or rotating it about a point – but triangles created in this way (they are sometimes called **congruent triangles**, see Fig. 4.14) will always have the same shape as triangle. So I first state that two triangles are congruent if the lengths of two of their sides and the interior angle between these sides are equal.

At this point, one might suspect that there are other situations in which triangles are congruent. In fact, two triangles, for example, are also congruent if the lengths of their three sides match. How can I see this? Well, the answer is also rooted in the cosine theorem. I will do another example for this:

Example 4.9 In this example, I give myself the three side lengths of a triangle – say $a = 5$, $b = 3$, and $c = 4$. By rearranging the formula from the cosine theorem, I can now calculate the three (interior) angles of the triangle. For example, $\cos(\alpha) = (b^2 + c^2 - a^2)/(2bc)$. So I calculate $\cos(\alpha) = (9 + 16 - 25)/24 = 0$ and so $\alpha = 90°$. So this happens to be a right triangle. I calculate the second angle β according to $\cos(\beta) = (a^2 + c^2 -$

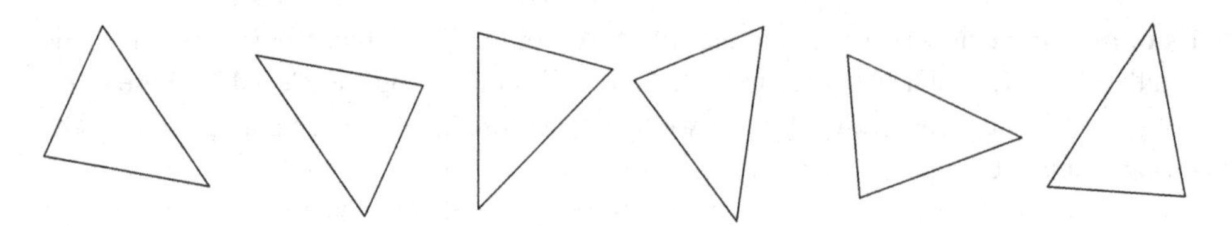

Fig. 4.14 Congruence of triangles: The three side lengths and the three (interior) angles are congruent

$b^2)/(2\ ac) = (25 + 16 - 9)/40 = 0.8$. This gives $\beta \approx 36.87°$ and so the remaining angle $\gamma \approx 53.13°$.

Exercise 4.11 Let a given triangle have side lengths $2y, 2y$ *and* $3y$ for a real number $y \neq 0$. Show that the three (interior) angles of the triangle are independent of y and determine these angles.⌧

It is easy to see that, given three angles α, β and γ (with $\alpha + \beta + \gamma = 180°$), a triangle in the sense of congruence is not yet fixed, since I can increase and decrease such a triangle at will. Another question is similarly easy for you to answer:

Exercise 4.12 Given lengths *a, b,* and *c,* can you actually always construct a triangle with these side lengths? Give reasons for your answer.

Quite at the beginning of this section I promised you that I would show you two statements concerning **isosceles** and **equilateral** triangles quite easily at a later time. I am now happy to keep this promise. If there is an isosceles triangle, then two sides of the triangle have the same length – so, for example, $b = a$. The cosine theorem then yields

$$\cos(\alpha) = \frac{b^2 + c^2 - a^2}{2bc} = \frac{a^2 + c^2 - a^2}{2ac} = \frac{c}{2a}$$

and

$$\cos(\beta) = \frac{a^2 + c^2 - b^2}{2ac} = \frac{a^2 + c^2 - a^2}{2ac} = \frac{c}{2a}$$

and I realize that the adjacent interior angles α, β of the remaining side match: $\alpha = \beta$. If there is an equilateral triangle, then $a = b = c$ and I can simplify the last two equations even further: $\cos(\alpha) = c/2\ a = a/2\ a = 1/2$ and $\cos(\beta) = 1/2$. Thus $\alpha = \beta = 60°$ and the remaining angle is also $\gamma = 60°$.

Another useful relationship of the lengths of the sides with the (interior) angles of any given triangle lltriangle can be derived by looking again at Fig. 4.13. There, the two partial triangles with vertices *A, D,* and *C, and B, D,* and *C*, respectively, are right angles. The sine of the angle at *A* and *B* is therefore the quotient of the lengths of the opposite cathetus and hypotenuse of these triangles. So I get

$$\sin(\alpha) = \frac{h}{b} \text{ and } \sin(\beta) = \frac{h}{a}.$$

So I can write the length of the height h on the side of the triangle with length c in two different ways:

$$h = b \cdot \sin(\alpha) \text{ and } h = a \cdot \sin(\beta).$$

But since this h must have a unique value, I get the relation

$$b \cdot \sin(\alpha) = a \cdot \sin(\beta)$$

for the given triangle. If I divide here by sin(α) and sin(β), I can also write this down as follows:

$$\frac{a}{\sin(\alpha)} = \frac{b}{\sin(\beta)}.$$

By exactly the same procedure, using the height on the side of the length b, I can see in the same way that also

$$\frac{a}{\sin(\alpha)} = \frac{c}{\sin(\gamma)}$$

applies.

Sine Theorem

For any triangle with side lengths a, b, and c and corresponding opposite interior angles α, β, and γ, $a/\sin(\alpha) = b/\sin(\beta) = c/\sin(\gamma)$ holds.

In other words, the sine theorem states that the quotient of two side lengths of a triangle gives the same value as the quotient of the sines of the opposite interior angles (also: **opposite angles**):

$$\frac{a}{b} = \frac{\sin(\alpha)}{\sin(\beta)}, \frac{b}{c} = \frac{\sin(\beta)}{\sin(\gamma)} \text{ and } \frac{c}{a} = \frac{\sin(\gamma)}{\sin(\alpha)}.$$

I would like to remark here that the sine theorem also follows from the fact that the peripheral angle α is always the same as in Fig. 4.10 over a fixed arc of the circumcircle of a triangle. Indeed, the determination of the sine shows me that for the right triangle triangle$_1$ in Fig. 4.10 (and thus for each of the triangles there), the relation

$$\sin(\alpha) = \frac{a}{2r}$$

where r is the radius of the circumcircle of triangle = triangle$_2$ and triangle = triangle$_1$ respectively. Thus also

$$\sin(\beta) = \frac{b}{2r} \text{ and } \sin(\gamma) = \frac{c}{2r}$$

is satisfied for each triangle lltriangle as in Fig. 4.13. Solving these three formulas for r and equating gives the sine theorem.

The sine theorem is especially useful when two (interior) angles – say α and β – of a triangle and one side length – say a – are given. In this case, I can easily determine the remaining angle by $\gamma = 180° - \alpha - \beta$, and then the missing side lengths b and c are obtained by rearranging the formulas of the sine theorem:

$$b = a\frac{\sin(\beta)}{\sin(\alpha)} \text{ and } c = a\frac{\sin(\gamma)}{\sin(\alpha)}.$$

So by applying the sine theorem, I see that by specifying the angles and one side length of a triangle, the lengths of the remaining two sides are fixed. In particular, two triangles are congruent if one of their side lengths and all of their interior angles agree.

Example 4.10 I consider the situation where the length of one side – say $a = 3$ – and the two interior angles adjacent to this side – say $\beta = 120°$ and $\gamma = 20°$ – are given. Of course, then the remaining angle is $\alpha = 180° - 120° - 20° = 40°$. I now calculate the two remaining side lengths by $b = a\,(\sin(\beta)/\sin(\alpha)) = 3\,(\sin(120°)/\sin(40°)) \approx 4.0419$ and $c = 3\,(\sin(20°)/\sin(40°)) \approx 1.5963$.

Above I showed you that the cosine theorem is directly applicable when the lengths of two sides of a triangle and the angle enclosed by them are given. Now, if the lengths of two sides of a triangle are given but one of the other two angles is given, the sine theorem *often* turns out to be a useful tool.

Example 4.11 For a triangle like the one in Fig. 4.13, let the lengths of two sides – say $a = 3$ and $b = 1$ – and one of the angles *not* included by those sides – say $\alpha = 50°$ – be given. Then, using the sine theorem, I first calculate the remaining angle of this kind: $\sin(\beta) = \sin(\alpha)(b/a) = \sin(50°)/3 \approx 0.2553$. The calculator now gives me the value $\beta \approx 14.79°$ after pressing INV and SIN. It now gives $\gamma \approx 115.21°$. For the remaining length I have free choice, because I can now apply both the sine and cosine theorems. I calculate for this $c \approx 3.5401$.

I have just shown you that in this example, by giving the length of two sides and an angle, the third side length and the two remaining angles can be calculated by applying the sine theorem. I have to be a little careful here in general, though, because if the lengths of the sides a and b and α, the angle opposite the side with length a, *are* given, then it is *not* necessary that

$$\left| \sin(\alpha) \frac{b}{a} \right| \leq 1$$

hold. But since this should always be satisfied because of $|\sin(\beta)| \leq 1$, I can see that something goes wrong for certain specifications of a, b and α, that is, a triangle cannot be constructed for some choices of these quantities.

Example 4.12 For example, if I consider a right triangle $\alpha = 90°$, that is $\sin(\alpha) = 1$, then b must not be greater than a. Otherwise, the length of the hypotenuse a would be smaller than the length b *of* one of the cathets, which is not possible. So for $\alpha = 90°$, for example, $b = 2$ and $a = 1$ is not a valid choice for side lengths of a triangle.

If, on the other hand, the calculation succeeds, as is the case in Example 4.11, then the triangle is already completely fixed in its shape by the specification of the quantities a, b and α. Two such triangles are then called to be congruent.

Exercise 4.13 In the triangle as in Fig. 4.13, let $a = 2$ and $b = 8$ and $\alpha = 10°$. Determine the remaining side length c and the two remaining interior angles β and γ. Now, consider the same given side lengths, but $\alpha = 40°$. What do you observe?

Exercise 4.14 Derive the formula $F = (a\ b\ c)/(4\ r)$ for the area of a triangle with side lengths a, b, c and perimeter radius r.

At the end of this section, for the sake of completeness, I will formulate the tangent theorem, which can sometimes be used to shorten some calculations. I use the terms from Fig. 4.13 to formulate it. I could derive this statement from the sine theorem using the addition theorems and the relation $\tan = \sin / \cos$.

Tangent Theorem
For any triangle with side lengths a and b and interior angles α and β opposite these sides, $\tan((\alpha - \beta)/2) = \tan((\alpha + \beta)/2)(a - b)/(a + b)$.

4.2 Plane Geometric Figures

In this section I want to take a look at geometric figures in the plane that are not quite as simple as triangles. Some of these figs. I get by putting triangles together in an appropriate way – these are **quadrilaterals and** more general figures called **polygons**. Polygons are bounded by lines. This distinguishes them from plane figures, which have a curved edge.

Such plane figures – I treat here the circle and the ellipse – cannot be constructed perfectly by composition of triangles. But I can at least approximate them with arbitrary accrurancy by using enough triangles.

The simplest plane figures that differ from triangles are quadrilaterals. They are defined by four vertices *A*, *B*, *C* and *D*. Unlike the triangle, however, I have to be a little careful when describing a quadrilateral: The vertices *A*, *B*, *C* and *D should be* arranged in such a way, that I can draw the lines from *A* to *B*, *B* to *C*, *C to D* and finally from *D* to *A* again in such a way, that they do not intersect and thus actually lead to the outline of a two-dimensional figure – the quadrilateral.

In Fig. 4.15 I show you various types of quadrilaterals square together with a few designations, some of which will play a role below. The classification made there distinguishes between **general** quadrilateral, **trapezoid**, **kite** as well as **(true) parallelogram** (also: **rhomboid**), **rectangle and square.**

Simple quadrilaterals are certainly rectangles – here all sides are perpendicular to each other and I have exactly two different side lengths. With squares, these lengths are even equal. Parallelograms are, to put it casually, *oblique* rectangles. These are formed from two pairs of parallel sides, and the lengths of opposite sides are thus equal in each case. They form a more general class of quadrilaterals than rectangles, since the sides do not necessarily have to be perpendicular to each other. If all the sides of a parallelogram are equal in length, it is called a **rhombus**. For kites, any two adjacent sides have equal length. For trapezoids, only two sides are parallel. If the legs (in the picture of Fig. 4.15 the distances from *A* to *D* and *B* to *C*) are equal in length, such a trapezoid **is** also called an **isosceles trapezoid**. General quadrilaterals are further divided into **convex quadrilaterals** and **concave quadrilaterals**. In convex quadrilaterals, each interior angle is less than or equal to 180° and I can draw both **diagonals** (that is, lines connecting opposite points in the quadrilateral), while in concave quadrilaterals only one diagonal runs inside the quadrilateral.

I can divide a quadrilateral square into two triangles by adding a diagonal in it interior. I have illustrated this, for example, for the quadrilaterals in Fig. 4.15. The sum of the interior angles of the two resulting triangles is 180° – as is well known, this is the case for every triangle. With the relations at the general quadrilateral in Fig. 4.15, I get thus

$$\alpha' + \beta + \gamma' = 180^\circ \text{ and} \alpha'' + \gamma'' + \delta = 180^\circ .$$

If $\alpha = \alpha' + \alpha''$ and $\gamma = \gamma' + \gamma''$ are the angles in square occurring besides β and δ, I use them to calculate the sum of the angles in square as

$$\begin{aligned} \alpha + \beta + \gamma + \delta &= (\alpha' + \alpha') + \beta + (\gamma' + \gamma') + \delta \\ &= (\alpha' + \beta + \gamma') + (\alpha'' + \gamma'' + \delta) = 2 \cdot 180^\circ = 360^\circ . \end{aligned}$$

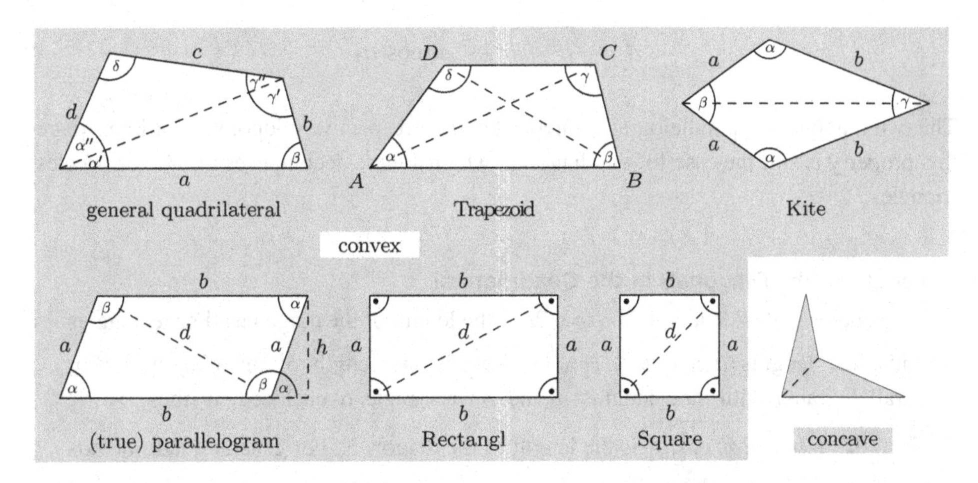

Fig. 4.15 Classification of quadrilaterals

> **Sum of the Interior Angles of a Quadrilateral**
> The sum $\alpha + \beta + \gamma + \delta$ of the interior angles $\alpha, \beta, \gamma, \delta$ of a quadrilateral is always 360°.

For some classes of quadrilaterals still apply certain properties and relationships of the angles. Above I already told you that in rectangles all angles are **right angles**, i.e. 90°. I also note that in kites there are always two angles of the same size-for the kite in Fig. 4.15 I have called this angle α. Furthermore, in a parallelogram there are only two different angles α and β and these add up to $\alpha + \beta = 180°$.

It is often useful to have knowledge of the **length of the diagonal** d in a quadrilateral square. For rectangles I get this easily by applying the **Pythagorean theorem**

$$d = \sqrt{a^2 + b^2},$$

because adding a diagonal leads to two **right triangles** with cathetus lengths a and b. I observe that in this case the two diagonals are of equal length. The same is true for squares, of course – here the diagonals have length

$$d = \sqrt{a^2 + a^2} = \sqrt{2a^2} = \sqrt{2}a.$$

The determination of d for parallelograms and the other more general quadrilaterals becomes somewhat more difficult. Here I often have to remember the generalization of Pythagoras' theorem from the last section: Using the **cosine** theorem, I calculate, for example, for the parallelogram in Fig. 4.15:

$$d = \sqrt{a^2 + b^2 - 2ab\cos(\alpha)}$$

The two diagonals in parallelograms intersect at their respective midpoints. For kite square this property is still the case for the diagonal, which square decomposes into two isosceles triangles.

Length of the Diagonals in the Quadrilateral

For squares with side length a, $d = \sqrt{2}a$ is the length of the diagonals. For rectangles with side lengths a and b, $d = \sqrt{a^2 + b^2}$ is the length of the diagonals. For parallelograms with side lengths a and b and angle α enclosed by these sides, $d = \sqrt{a^2 + b^2 - 2ab\cos(\alpha)}$ is the length of the diagonals. For general quadrilaterals square the lengths of the diagonals d are also determinable with the cosine theorem, if besides the side lengths also the corresponding included angles in square are known.

Have you ever installed a carpet before? If so, you have certainly thought about the area of squares and more general figures. If you may have neglected to do so, you may know the fatal consequences of ignorance of what I am about to begin describing to you. I am, in fact, concerned with computing areas. I will start with quadrilaterals and later I consider this for more general geometric figures.

The area F of squares and rectangles can be calculated very easily: This is $F = a^2$ *and* $F = ab$, respectively. I have already shown you how to calculate the area F *of* a parallelogram in the last section: You need the length of a height h *for* this. If the angle α (see Fig. 4.15) in the parallelogram is known, I can *use* the formula $h = a\sin(\alpha)$. With the help of the length of the remaining side I get $F = \sin(\alpha)\,ab$. Below, I will show you how to calculate the area of a trapezoid based on this formula.

Area of Parallelograms

The area F of a parallelogram with sides of length a and angle α enclosed by these sides is $F = \sin(\alpha)\,ab$. More specifically, the area F *of* a rectangle ($\alpha = 90°$) is $F = ab$ and the area of a fine square ($\alpha = 90°$ and $a = b$): $F = a^2$.

Exercise 4.15 Given a parallelogram square as in Fig. 4.15, let the side length $a = 5$ and $b = 3$ and the angle $\beta = 120°$. Calculate the length of the diagonal and the area of square.

But if the given quadrilateral square is of a more general or different (for example kite of type), I sometimes have to try a bit harder to calculate the area of square. The *main trick* is to split the square into two (or even four) triangles by adding one (or both) diagonal(s) of square. Depending on the given sizes, I can then choose the most convenient formula for calculating the area of the partial triangles from the previous section.

Example 4.13 In a kite square let the lengths a and b be given and the angle α be known as in Fig. 4.15. Then, using the **cosine theorem from** the last section, I can first calculate the length d of the diagonal in square connecting the two vertices with the remaining interior angle $d = \sqrt{a^2 + b^2 - 2ab\cos(\alpha)}$. The two triangles created by adding this diagonal from square, triangle$_1$ and triangle$_2$ are **congruent** – so in particular they have the same area. Since I now know the three side lengths of triangle$_1$, I can **apply Heron's formula** to determine the area of triangle$_1$. Twice that is then the value of the area of square.

Example 4.14 In a kite square let the lengths a and b and the angle β be given as in Fig. 4.15. I then add the diagonal to square, which connects the two vertices with the same interior angle α, and denote its length by d. Two **isosceles triangles** are then formed, whose base side is this diagonal. To determine their length d, I can again apply the **cosine theorem**: $d = \sqrt{2a^2 - 2a^2\cos(\beta)} = a\sqrt{2(1 - \cos(\beta))}$. After that I could use Heron's formula. However, an alternative possibility arises from Exercise 4.4, in which I asked you to derive a general formula for the area of isosceles triangles. The sum of the areas of the two triangles involved then gives the area of

$$F = \frac{d}{4}\left(\sqrt{4a^2 - d^2} + \sqrt{4b^2 - d^2}\right)$$

As is so often the case in life, the simplicity or difficulty of a task depends on the given basic conditions. It is no different with the calculation of the area of a kite – please have a look at the next task:

Exercise 4.16 Justify why the area F of a kite square is half the product of the lengths of the diagonals in square.

If for a general quadrilateral the side lengths a, b, c and d are known, as *well as* the opposite angles β and δ, as in Fig. 4.15, I can use the same trick as in the above two examples. Adding the diagonal as in Fig. 4.15 divides the quadrilateral into two triangles, and by applying the **cosine theorem**, after some laborious calculation, which I will spare us, we arrive at the following formula for the area F of the general quadrilateral, which is related in construction to **Heron's formula**:

$$F = \sqrt{(s-a)(s-b)(s-c)(s-d) - abcd\cos(\sigma)},$$

where $s = (a + b + c + d)/2$ and $\sigma = (\beta + \delta)/2$. If the quadrilateral is concave, the angles β and δ given here should not be shared by the diagonal.

So you can see that it is usually successful to calculate sizes of somewhat complicated figures using known relations for simpler figures. As a final example in the context of calculating the area of quadrilaterals, I will consider the **trapezoid**. Here I can proceed as in Fig. 4.16. There I assume that the lengths a and b *of* the two parallel sides as well as the

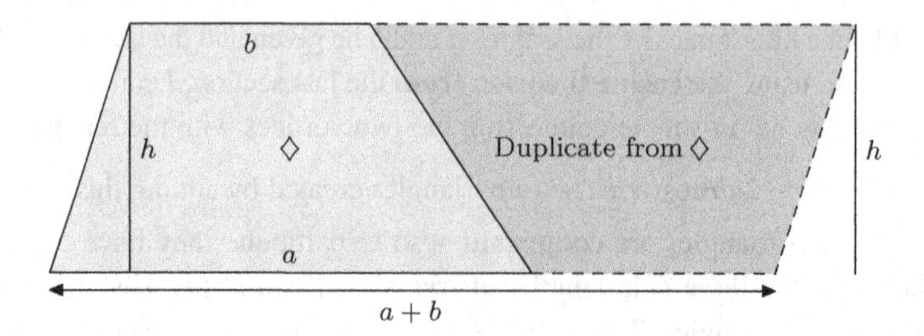

Fig. 4.16 The area of a trapezoid llsquare is half the area of a suitable parallelogram obtained by duplicating llsquare

length of the height h of a trapezoid square are given. One idea now is to attach an inverted duplicate of the trapezoid to square, so that a simpler figure – a parallelogram – is created altogether. This parallelogram has a base side of length $a + b$ and the corresponding height of length h. Furthermore, this parallelogram has twice the area of square, so I arrive at the following statement:

Area of the Trapezoid

The area F of a trapezoid with side lengths a and b *of* parallel sides and height length h is $F = (a + b)\, h/2$.

Exercise 4.17 For a trapezoid square the lengths $a = 6$ and $b = 4$ of the two parallel sides are given. Further, let the side of length *a enclose* an angle of $\alpha = 45°$ with a side of square of length $c = \sqrt{2}$. Determine the area of square.⍰

Exercise 4.18 Determine *all the* side lengths and the angles of an isosceles trapezoid of area 1 in which one of the two parallel sides has the same length as the height to that side and the remaining parallel side is twice as long.

If I now want to consider more general plane figures than triangles and quadrilaterals, which I can nevertheless draw in the plane in a simple way, I arrive at **polygons**, which are also called ***N*-corners**, because their number of corners is N – a natural number greater than 2. To construct an *N-corner*, I give myself N points in the plane and connect them with N lines in such a way that they do not intersect and the starting point of the first line coincides with the ending point of the last line (see Fig. 4.17). These two properties guarantee that such a **closed polygon** (all lines) is actually the boundary of a two-dimensional figure – the *N-corner.*

I now realize, by briefly considering what *N vertices* look like for $N = 3$ and $N = 4$, that triangles and quadrilaterals are just special cases of polygons. I explained to you that the angles in triangles are always less than 180°, and that this need not be true for quadrilaterals, because they can be concave. Now more generally, for $N \geq 4$, an *N-corner*

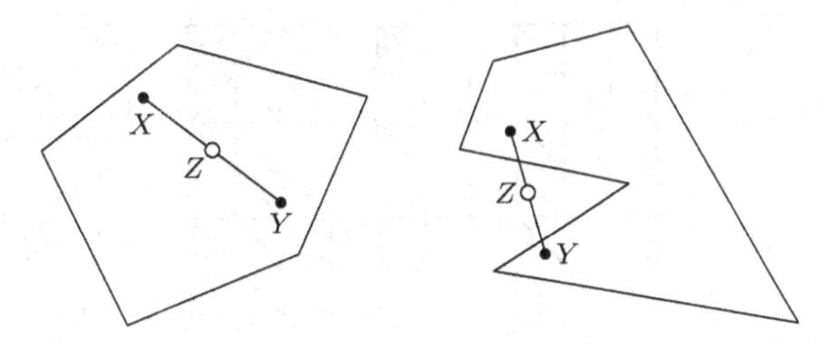

Fig. 4.17 A convex pentagon (five-corner) and a concave hexagon (six-corner)

Λ is called **convex** if all the angles in Λ are less than or equal to 180° and otherwise it is called **concave** *N-corner*. In other words, this means that in convex *N-corners* Λ with two points X, Y each, also every point Z *of* the connecting line between these points, i.e., all points Z of the form

$$Z = tX + (1 - t)Y \text{ for ein } t \in [0, 1],$$

lies in Λ (see Fig. 4.17). On the other hand, in concave polygons Λ, points X *and* Y exist in Λ, so at that point Z on their connecting line is not in Λ. The addition of points in the plane and multiplication by a real scalar is to be understood here as I explained it to you in the last section before calculating the intersection of the side bisectors in the triangle.

Sometimes it is necessary to do similar calculations as I showed you above for the quadrilateral, also for given *N-corners.* You may remember this from laying your carpet: it is rare that the part of the floor you want to carpet actually corresponds to a rectangular footprint. Usually there is a cupboard in the way, the room goes around a corner, the walls in your old flat are crooked or something similar. In other words, the area to be covered often forms a more general *N-corner*, like the outlined area in Fig. 4.18. Now, when you are standing in the carpet shop, you are usually faced with the unpleasant question of how large a rectangular piece of carpet offered there should be. On the one hand, you are aiming to cover the entire floor – on the other hand, you do not want to have too much carpet residue after laying, because you do not that much money – otherwise you would hardly lay the carpet yourself.

Fortunately, it is now the case that I can always decompose *N-corners* into triangles – and I know about triangles relatively well after browsing the last section. So this is still my *main trick*, so it is now applicable to *N-corners as* well. So in principle, this works the same way I showed you above for quadrilaterals, that is, I add certain interior lines to a given *N-corner*. Figure 4.19 (left) illustrates the procedure. For a given *N-corner* Λ, I can peel off step by step the triangles with the numbering there, or draw in the corresponding dashed stretches, until finally the whole *N-corner* is decomposed into triangles. The resulting **triangles** form a **triangulation** $\mathcal{T}$ of Λ and one even knows the number of triangles: $N - 2$. This procedure works for any *N-corner* – I show a slightly larger example in Fig. 4.19 (right). Sometimes it makes sense to look at other triangulations $\mathcal{T}$ – for example, you can cover the floor of your living room (see Fig. 4.18 (left)) with regular triangles by adding a few inner points – this has the advantage that you only need to know, for example, the

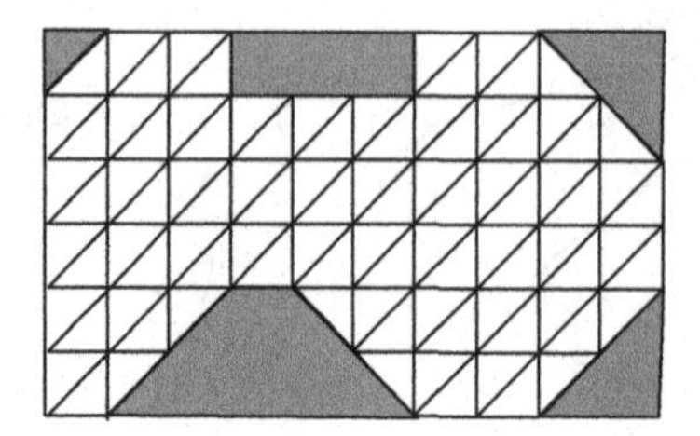

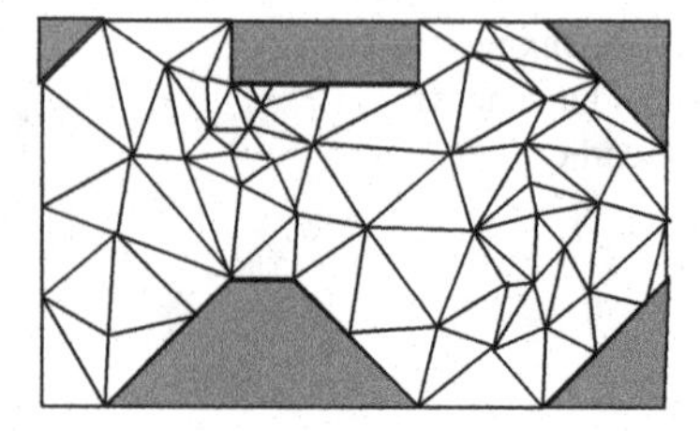

Fig. 4.18 Even and uneven triangulations of the floor of your living room. *Grey areas*: cupboards, glass cabinet, TV cabinet, etc.

number of triangles and their area to see how much carpeting you need. So, for example, if 0.5 m^2 is the area of each of the 93 triangles in Fig. 4.18, the area to be carpeted is 46.5 m^2. If your carpet shop sells carpet on rolls of width 2 meters, you should buy 24 m of carpet. In this case you will have 1.5 m^2 left over – which is not a mistake, because the next red wine glass is already waiting to be tipped over. You can, of course, consider more general triangulations like the one in Fig. 4.18 (right). Certain mathematicians deal with such objects – but from the practical point of view of carpet laying this is less recommendable as far as I know.

A typical example where I can apply the decomposition of an *N-corner* into triangles is to determine the sum of all N interior angles α of any *N-corner* Λ. By decomposing these angles into angles such as α' and α'' in Fig. 4.19 (left) according to all the dashed vertices ending in a vertex, and using that the sum of the angles of each of the $N - 2$ triangles in Λ is always 180°, I arrive at the following result:

Sum of the Interior Angles of an *N-corner*
The sum of the N interior angles of an *N-corner* is always $(N - 2)$ 180°.

Of particular interest are regular *N-corners*. These are *N-corners* as in Fig. 4.20, where the N sides all have the same length b and the N interior angles α always coincide, i.e.

$$\alpha = \frac{(N-2)180^\circ}{N} = 180^\circ - \frac{1}{N}360^\circ$$

Holds true. To calculate the area F_N of such an *N-corner* Λ, I can make life easy for myself by choosing another point M at the center of Λ and considering the uniform triangulation $\mathcal{T}$ of Λ that arises when I connect M to all vertices of Λ. The N vertices of the *N-corner* then lie on a circle with center M and radius a (see Fig. 4.20 (left)). The triangulation $\mathcal{T}$ obviously consists of N **isosceles** triangles that are **congruent.** For each of these triangles Δ, I can immediately see that the angle in Δ *opposite* the side with length b has value $\beta = 360°/N$. Using the sine therefore gives

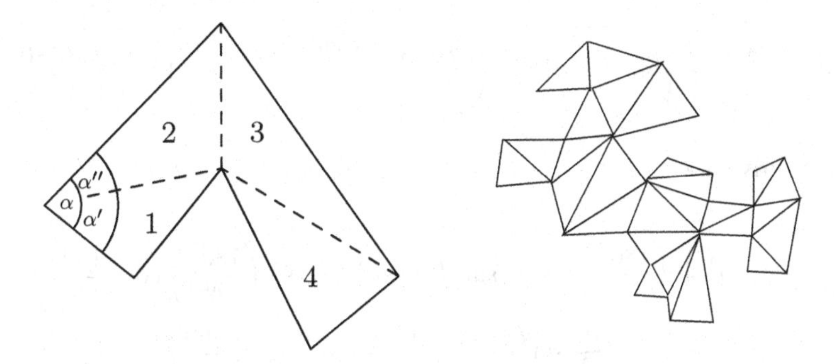

Fig. 4.19 *N-corners* can always be decomposed into triangles, resulting in a triangulation

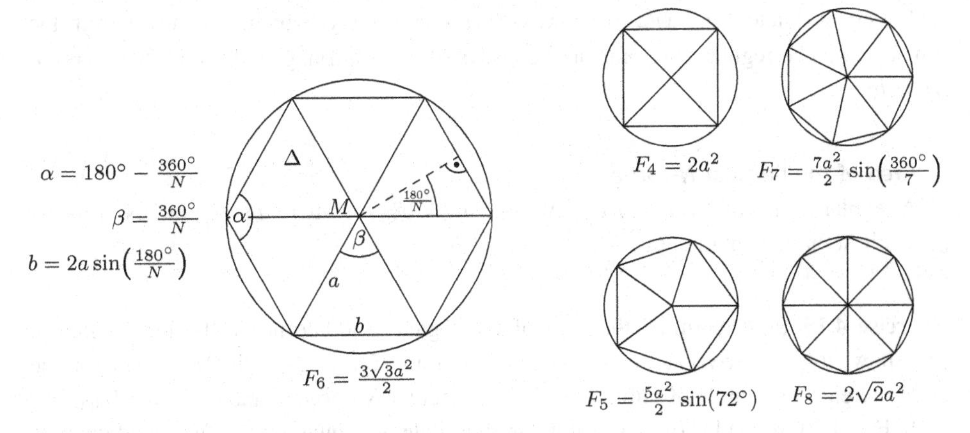

Fig. 4.20 The area F_N of regular *N-corners* is determined according to the formula $F_N = (Na^2/2)\sin(360°/N)$. On the left, one can see a regular hexagon together with the symbols and contexts used in the derivation of the general formula. Here, exceptionally, $a = b$ holds. On the right one sees a regular quadrilateral, pentagon, heptagon and octagon together with the formulas for their area for these choices of N, that is $N = 4, 5, 7, 8$

$$\sin\left(\frac{180°}{N}\right) = \frac{\frac{b}{2}}{a}, \text{respectively } b = 2a\sin\left(\frac{180°}{N}\right),$$

where a is the length of the legs of $\triangle$. Now, remembering the area formula for isosceles triangles from Exercise 4.4, I can find the area $F\triangle$ *of* the triangles $\triangle$ by substituting b

$$F_{\triangle} = \frac{b}{4}\sqrt{4a^2 - b^2} = \frac{2a\sin\left(\frac{180°}{N}\right)}{4}\sqrt{4a^2 - \left(2a\sin\left(\frac{180°}{N}\right)\right)^2}.$$

By cancelling, squaring and taking the square root of $4\,a^2$ I get

$$F_{\triangle} = \frac{a\sin\left(\frac{180^\circ}{N}\right)}{2}\sqrt{4a^2 - 4a^2\sin^2\left(\frac{180^\circ}{N}\right)} = \frac{a\sin\left(\frac{180^\circ}{N}\right)}{2}2a\sqrt{1-\sin^2\left(\frac{180^\circ}{N}\right)}.$$

Finally, we arrive at

$$\begin{aligned} F_{\triangle} = & \; a^2\sin\left(\frac{180^\circ}{N}\right)\sqrt{1-\sin^2\left(\frac{180^\circ}{N}\right)} = a^2\sin\left(\frac{180^\circ}{N}\right)\cos\left(\frac{180^\circ}{N}\right) \vdash \\ = & \; \frac{a^2}{2}\sin\left(\frac{180^\circ}{N}+\frac{180^\circ}{N}\right) = \frac{a^2}{2}\sin\left(\frac{360^\circ}{N}\right), \end{aligned}$$

where for the second and the third last equality I exploited the relations $\sin(x) = \sqrt{1-\cos^2(x)}$ and $\sin(2\,x)/2 = \sin(x)\cos(x)$, respectively, which are known from the last section. The regular *N-corner* now consists of *N* such triangles lltriangle with area *F triangle*.

Area of the Regular *N-corner*
A regular *N-corner* has area $F_N = N\,(a^2/2\,\sin(360°/N))$, where *a* is the radius of the smallest circle containing it.

Exercise 4.19 (a) Determine the area of the regular 3600-corner with side lengths *b*. (b) Show how to arrive at the approximate formula $U_N = 2\,N\,\sin(180°\,/\,N)$ for the circumference of a circle of radius $a = 1$ using regular *N-corners*, and calculate U_{3600}.

In Fig. 4.20 (right) I show you a regular quadrilateral, pentagon, heptagon and octagon together with the special formulas for their areas. You can see from these special cases – but also from the general formula – that the area of these *N-corners* depends not only on the radius *a of* the circle with center *M*, which I have **circumscribed to** these figures, but also only on the angle 360°/N, which is determined by *N*.

Speaking of circles. Of course, you already know that I form a general **circle** in the plane by considering, as in Fig. 4.21, the set of all points $(x|y)$ that have the same fixed distance *r* - called the radius - from a fixed **center** $M = (x_1|y_1)$. The double length of the radius is called the **diameter of** the circle. The distance of two points $(x|y)$ and $(x_1|y_1)$ in the plane I can calculate according to what I told you in the last section, using the **Pythagorean theorem** by the formula $r = \sqrt{(x-x_1)^2 + (y-y_1)^2}$. Thus, the set of all points $(x|y)$ on the circle with center $M = (x_1|y_1)$ and radius *r is given by* the **circle equation**

$$(x-x_1)^2 + (y-y_1)^2 = r^2$$

described. Some people also write this down as follows:

$$\left(\frac{x-x_1}{r}\right)^2+\left(\frac{y-y_1}{r}\right)^2=1.$$

Remembering the introduction of the sine and cosine from the last section and taking a quick look at Fig. 4.21, I can see the validity of the alternative **parameter representation of the circle:**

$$x=x_1+r\cos(t)\,\text{und}\,y=y_1+r\sin(t), t\in[0,2\pi).$$

For orientation I may consider that the N vertices $(x_k|y_k)$, $k = 0, \ldots, N-1$, of a regular *N-corner* as in Fig. 4.20 for example are given by

$$(x_k|y_k)=\left(x_1+r\cos\left(\frac{k360^\circ}{N}\right)|y_1+r\sin\left(\frac{k360^\circ}{N}\right)\right), k=0,\ldots,N-1,$$

(where $r = a$) are given. Already in the last section I explained to you that three points forming the vertices of a triangle already completely define a circle – the **circumcircle of triangle**. Now we can explicitly calculate such circles. Here I consider a slightly simplified example – because of the simple choice of points.

Example 4.15 I determine the circle through the points (0|0), (1|0), and (1|1) by substituting them into the general equation of the circle:

$$x_1^2+y_1^2=r^2, (1-x_1)^2+y_1^2=r^2\,\text{und}\,(1-x_1)^2+(1-y_1)^2=r^2.$$

Subtracting the first equation from the second equation, I get $(1-x_1)^2-x_1^2=0$, that is, after applying the **binomial formula** $1 - 2x_1 = 0$, so $x_1 = 1/2$. Subtracting the third equation from the second equation leads to $y_1^2-(1-y_1)^2=0$, and I calculate just as I just did $y_1 = 1/2$. The center point is thus $M = (1/2|1/2)$. Furthermore, I get the radius r by substituting $x_1 = y_1 = 1/2$ into one of these three equations – say the first one: $1/4 + 1/4 = r^2$ – so $r=1/\sqrt{2}=\sqrt{2}/2$. The circle equation found is $(x - 1/2)^2 + (y - 1/2)^2 = 1/2$. Alternatively, of course, I could calculate the center point as the intersection of two perpendicular bisectors and then find the radius as I just did.

Definition of Circles

Three points that do not lie on a common straight line (i.e. they form a triangle) uniquely define a plane circle.

Circles differ from the plane figures discussed above primarily in that they are not bounded by lines – one also speaks of a **curved edge**. Now, at first glance, it seems difficult

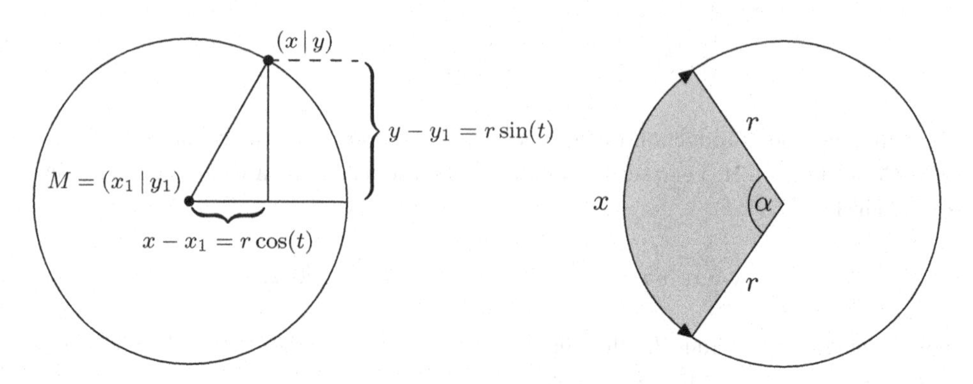

Fig. 4.21 The points $(x|y)$ on a circle with center $M = (x_1|y_1)$ and radius r can be expressed parametrically using the sine and cosine. The center angles α are proportional to the length x of the corresponding arc

to calculate the area of a circle using a convenient formula. After all, it is probably impossible to *exactly* represent a circle with a finite number of triangles or other plane figures bounded only by straight lines. Here I am helped by an observation I can make by looking closely at Fig. 4.20. With increasing number of corners N the area of the regular *N-corner* seems to be a better and better approximation to the area of the **unit circle** (that is the circle with radius $r = a = 1$) with center – say $M = (0|0)$. The area of the regular *N-corner* as in Fig. 4.20 is for all natural numbers N always smaller than the area of the circle – but in fact it is the case that the latter approximates itself arbitrarily exactly – one says *from below*. At this point I do not want to define a stretched plane figure approximating the area of the circle *from above, in order* to then carry out possible **limit value considerations**, but instead I offer you Table 4.2. There I have calculated the area of *N-corners* for large values N *on* a long winter evening in laborious domestic work:

I then chose a huge N and calculated the area of the unit circle – also called **the number** π – to 40 decimal places

$$\pi \approx 3.1415926535897932384626433832795028841971 6.$$

You can see that the decimal places of π are completely irregular and no schemes are discernible in their structure. This is in the nature of things, because π – as we now know – is not a **rational number**. What is more, π is also crucially different from roots – such as $\sqrt{2}$ – but we will spare these details about **transcendental numbers** and the **impossibility of squaring the circle** from algebra. More importantly, as you begin your study, I note the following:

Circular Area

The unit circle has the area $F = \pi$. Any circle with radius r has area $F = \pi\, r^2$.

Table 4.2 Approximations of the area π of the unit circle calculated with a computer algebra (in a few seconds) by using regular N *vertices* for different values of N

N	F_N
16	3.061467458920
64	3.136548490545
256	3.141277250932
1024	3.141572940367
4096	3.141591421511
16,384	3.141592576584
65,536	3.141592648776
262,144	3.141592653288
1,048,576	3.141592653570
4,194,304	3.141592653588

Now how did I come up with the area of any circle with radius r so quickly? Well – there is no mystery here. The area of the regular *N-corner* in the unit circle is $F_N = (N/2)$ sin (360°/N), while the area of the regular *N-corner* in any circle with radius $r = a$ *is* given by $F_N \cdot r^2$. However, the different multiplicative factor r^2 is just the factor that also appeared in the last statement.

To calculate the length of the distance I have to walk around a circle once, the so-called **circumference of a circle**, I can now proceed as follows, similar to the calculation of the area. Your experience from Exercise 4.19b) will help you here.

Circumference of the Circle

The unit circle has a circumference of 2π. Any circle with radius r has circumference $U = 2\pi r$.

At this point, if I ask myself how long is the arc x *of* a circle of radius r to a given central angle α as in Fig. 4.21, I can quickly see that for this $x = 2\pi\alpha/360° = \pi\alpha/180°$. A slight generalization of the value x *is* what I called **radians in** the previous section (where I also allowed negative values) – as I am sure you remember, since you are probably younger than I am.

If, on the other hand, I want to calculate the area F_α of a **circle section** (also: **circle sector**) with center angle α as in Fig. 4.21, this is no longer a problem for the experienced pizza eater – this area is given by

$$F_\alpha = \frac{\alpha}{360°}\pi r^2 = xr^2/2.$$

Exercise 4.20 Thursday is, as always, big pizza day and you order a family pizza with a diameter of 80 cm for you and your seven friends. How many cm^2 of pizza – assuming even distribution – does each of you get? Approximately how many cm of uncovered crust is available to each of you? This week, your pizza chef makes you a very special offer: he sells you a pizza in the shape of a circular ring with an outer diameter of 1 m and an inner diameter of 50 cm for the same price. You calculate briefly how many cm^2 of pizza and how many cm of uncovered edge each of your friends would now get. Do you accept the offer?⍰

Finally, let me remind you of a few things about **ellipses**: These are more general plane figures than circles, which also have curvilinear edges. They are formed from the set of all points $(x \mid y)$ for which the sum $r_1 + r_2$ of the distances to two fixed points $A = (a_1 \mid a_2)$ and $B = (b_1 \mid b_2)$ – called **focal points** – has a constant value $2r$. So if I remember how to calculate the distance between two points in the plane using the **Pythagorean theorem**, I get the **general elliptic equation**

$$\overbrace{\sqrt{(x-a_1)^2+(y-a_2)^2}}^{r_1}+\overbrace{\sqrt{(x-b_1)^2+(y-b_2)^2}}^{r_2}=2r.$$

In the following, for the sake of simplicity of description, I always assume that the two foci lie on the *x-axis* as in Fig. 4.22 and that the **center of** the ellipse (center of the foci) is the **origin** $(0 \mid 0)$: $A = (-c \mid 0)$ and $B = (c \mid 0)$, where $c > 0$. The ellipse equation is then of the following form:

$$\sqrt{(x+c)^2+y^2}+\sqrt{(x-c)^2+y^2}=2r.$$

In this case the two **major vertices of** the ellipse are the points $H_1 = (-r \mid 0)$ and $H_2 = (r \mid 0)$, the line of length $2r$ connecting these points is called the **major axis of** the ellipse and r is called the **major axis radius.** The two **minor vertices of** the ellipse are then the points of the form $N_1 = (0 \mid b)$ and $N_2 = (0 \mid -b)$ on the ellipse – that is, the points of intersection of such an ellipse with the *y-axis*. These points are the intersections of the circles of radius r about the foci, and I obtain from the **Pythagorean theorem** $b^2 = r^2 - c^2$. Thus $N_1 = \left(0 \middle| \sqrt{r^2-c^2}\right)$ and $N_2 = \left(0 \middle| -\sqrt{r^2-c^2}\right)$. The minor vertices form the minor **axis of** the ellipse and $b = \sqrt{r^2-c^2}$ denotes the **minor axis radius.**

From the triangle inequality at the beginning of the previous section it follows that the distance of the foci $2c$ must be smaller than $r_1 + r_2 = 2r$. This is the criterion which already shows me how large I should choose the major axis radius r at least in order to be able to construct an ellipse with given foci. But if points are given on an ellipse to be constructed, the situation is more difficult than with circles, because three points are generally not enough to define an ellipse. For your information, I will tell you here that four *suitable points* are needed. Further, I can observe that ellipses are symmetrical figures with respect to the major and minor axes.

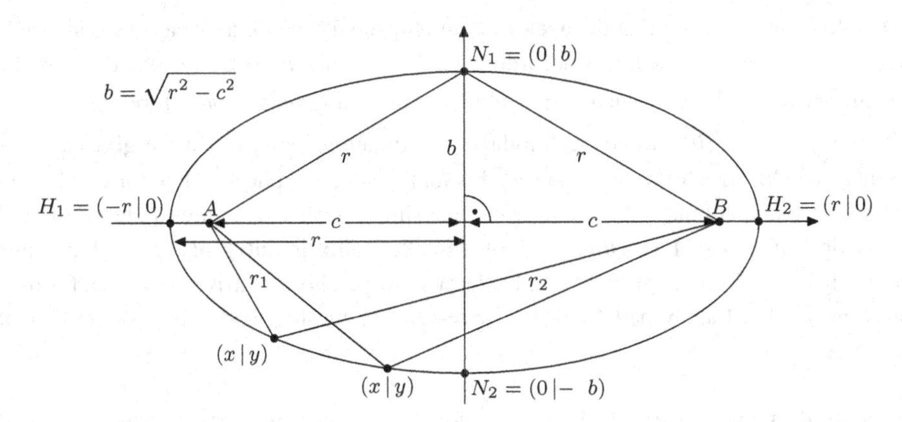

Fig. 4.22 The points (x|y) on an ellipse have the property that the sum $r_1 + r_2$ of the distances r_1 and r_2 to the foci A and B is constant $2r$

If I consider the special case that the two foci coincide $A = B = (0 \mid 0)$, respectively $c = 0$, I can first write the elliptic equation in the form $\sqrt{x^2 + y^2} + \sqrt{x^2 + y^2} = 2r$ and then as $2\sqrt{x^2 + y^2} = 2r$. Cancelling 2 and then squaring both sides gives $x^2 + y^2 = r^2$ – so this special case is a circle with radius r and center (0|0). At this point the question arises, if I could not write the above form of the elliptic equation much simpler. Indeed: By relatively tricky transforming one arrives at the alternative representation

$$\frac{x^2}{r^2} + \frac{y^2}{b^2} = 1,$$

where $b^2 = r^2 - c^2$ still applies. This is also called the **midpoint equation of** the ellipse. I notice that the squares of the lengths of the major and minor axes occur here in the denominator, respectively, and the special case $c = 0$ (thus $b = r$) – as already seen above – represents the circle with radius r and center $(0 \mid 0)$. Also the parameter representation of the circle can be generalized for ellipses:

$$x = r\cos(t) \quad \text{and} \quad y = b\sin(t), \quad t \in [0, 2\pi).$$

Example 4.16 I can calculate the ellipse $\mathcal{E}$ through the point $X = (6|4)$ with center (0|0) and major axis radius $r = 10$ as follows: Since X lies on $\mathcal{E}$, $6^2/10^2 + 4^2/b^2 = 1$ and $16/b^2 = 1 - 36/100 = 16/25$, respectively. Thus: $b^2 = 25$ and so $c = \sqrt{r^2 - b^2} = \sqrt{75} = 5\sqrt{3}$. Thus, the ellipse has the equation $x^2/100 + y^2/25 = 1$ and its parameter representation is as follows: $x = 10\cos(t)$ and $y = 5\sin(t)$, $t \in [0,2\pi]$. The foci of $\mathcal{E}$ are $(5\sqrt{3}|0)$ and $(-5\sqrt{3}|0)$, while the minor vertices of $\mathcal{E}$ are the points $(0 \mid 5)$ and $(0 \mid -5)$.

Finally, I will tell you that the **area** F *of* an **ellipse** with major axis radius r and minor axis radius b can be calculated according to the formula $F = \pi\ r\ b$ *and* that for the **circumference** U **of** such an **ellipse** the approximation formula $U \approx \pi\left(1.5 \cdot (r + b) - \sqrt{rb}\right)$ is known. Similar approximation formulas can be **given** for the length of an **elliptic arc**. As you can see, this looks more complicated than for circles – in fact, the derivation of these statements is not that simple at all and requires quite a profound knowledge of **integral calculus** and of so-called **elliptic integrals**. I do have this knowledge – and in Example 6.43 I will at least show you how to arrive at the area formula for ellipses – but for my part I will now interrupt the writing work and prepare a small barbecue.

Exercise 4.21 At a summer barbecue, you notice that the 20-cm-diameter round hamburgers, when frozen, meld together into an ellipse two-thirds the size (area), with a major-axis radius twice the minor-axis radius. The problem would not let you go: you are dying to know what the major and minor axis radii of the resulting Burger ellipses are.

Introduction to Linear Algebra

5

By now you have made quite a bit of progress, have not you? You have remembered the old unpleasant things like parentheses, multiplying out or logarithms, you have seen how to deal with certain functions, you have been maltreated with different kinds of equations and even a little geometry has been offered to you. That was not so little, and once you take a closer look, you will notice that the first three of the chapters so far have one distinct thing in common: Everywhere, even in the discussion of elementary arithmetic methods, we have had to leave the realm of pure numerical arithmetic and resort instead to arithmetic with letters. Sure, calculating with concrete numbers may be more convenient, but when it comes to formulating general rules or simplifying any formulas, for example, there is simply no getting around calculating with the so-called variables. In general, this alphabetic arithmetic is referred to by the old term "algebra".

Now, however, you could find that this algebra has led you down a wide variety of paths. After all, we gather under this term such different things as calculating with logarithms and simple addition, such different types of equations as the simple linear equation and the incomparably more complicated root or even exponential equation. So we were dealing with a pretty broad area, and in this chapter I want to narrow it down a bit and just deal with "linear algebra". If you recall linear equations for a moment, you will have a good outlook, because such troublesome operations as squaring or logarithmizing were not to be found there by any stretch of the imagination: Linear equations look essentially like $5\,x + 3 = 17$, and no x^2 or $\log_2 x$ in the world has the right to stray there. Everything I consider in this chapter will be **linear**, and that means that the arithmetic operations that occur do not go beyond basic arithmetic.

Sounds good, does not it? But in life, nothing is free (though some things are in vain). While I will subject myself to very strong restrictions in the permissible arithmetic operations, I may on the other hand, so to speak through the back door, bring a little

G. Walz et al., *Foundations of Mathematics: A Preparatory Course*,
https://doi.org/10.1007/978-3-662-67809-1_5

more excitement into the matter, otherwise I could namely give this chapter to myself. When solving linear equations, and other equations as well, there was always *a single* unknown whose value you had to figure out. Occasionally there were multiple solutions, but those were always solutions for the one unknown you were looking for. Unfortunately, it happens more often in reality that not only one quantity is unknown, but two, three or seventeen at once. We have not even talked about such situations here, and by now you have learned enough algebra to tackle such problems as well. So in this chapter on linear algebra we make a small deal: On the one hand, I refrain from bringing unpleasant arithmetic operations into play and restrict myself strictly to the linear, on the other hand, I may increase the number of quantities used, which so far has mostly been one, at will – otherwise linear algebra would also be too boring and you could safely skip the chapter.

After this long preface, let us look at what I actually mean using vectors as an example.

5.1 Vectors

In linear algebra, we are always dealing with vectors, and I do not want to deprive you of them either. You will see in Example 5.1 that you can use them to combine different pieces of information into a total piece of information:

Example 5.1 Imagine that you are the happy owner of a company *MyCompany* that produces three different products A_1, A_2 and A_3, leaving the design of the products to your imagination and preferences. Now, *MyCompany surely* will not produce the same quantity of each of the three products, because you cannot expect the sales market to be able to handle, say, as many toilet lids as desk lamps. So your company will have a certain sales volume per product, and let us assume that the sales volume for the first product is 18 units, for the second product it is 34 units, and for the third product it is 17 units. Of course, units can also mean packs of thousands or something else, but we do not need to know that now. The question is: How can I represent this situation in a meaningful formula, preferably in a data structure that I can also save with a computer?

But there are not only sales quantities; to calculate the turnover, of course, you still need the unit prices of the three products. Let us say the prices are 12 monetary units for one unit of the first product, 9 monetary units for the second product, and 13 for the third. Now, how can you combine these three unit prices into a single data structure?

You can probably already see where this is going. In order to combine three numerical values that should somehow form a unit without losing the individual values, I simply make a so-called three-dimensional vector out of them. We will take a look at what something like that looks like in a moment:

Three Dimensional Vector

A three-dimensional vector is a quantity of the form

$$\mathbf{x} = \begin{pmatrix} x_1 \\ x_2 \\ x_3 \end{pmatrix},$$

where $x_1 \in \mathbb{R}, x_2 \in \mathbb{R}, x_3 \in \mathbb{R}$. The set of all three-dimensional vectors is denoted by $\mathbb{R}^3$.

In a three-dimensional vector, one summarizes only three real numbers, no more and no less. Often such vectors are named with bold lower case letters, and I do not want to deviate from this rule here, so the vector is called $\boldsymbol{x}$ and not just x. I can now solve the two problems from Example 5.1 quite easily with the help of two three-dimensional vectors, namely the "heel vector"

$$\mathbf{a} = \begin{pmatrix} 18 \\ 34 \\ 17 \end{pmatrix}$$

and a "price vector"

$$\mathbf{p} = \begin{pmatrix} 12 \\ 9 \\ 13 \end{pmatrix}.$$

So it does not matter whether you enter sales quantities, prices or anything else in the vector; the structure always remains the same.

You are probably wondering what to do with a company that manufactures only two products or four or five products at once. It would make little sense to write down a new definition for each dimension, especially since there are an infinite number of natural numbers and I do not have an infinite number of pages available. We therefore need a vector term that takes care of all possible cases, i.e. all conceivable dimensions, in one fell swoop, and this is what one does by talking about *n-dimensional* vectors. If one does not want to specify how many entries the vector should have, then one takes the liberty to denote the number of entries with the variable n, and depending on the need then $n = 3$ or $n = 2$ or $n =$ something else.

Vector

An *n-dimensional* vector is a quantity of the form

$$\mathbf{x} = \begin{pmatrix} x_1 \\ x_2 \\ \vdots \\ x_n \end{pmatrix},$$

where $x_1 \in \mathbb{R}, \ldots, x_n \in \mathbb{R}$ holds. The set of all *n-dimensional* vectors is denoted by $\mathbb{R}^n$.

Mostly, one simply speaks of a vector and omits the addition *n-dimensional,* because the vector must have some dimension anyway. If you are irritated by the three dots standing around in the middle of the vector: They just symbolize that there should be a total of n numbers below each other here, and since you do not know the value of n in this general case, you are comfortable with this simple three-dot notation. Two- and three-dimensional vectors, by the way, allow a completely different representation, which has something to do with geometry; but we will come to that in the fourth section of this chapter.

So now we can write vectors down, but that is about it. Of course, you could also throw a vector around or paint it green, but who cares about vectors painted green? We were dealing with objects made of numbers here, and it should be possible to do math with them. In the next example, you will see how to add vectors in a meaningful way.

Example 5.2 Again, I use the sales vector $\boldsymbol{a}$ from Example 5.1. Assuming that it describes sales in the first quarter of the current year, and further assuming that you have a vector $\boldsymbol{b}$ that stores production in the second quarter, you can ask yourself how much your company *MyCompany* probably produced during the first half of the year. With the vectors

$$\mathbf{a} = \begin{pmatrix} 18 \\ 34 \\ 17 \end{pmatrix} \quad \text{and} \quad \mathbf{b} = \begin{pmatrix} 19 \\ 33 \\ 21 \end{pmatrix}$$

for the first half of the year, this results in an output volume of

$$\begin{pmatrix} 18+19 \\ 34+33 \\ 17+21 \end{pmatrix} = \begin{pmatrix} 37 \\ 67 \\ 38 \end{pmatrix}.$$

And already you know how to add vectors.□

According to the scheme of Example 5.2 one can always proceed, if one wants to add two vectors: One simply adds their **components** fitting to each other. This can only work, however, if you are dealing with vectors of the same dimension, otherwise you will end up with a vector with one component left over, and no one can tell it where to lay its weary head. So we only add vectors of the same dimension, better leave mixed combinations alone.

Addition and Subtraction of Vectors

Are

$$\begin{pmatrix} x_1 \\ \vdots \\ x_n \end{pmatrix} \text{ and } \begin{pmatrix} y_1 \\ \vdots \\ y_n \end{pmatrix}$$

two *n-dimensional* vectors, then set

$$\begin{pmatrix} x_1 \\ \vdots \\ x_n \end{pmatrix} \pm \begin{pmatrix} y_1 \\ \vdots \\ y_n \end{pmatrix} = \begin{pmatrix} x_1 \pm y_1 \\ \vdots \\ x_n \pm y_n \end{pmatrix}.$$

Thus, one adds or subtracts two vectors of the same dimension by adding or subtracting the respective components of the vectors, that is, their entries.

As you can see, I did the subtraction at the same time, which was probably not very surprising. You have already seen an example of addition above. Nevertheless, I will show you more examples in a moment, but first I want to go one step further and demonstrate another calculation method: multiplication by a number. But there is nothing complicated behind this either, as you can easily figure out. If we take, for example, the two-dimensional vector $\begin{pmatrix} 1 \\ 2 \end{pmatrix}$, then of course $\begin{pmatrix} 1 \\ 2 \end{pmatrix} + \begin{pmatrix} 1 \\ 2 \end{pmatrix} = \begin{pmatrix} 2 \\ 4 \end{pmatrix}$ applies according to the rule for addition that you have just learned. But a thing plus the same thing is still the twofold thing, and consequently I multiplied the vector $\begin{pmatrix} 1 \\ 2 \end{pmatrix}$ by the number 2 to get the vector $\begin{pmatrix} 2 \\ 4 \end{pmatrix}$. I guess you see where I am going with this, because it holds:

$$\begin{pmatrix} 2 \\ 4 \end{pmatrix} = \begin{pmatrix} 2 \cdot 1 \\ 2 \cdot 2 \end{pmatrix}.$$

So you multiply a vector by a number by multiplying each of its components by that number.

Multiplication of a Vector by a Number

For a real number λ and a vector

$$\begin{pmatrix} x_1 \\ \vdots \\ x_n \end{pmatrix}$$

one sets

$$\lambda \cdot \begin{pmatrix} x_1 \\ \vdots \\ x_n \end{pmatrix} = \begin{pmatrix} \lambda \cdot x_1 \\ \vdots \\ \lambda \cdot x_n \end{pmatrix}.$$

Do not mind the ominous λ, that is just the Greek version of an *l*, and it is become a bit common, in the context of vectors, to refer to the simple numbers with Greek letters so that they can be easily distinguished from vectors. It is pronounced "lambda", by the way.

But now it is time for an example or two:

Example 5.3 It applies:

$$\begin{pmatrix} -5 \\ 3 \\ 2 \\ 8 \end{pmatrix} + \begin{pmatrix} 9 \\ 7 \\ 1 \\ 17 \end{pmatrix} = \begin{pmatrix} -5+9 \\ 3+7 \\ 2+1 \\ 8+17 \end{pmatrix} = \begin{pmatrix} 4 \\ 10 \\ 3 \\ 25 \end{pmatrix}.$$

Furthermore

$$3 \cdot \begin{pmatrix} 17 \\ 8 \\ 12 \end{pmatrix} = \begin{pmatrix} 3 \cdot 17 \\ 3 \cdot 8 \\ 3 \cdot 12 \end{pmatrix} = \begin{pmatrix} 51 \\ 24 \\ 36 \end{pmatrix}.$$

In contrast, the addition

$$\begin{pmatrix} 2 \\ 3 \end{pmatrix} + \begin{pmatrix} 1 \\ 2 \\ 4 \end{pmatrix}$$

is not possible, because the two vectors have different dimensions.

Would you like a little practice? That is not where it should fail:

Exercise 5.1 Calculate

$$\begin{pmatrix} 1 \\ -2 \\ 5 \end{pmatrix} - 4 \cdot \begin{pmatrix} -2 \\ 1 \\ 4 \end{pmatrix} + 2 \cdot \begin{pmatrix} 12 \\ 0 \\ 1 \end{pmatrix} \quad \text{and} \quad -\begin{pmatrix} 3 \\ -4 \end{pmatrix} + 3 \cdot \left[\begin{pmatrix} 2 \\ 1 \end{pmatrix} - 2 \cdot \begin{pmatrix} -1 \\ 4 \end{pmatrix} \right]. \quad \blacksquare$$

The basic arithmetic operations should already be done with this, because what else should you do apart from adding, subtracting and multiplying? The answer is obvious: division would not be bad either. I am sorry to tell you, however, that it is not possible to divide one vector by another; nothing sensible can come out of that. What does work, though, is dividing a vector by a number, and that is hardly worth mentioning, strictly speaking, because we have actually already discussed it. A small example shows what I mean.

Example 5.4 What do you get when you divide $\begin{pmatrix} 2 \\ 4 \end{pmatrix}$ by 2? As you know, dividing by 2 is equivalent to multiplying by 1/2, and therefore the result is:

$$\frac{1}{2} \cdot \begin{pmatrix} 2 \\ 4 \end{pmatrix} = \begin{pmatrix} 1 \\ 2 \end{pmatrix}. \quad \blacksquare$$

So you divide a vector by a number by multiplying it by the inverse of the number – and multiplying by a number is old hat to you by now. I told you that you basically already know division by a number.

But now to something really new. If you multiply a vector with a number, the result is a vector again. But there is a kind of multiplication between vectors, which leads as a result to a number, which so to speak turns the usual upside down. Let us take a look at how something like this can happen.

Example 5.5 Again, I assume that your company *MyCompany* sells three items A_1, A_2 and A_3. There are 18 units sold of A_1, 34 units sold of A_2 and 17 units sold of A_3. The corresponding unit prices are 12 Euros for a unit of A_1, 9 Euros for a unit of A_2 and 13 Euros for a unit of A_3. The turnover of your emerging company is then calculated as:

$$turnover = 18 \cdot 12 + 34 \cdot 9 + 17 \cdot 13 = 743.$$

But now the given information can be written down again with the help of vectors, because I have here a sales vector ***a*** and a price vector ***p*** with

$$\mathbf{a} = \begin{pmatrix} 18 \\ 34 \\ 17 \end{pmatrix} \quad \text{and} \quad \mathbf{p} = \begin{pmatrix} 12 \\ 9 \\ 13 \end{pmatrix}.$$

I calculated the turnover by multiplying the first component of one vector with the first component of the other, then the second component of the first with the second component of the other and finally the third component of the first with the third component of the other – after that the individual products were added. Because some things are multiplied here, one speaks of a product, and because a number results at the end, one could call the whole thing a number product. But you do not. The more polite term for "number" is "scalar", and that is why the operation I just performed is called the scalar product of the two vectors ***a*** and ***p***. Written in formulas, it looks like this:

$$\begin{pmatrix} 18 \\ 34 \\ 17 \end{pmatrix} \cdot \begin{pmatrix} 12 \\ 9 \\ 13 \end{pmatrix} = 18 \cdot 12 + 34 \cdot 9 + 17 \cdot 13 = 743. \quad ■$$

Of course, this procedure works not only for two three-dimensional vectors, but whenever you are dealing with two vectors of the same dimension.

Scalar Product

Are

$$\begin{pmatrix} x_1 \\ \vdots \\ x_n \end{pmatrix} \text{ and } \begin{pmatrix} y_1 \\ \vdots \\ y_n \end{pmatrix}$$

two *n-dimensional* vectors, then set

$$\begin{pmatrix} x_1 \\ \vdots \\ x_n \end{pmatrix} \cdot \begin{pmatrix} y_1 \\ \vdots \\ y_n \end{pmatrix} = x_1 \cdot y_1 + \cdots + x_n \cdot y_n.$$

This operation is called the scalar product of the two vectors.

The scalar product of two vectors thus serves to combine the contents of these vectors in a certain way. For this, however, both vectors must be of the same length, i.e. have the same dimension. To get you used to this construction, let us look at some more examples.

Example 5.6 It applies:

$$\begin{pmatrix} 2 \\ 5 \\ -1 \end{pmatrix} \cdot \begin{pmatrix} 1 \\ -3 \\ 2 \end{pmatrix} = 2 \cdot 1 + 5 \cdot (-3) + (-1) \cdot 2 = -15.$$

In contrast, the calculation of

$$\begin{pmatrix} 1 \\ 2 \end{pmatrix} \cdot \begin{pmatrix} 3 \\ 4 \\ 5 \end{pmatrix}$$

is not possible, because the two vectors have different lengths, that is, different dimensions.

So now you know how to multiply two vectors together and get a number out of it. However, sometimes life is a little more complicated and it is not enough to be satisfied with just two vectors. Imagine, for example, that you not only have a vector describing the sales quantity and another vector containing the unit prices, but the price vector is split into a net price vector and a VAT vector; a look at your last invoice from the garage shows that this can happen in real life at any time. In this case, you need to scalar multiply the sales volume vector by the sum of the other two vectors, because the two together only give you the unit price. So it may sometimes be necessary to work out scalar products in more complicated situations by multiplying not just one vector by another, but one vector by the sum of two or even more others. However, as soon as such entanglements arise, where several arithmetic operations are involved, one likes to write down arithmetic rules so that one knows at all times what is allowed and what is not. Let us first look at the basic situation with an example:

Example 5.7 On the one hand

$$\begin{pmatrix} 1 \\ 2 \\ 3 \end{pmatrix} \cdot \left[\begin{pmatrix} 0 \\ 1 \\ -1 \end{pmatrix} + \begin{pmatrix} 2 \\ -1 \\ 0 \end{pmatrix} \right] = \begin{pmatrix} 1 \\ 2 \\ 3 \end{pmatrix} \cdot \begin{pmatrix} 2 \\ 0 \\ -1 \end{pmatrix} = 2 + 0 - 3 = -1,$$

where I added the two vectors inside the square brackets after the first equal sign. On the other hand, you could also try multiplying out the parentheses, as you learned in the first chapter in the usual rules for number crunching. This yields:

$$\begin{aligned} \begin{pmatrix} 1 \\ 2 \\ 3 \end{pmatrix} \cdot \left[\begin{pmatrix} 0 \\ 1 \\ -1 \end{pmatrix} + \begin{pmatrix} 2 \\ -1 \\ 0 \end{pmatrix} \right] &= \begin{pmatrix} 1 \\ 2 \\ 3 \end{pmatrix} \cdot \begin{pmatrix} 0 \\ 1 \\ -1 \end{pmatrix} + \begin{pmatrix} 1 \\ 2 \\ 3 \end{pmatrix} \cdot \begin{pmatrix} 2 \\ -1 \\ 0 \end{pmatrix} \\ &= 0 + 2 - 3 + 2 - 2 + 0 = -1. \end{aligned}$$

So I get the same result both ways, and that is not a coincidence at all.□

Multiplying out goes just as well with the scalar product as it does with the familiar multiplication of numbers. And that is not all, because in truth *all the* rules you know about multiplying numbers also apply to the scalar product.

Calculation Rules for the Scalar Product

For $\boldsymbol{x}, \boldsymbol{y}, \boldsymbol{z} \in \mathbb{R}^n$ and $\lambda \in \mathbb{R}$ the following rules hold:

(a) $\boldsymbol{x} \cdot \boldsymbol{y} = \boldsymbol{y} \cdot \boldsymbol{x}$;
(b) $\boldsymbol{x} \cdot (\boldsymbol{y} + \boldsymbol{z}) = \boldsymbol{x} \cdot \boldsymbol{y} + \boldsymbol{x} \cdot \boldsymbol{z}$;
(c) $\lambda \cdot (\boldsymbol{x} \cdot \boldsymbol{y}) = \lambda \cdot \boldsymbol{x} \cdot \boldsymbol{y}$.

Note that in rule (b) the two plus signs have two different meanings. On the left side of the equation it is the addition of vectors, on the right side it is the simple addition of ordinary numbers. Nevertheless, you can do the calculation one way or the other according to your personal taste and get the same result both times. In rule (c), on the other hand, the multiplication point has different meanings: First, the multiplication of a number by a vector ($\lambda \cdot x$), second, the scalar product of two vectors ($x \cdot y$), and third, the plain multiplication of two numbers (product of λ and $x \cdot y$). But the rule shows that there is no harm in naming such fundamentally different operations with the same sign, because in the end everything turns out all right.

Now it is time for some exercise again.

Exercise 5.2

(a) Calculate

$$\begin{pmatrix} 3 \\ -4 \\ 1 \end{pmatrix} \cdot \begin{pmatrix} 1 \\ 0 \\ -12 \end{pmatrix} \quad \text{and} \quad \begin{pmatrix} 2 \\ 1 \\ -1 \end{pmatrix} \cdot \begin{pmatrix} -1 \\ 5 \\ 6 \end{pmatrix}.$$

(b) Calculate the scalar product

$$\begin{pmatrix} -1 \\ 2 \end{pmatrix} \cdot \left(\begin{pmatrix} 3 \\ 4 \end{pmatrix} + \begin{pmatrix} -2 \\ 5 \end{pmatrix} \right)$$

in two ways.

The current state is not bad at all, because we now have the essential arithmetic operations for vectors together. On their basis, you can also tackle somewhat more complex tasks, as the following example shows.

Example 5.8 Your old familiar company *MyCompany* produces three products each A_1, A_2 and A_3 in two halls H_1 and H_2. Due to quality differences, the products from H_1 have a lower price than the products from H_2. For example, product A_1 from hall H_1 costs 12 Euros, product A_2 costs 8 Euros and product A_3 costs 17 Euros. In contrast, if the three products were produced in Hall H_2, they cost 17, 10, and 20 Euros, respectively. What is the total turnover if 10 units of each product are manufactured in H_1 and 15 units of each product are manufactured in H_2?

The price vector of H_1 is $\begin{pmatrix} 12 \\ 8 \\ 17 \end{pmatrix}$ the price vector of H_2 is $\begin{pmatrix} 17 \\ 10 \\ 20 \end{pmatrix}$. We obtain the results per item as a combination of both vectors, namely:

$$10 \cdot \begin{pmatrix} 12 \\ 8 \\ 17 \end{pmatrix} + 15 \cdot \begin{pmatrix} 17 \\ 10 \\ 20 \end{pmatrix} = \begin{pmatrix} 120 \\ 80 \\ 170 \end{pmatrix} + \begin{pmatrix} 255 \\ 150 \\ 300 \end{pmatrix} = \begin{pmatrix} 375 \\ 230 \\ 470 \end{pmatrix}.$$

The result vector describes in its first component the turnover for product A_1, then the turnover for product A_2 and finally the turnover for product A_3. The total turnover is then obtained from

$$\text{turnover} = 375 + 230 + 470 = 1075. \quad \blacksquare$$

What have I done here with the two vectors that appear? I multiplied each by a number and then added the resulting vectors. This is a very simple thing, but because it is incredibly common in vector calculus, it has been given its own name and is called a linear combination: obviously, the vectors are being combined here, and the nature of the combination is based on innocuous linear operations such as addition and multiplication by a number. Of course, such linear combinations can be formed not only from two vectors, but from any number of them. But the principle is always the same.

Linear Combination

If $\boldsymbol{x}_1, \ldots, \boldsymbol{x}_m \in \mathbb{R}^n$ are vectors and $c_1, \ldots, c_m \in \mathbb{R}$ are real numbers, then is called

$$c_1 \cdot \mathbf{x}_1 + \cdots + c_m \cdot \mathbf{x}_m$$

is called a linear combination of the vectors $\boldsymbol{x}_1, \ldots, \boldsymbol{x}_m$.

This is just a fancy name for a matter of course, which the next example will show once again:

Example 5.9

$$2 \cdot \begin{pmatrix} 5 \\ 3 \\ 1 \end{pmatrix} - 3 \cdot \begin{pmatrix} -1 \\ 2 \\ 4 \end{pmatrix} + \begin{pmatrix} 0 \\ 8 \\ 9 \end{pmatrix} = \begin{pmatrix} 10 \\ 6 \\ 2 \end{pmatrix} - \begin{pmatrix} -3 \\ 6 \\ 12 \end{pmatrix} + \begin{pmatrix} 0 \\ 8 \\ 9 \end{pmatrix} = \begin{pmatrix} 13 \\ 8 \\ -1 \end{pmatrix}.$$

Here a linear combination of three vectors is calculated, where the third vector has no prefactor, but since you can of course always write the number 1 in front of a vector, the calculation still fits the linear combination scheme.

The last term I have to plague you with in this section is the term linear dependence. It too has a rather practical background. Imagine that you want to bake a cake consisting of three ingredients, and of course the ingredients must be in a certain mixing ratio. So, for example, if you have two units of the first ingredient, three units of the second ingredient, and one unit of the third ingredient to mix, you can obviously describe the ingredient scheme by the three-dimensional vector $\begin{pmatrix} 2 \\ 3 \\ 1 \end{pmatrix}$. Now what about a cake for which you need four units of the first ingredient, six units of the second ingredient, and two units of the third ingredient? At first glance, this looks like a completely different cake, but it is actually the same cake, just twice the amount, because you have only doubled the respective units, but left the mixing ratio the same. Therefore, the equation $\begin{pmatrix} 4 \\ 6 \\ 2 \end{pmatrix} = 2 \cdot \begin{pmatrix} 2 \\ 3 \\ 1 \end{pmatrix}$ also applies to the new ingredient vector. Since the new vector can be easily calculated from the old one in this way, i.e. it does not actually offer any new information, the two vectors are said to be **linearly dependent**.

This also works with more than two vectors. For example, if we assume that your frequently used company *MyCompany* produces fruit juices, then one juice could be based on mixing two parts orange juice with one part lemon juice, and the second juice in the range does exactly the opposite: for it, one part orange juice is mixed with two parts lemon juice. The corresponding blend vectors are then $\begin{pmatrix} 2 \\ 1 \end{pmatrix}$ and $\begin{pmatrix} 1 \\ 2 \end{pmatrix}$. For both juice blends you need a machine, which unfortunately also costs money, and it gets really bad when you realize that many customers would also like a juice blend where you mix orange and lemon juice in a ratio of one to one. But before you start calculating the cost of a new machine in the dark of night, you should think back to the vector calculation and remember that

$$\begin{pmatrix} 1 \\ 1 \end{pmatrix} = \frac{1}{3} \cdot \begin{pmatrix} 2 \\ 1 \end{pmatrix} + \frac{1}{3} \cdot \begin{pmatrix} 1 \\ 2 \end{pmatrix}$$

applies. So as soon as you mix the juice blends already in the range in equal parts, you have already fulfilled the new customer request, and that is probably easier to manage than building a new factory building. The situation is similar with a desired mixing ratio of five to four, because it applies:

$$\begin{pmatrix} 5 \\ 4 \end{pmatrix} = 2 \cdot \begin{pmatrix} 2 \\ 1 \end{pmatrix} + \begin{pmatrix} 1 \\ 2 \end{pmatrix}.$$

You see what it comes down to. One will also here the vectors

$$\begin{pmatrix}1\\1\end{pmatrix}, \begin{pmatrix}2\\1\end{pmatrix}, \begin{pmatrix}1\\2\end{pmatrix}$$

call linearly dependent, because you can represent the first vector as a linear combination of the other two. And of course, for the same reason

$$\begin{pmatrix}5\\4\end{pmatrix}, \begin{pmatrix}2\\1\end{pmatrix}, \begin{pmatrix}1\\2\end{pmatrix}$$

are linearly dependent.

Now that you know what linear dependence means, I can formulate the concept officially.

Linear Dependence and Independence

The vectors $\boldsymbol{x}_1, \ldots, \boldsymbol{x}_m \in \mathbb{R}^n$ are called *linearly dependent* if one of them can be represented as a linear combination of the remaining $m-1$ vectors. Vectors that are not linearly dependent are called *linearly independent*.

This is again the same as before, only this time formulated in general terms for any number of vectors. Let us take a look at some examples right away:

Example 5.10

(a) $\begin{pmatrix}1\\0\end{pmatrix}$ and $\begin{pmatrix}0\\1\end{pmatrix}$ are linearly independent, because it is not possible to represent $\begin{pmatrix}1\\0\end{pmatrix}$ as $c \cdot \begin{pmatrix}0\\1\end{pmatrix}$ or $\begin{pmatrix}0\\1\end{pmatrix}$ as $c \cdot \begin{pmatrix}1\\0\end{pmatrix}$ with a real number c.

(b) The vectors

$$\begin{pmatrix}1\\2\end{pmatrix}, \begin{pmatrix}2\\3\end{pmatrix}, \begin{pmatrix}4\\7\end{pmatrix}$$

are linearly dependent, because

$$\begin{pmatrix}4\\7\end{pmatrix} = 2 \cdot \begin{pmatrix}1\\2\end{pmatrix} + \begin{pmatrix}2\\3\end{pmatrix}.$$

The vector $\begin{pmatrix}4\\7\end{pmatrix}$ does not offer any new information that is not already contained in the other two vectors. That is why we speak of dependence. However, the question arises here how I actually came up with the prefactors of the other two vectors that I need for the linear combination. As things stand, they still seem to be falling from the sky, and we will have to figure out how to get them.

(c) The vectors

$$\begin{pmatrix} 1 \\ 2 \end{pmatrix}, \begin{pmatrix} 2 \\ 1 \end{pmatrix}$$

are linearly independent, because it is not possible to represent $\begin{pmatrix} 1 \\ 2 \end{pmatrix}$ as $c \cdot \begin{pmatrix} 2 \\ 1 \end{pmatrix}$ or $\begin{pmatrix} 2 \\ 1 \end{pmatrix}$ as $c \cdot \begin{pmatrix} 1 \\ 2 \end{pmatrix}$ with a real number c.

(d) The vectors

$$\begin{pmatrix} 5 \\ 1 \\ 3 \end{pmatrix}, \begin{pmatrix} 1 \\ 4 \\ 2 \end{pmatrix}, \begin{pmatrix} 1 \\ 2 \\ 6 \end{pmatrix}$$

are linearly independent. To test for linear dependence, one would apply, for example:

$$\begin{pmatrix} 5 \\ 1 \\ 3 \end{pmatrix} = c_1 \cdot \begin{pmatrix} 1 \\ 4 \\ 2 \end{pmatrix} + c_2 \cdot \begin{pmatrix} 1 \\ 2 \\ 6 \end{pmatrix},$$

because a vector must be representable as a linear combination of the other vectors in case of linear dependence. If one now writes down the equations for the individual components, then this means:

$$\begin{aligned} c_1 \;&+\; c_2 &= 5 \\ 4c_1 \;&+\; 2c_2 &= 1 \\ 2c_1 \;&+\; 6c_2 &= 3. \end{aligned}$$

So I have a so-called linear system of equations with three equations and two unknowns. If this system has any solution, i.e. if there are two real numbers c_1 and c_2, so that when I insert them into the three equations everything works out, then the vectors are linearly dependent, because in this case I can actually represent one vector as a linear combination of the other vectors. But if there is no solution, then they are linearly independent, because in that case you just cannot represent one vector as a linear combination of the other vectors. So linear dependence obviously has something to do with the solvability of systems of linear equations, and this is one of the reasons why I will devote myself to such systems of equations in the third section. As things stand, the problem at hand is not yet solvable; you will have to be patient until the systems of linear equations, and take my word for the linear independence of the vectors for now.

Linear dependence or independence can also be written down in a slightly different way. I come back to the vectors in Example 5.10(b), where I use the equation

$$\begin{pmatrix} 4 \\ 7 \end{pmatrix} = 2 \cdot \begin{pmatrix} 1 \\ 2 \end{pmatrix} + \begin{pmatrix} 2 \\ 3 \end{pmatrix}$$

to show the linear dependence of the three vectors involved. Bringing everything in this equation to one side, we get

$$2 \cdot \begin{pmatrix} 1 \\ 2 \end{pmatrix} + \begin{pmatrix} 2 \\ 3 \end{pmatrix} + (-1) \cdot \begin{pmatrix} 4 \\ 7 \end{pmatrix} = \begin{pmatrix} 0 \\ 0 \end{pmatrix}.$$

Thus there is a way to combine the so-called zero vector, which has the value zero in each component, from the given three vectors without each vector having to put up with the prefactor zero. This is by no means obvious; try combining the zero vector together from the two vectors $\begin{pmatrix} 1 \\ 0 \end{pmatrix}$ and $\begin{pmatrix} 0 \\ 1 \end{pmatrix}$. To do this, you would need to find a linear combination of the two vectors that yields the zero vector, thus:

$$c_1 \cdot \begin{pmatrix} 1 \\ 0 \end{pmatrix} + c_2 \cdot \begin{pmatrix} 0 \\ 1 \end{pmatrix} = \begin{pmatrix} 0 \\ 0 \end{pmatrix}.$$

But if you now calculate the linear combination on the left side, you will find as result the vector $\begin{pmatrix} c_1 \\ c_2 \end{pmatrix}$, and this means that $\begin{pmatrix} c_1 \\ c_2 \end{pmatrix} = \begin{pmatrix} 0 \\ 0 \end{pmatrix}$ and therefore also $c_1 = c_2 = 0$.

So some vectors you can combine to the zero vector without simply multiplying all vectors by the number zero, others you cannot. And you have already seen which vectors are the well-behaved ones: Those were exactly the linearly dependent ones. This is how you get to the following description of linear dependence and independence:

Linear Dependence and Independence

The vectors $\boldsymbol{x}_1, \ldots, \boldsymbol{x}_m \in \mathbb{R}^n$ are linearly independent if and only if from

$$c_1 \cdot \mathbf{x}_1 + c_2 \cdot \mathbf{x}_2 + \cdots + c_m \cdot \mathbf{x}_m = \mathbf{0}$$

always follows:

$$c_1 = c_2 = \cdots = c_m = 0,$$

where 0 is understood to be the zero vector. If, on the other hand, the vectors can be combined to form the zero vector without each of the prefactors $c_1, c_2, \ldots, c_m$ having to become zero, then the vectors are linearly dependent.

Let us look at two examples again:

Example 5.11

(a) The vectors

$$\begin{pmatrix} 1 \\ 0 \\ 0 \end{pmatrix}, \begin{pmatrix} 0 \\ 1 \\ 0 \end{pmatrix}, \begin{pmatrix} 0 \\ 0 \\ 1 \end{pmatrix}$$

are linearly independent. From

$$c_1 \cdot \begin{pmatrix} 1 \\ 0 \\ 0 \end{pmatrix} + c_2 \cdot \begin{pmatrix} 0 \\ 1 \\ 0 \end{pmatrix} + c_3 \cdot \begin{pmatrix} 0 \\ 0 \\ 1 \end{pmatrix} = \begin{pmatrix} 0 \\ 0 \\ 0 \end{pmatrix}$$

follows immediately

$$\begin{pmatrix} c_1 \\ c_2 \\ c_3 \end{pmatrix} = \begin{pmatrix} 0 \\ 0 \\ 0 \end{pmatrix} \quad \text{and so} \quad c_1 = 0, \ c_2 = 0, \ c_3 = 0.$$

(b) The corresponding system of equations for the vectors

$$\begin{pmatrix} 5 \\ 1 \\ 3 \end{pmatrix}, \begin{pmatrix} 1 \\ 4 \\ 2 \end{pmatrix}, \begin{pmatrix} 1 \\ 2 \\ 6 \end{pmatrix}$$

is

$$\begin{aligned} 5c_1 &+ c_2 &+ c_3 &= 0 \\ c_1 &+ 4c_2 &+ 2c_3 &= 0 \\ 3c_1 &+ 2c_2 &+ 6c_3 &= 0. \end{aligned}$$

This system of equations naturally has the solution $c_1 = c_2 = c_3 = 0$, because if you insert the value zero for c_1, c_2 and c_3 respectively, then the overall result of each left-hand side is zero again. The only question is whether the system of equations also has other solutions, because in the case of other solutions that are not just zeros, you would have proved linear dependence. I will show you how to calculate the different solutions of a linear system of equations in the third section, and I will come back to this example there.

At the end of this section I will tell you a little trick, with which you can sometimes spare yourself unpleasant calculations. If you have at least four three-dimensional vectors in your hand, then these four vectors are automatically linearly dependent without you having to do any calculations. And that is how it always is: If the number of vectors in question exceeds the dimension of these vectors, then the vectors are linearly dependent all by themselves, and no one can ask you to lift a finger to calculate.

Linear Dependence
If one has m *n-dimensional* vectors and if $m > n$ holds, then these vectors are linearly dependent.

So for example $\begin{pmatrix} 1 \\ 17 \end{pmatrix}, \begin{pmatrix} -12 \\ 38 \end{pmatrix}, \begin{pmatrix} 123 \\ -125 \end{pmatrix}$ are definitely linearly dependent because I have three vectors of dimension two here.

With this practical trick for avoiding arithmetic, I end the section on vectors. In the next section, I will show you how to deal with matrices and what they have to do with vectors – but not without giving you another opportunity to practice.

Exercise 5.3 Test the following vectors for linear dependence or independence.

(a) $\begin{pmatrix} 1 \\ 2 \\ 3 \end{pmatrix}, \begin{pmatrix} 2 \\ 4 \\ 7 \end{pmatrix}$

(b) $\begin{pmatrix} -2 \\ 5 \end{pmatrix}, \begin{pmatrix} 12 \\ 17 \end{pmatrix}, \begin{pmatrix} 234 \\ 1 \end{pmatrix}$

(c) $\begin{pmatrix} 2 \\ -1 \\ 2 \end{pmatrix}, \begin{pmatrix} -4 \\ 2 \\ -4 \end{pmatrix}$?

5.2 Matrices

There is nothing mysterious about matrices, even if the movies in the "Matrix" series make it seem so. If the main characters there had studied linear algebra a bit, they could easily answer their repeated question "What is the matrix?" with the sentence that every matrix is actually just a vector that is too wide. I admit: You could not have made a movie out of that, and maybe that is why the problem was solved a little differently in the movies.

Now you are not in a movie theater here, so as usual, let us look at an introductory example.

Example 5.12 Your now well-known company, *MyCompany,* has undergone a very pleasing development and is now so large that you have had to divide it into three departments, *A, B* and *C.* These departments are, of course, not isolated islands in the ocean of the company, but provide services to each other; what one department provides, the other receives. The service relationships are described by the following agreement:

> Division A gives 20 units to Division B and 30 units to Division C. Department B gives 30 units to Department A and 40 units to Department C. Division C gives 10 units to Division A and 20 units to Division B.

Whatever your company may produce and whatever the units may be, in any case the occurring power flows will have to be put into a clear schema, not only so that it can be read more easily, but also and above all so that the power allocation can be carried out mechanically. So I need a reasonable data structure to be able to represent the given situation decently. And this is where matrices come into play, which are nothing more than a kind of tabular listing of numbers. In our example, I can use the following matrix:

$$\begin{array}{c} \\ A \\ B \\ C \end{array}\begin{array}{c} \begin{array}{ccc} A & B & C \end{array} \\ \begin{pmatrix} 0 & 20 & 30 \\ 30 & 0 & 40 \\ 10 & 20 & 0 \end{pmatrix} \end{array}$$

So I only wrote the occurring numbers into a rectangular scheme and also labeled this scheme a little bit, so the reader knows what these numbers are about at all. But the labeling in the margin is just a convenience aid for the reader, the matrix itself consists only of the rectangular scheme filled with numbers.

So a matrix is only a rectangular scheme, containing numbers, and further it is nothing at all. Now I will save this result for the special type, which was in Example 5.12.

Matrix
A matrix with three rows and three columns is a rectangular scheme of the form

$$\begin{pmatrix} a_{11} & a_{12} & a_{13} \\ a_{21} & a_{22} & a_{23} \\ a_{31} & a_{32} & a_{33} \end{pmatrix},$$

where the entries in this scheme are real numbers.

Notice how I defined the entries in the matrix. With three rows and three columns, I need nine entries, and at first glance it would have seemed obvious to label them $a_1, a_2, \ldots, a_9$ just as I did with the vectors. But this would mean that some information would be lost, because I would not know at first sight at which position in the matrix the element a_5 is located, for example. It is better to use double numbering, and for example to designate the third entry of the second row as a_{23}, which is pronounced "a two three" and not, say, "a twenty-three". So the first number indicates the number of the current row, and the second the number of the current column, so that I can immediately tell each element which place in the matrix it feels at home.

Of course, there are not only matrices with three rows and three columns:

Matrix
If m and n are natural numbers, the rectangular number scheme consisting of m rows and n columns is called

$$\begin{pmatrix} a_{11} & a_{12} & \cdots & a_{1n} \\ a_{21} & a_{22} & \cdots & a_{2n} \\ \vdots & \vdots & & \vdots \\ a_{m1} & a_{m2} & \cdots & a_{mn} \end{pmatrix}$$

is an $m \times n$ *matrix*. The set of all $m \times n$ *matrices* is denoted by $\mathbb{R}^{m \times n}$.

Now this is exactly the same as in the case of the special matrix, except that here the number of rows and the number of columns are indeterminate. By the way, you can see that a matrix does not always have to be square, the number of rows and the number of columns are independent of each other. You can see this very clearly when you look at matrices with only one column or only one row: A matrix with only one column suspiciously resembles a vector, which can be more accurately called a column vector, and a matrix with only one row is just also a vector that had no more power to stand; such a thing is called a row vector.

So just like in the first section, we now have objects defined and need to see what can be done with them. The simplest operation is addition, which is done exactly as you probably imagine it.

Example 5.13 Your firm *MyCompany* once again produces three goods, which I now label I, II, III, and delivers them to distributors A, B, C, D. In the first and second quarter, respectively, the respective deliveries are described by the following matrices:

$$\begin{array}{c} \\ \text{I} \\ \text{II} \\ \text{III} \end{array} \begin{array}{c} \begin{array}{cccc} \text{A} & \text{B} & \text{C} & \text{D} \end{array} \\ \begin{pmatrix} 12 & 8 & 0 & 7 \\ 17 & 9 & 4 & 8 \\ 5 & 7 & 4 & 10 \end{pmatrix} \end{array}$$

for the first quarter and

$$\begin{array}{c} \\ \text{I} \\ \text{II} \\ \text{III} \end{array} \begin{array}{c} \begin{array}{cccc} \text{A} & \text{B} & \text{C} & \text{D} \end{array} \\ \begin{pmatrix} 13 & 7 & 2 & 9 \\ 15 & 10 & 6 & 10 \\ 6 & 6 & 5 & 9 \end{pmatrix} \end{array}$$

for the second quarter. So, for example, retailer A received five units of product III in the first quarter, but they apparently sold so well that they ordered one more unit in the second quarter. Now, to determine which retailer received how many units of each product in the entire first half of the year, the two matrices are added component-wise. This gives the matrix for the first half of the year:

$$\begin{array}{c} \\ \text{I} \\ \text{II} \\ \text{III} \end{array} \begin{array}{c} \begin{array}{cccc} \text{A} & \text{B} & \text{C} & \text{D} \end{array} \\ \begin{pmatrix} 25 & 15 & 2 & 16 \\ 32 & 19 & 10 & 18 \\ 11 & 13 & 9 & 19 \end{pmatrix} \end{array}$$

and thus we know, for example, that Dealer D received a total of 18 units of Product II in the first half of the year.□

Matrices are therefore simply added component by component.

Addition and Subtraction of Matrices

Let

$$A = \begin{pmatrix} a_{11} & a_{12} & \cdots & a_{1n} \\ a_{21} & a_{22} & \cdots & a_{2n} \\ \vdots & \vdots & & \vdots \\ a_{m1} & a_{m2} & \cdots & a_{mn} \end{pmatrix}$$

and

(continued)

$$B = \begin{pmatrix} b_{11} & b_{12} & \cdots & b_{1n} \\ b_{21} & b_{22} & \cdots & b_{2n} \\ \vdots & \vdots & & \vdots \\ b_{m1} & b_{m2} & \cdots & b_{mn} \end{pmatrix}$$

be matrices with m rows and n columns. Then one sets:

$$A \pm B = \begin{pmatrix} a_{11} \pm b_{11} & a_{12} \pm b_{12} & \cdots & a_{1n} \pm b_{1n} \\ a_{21} \pm b_{21} & a_{22} \pm b_{22} & \cdots & a_{2n} \pm b_{2n} \\ \vdots & \vdots & & \vdots \\ a_{m1} \pm b_{m1} & a_{m2} \pm b_{m2} & \cdots & a_{mn} \pm b_{mn} \end{pmatrix}.$$

The sum and the difference of the two matrices are then again matrices with m rows and n columns.

So you add and subtract matrices the same way you add and subtract vectors, component by component. So if you have to add two matrices, each with three rows and four columns, then obviously each of the two matrices has twelve entries and you will therefore do twelve additions. However, since you can only add components together that are present, it is clear that you must never add or subtract matrices that do not match in their number of rows or columns. Matrices can only be added and subtracted if their row and column numbers match.

Again, examples of this:

Example 5.14

(a)

$$\begin{pmatrix} 5 & 3 \\ 1 & 0 \end{pmatrix} - \begin{pmatrix} 2 & 7 \\ 3 & 5 \end{pmatrix} = \begin{pmatrix} 5-2 & 3-7 \\ 1-3 & 0-5 \end{pmatrix} = \begin{pmatrix} 3 & -4 \\ -2 & -5 \end{pmatrix}.$$

(b) The addition

$$\begin{pmatrix} 5 & 3 \\ 1 & 0 \end{pmatrix} + \begin{pmatrix} 1 & 2 & 3 \\ 4 & 5 & 6 \end{pmatrix}$$

cannot be performed because the first matrix has two columns, but the second has three columns.

So the addition of matrices is nothing exciting, and the subtraction does not promise any increased thrill either. For now, I cannot promise much more excitement either, because next I will show you how to multiply a matrix by a scalar, that is, by a number. Since I am sure you remember the corresponding operation for vectors, nothing here will surprise you.

Example 5.15 You also have to think outside the box sometimes and make an effort to see other countries and other customs. So I will leave your good old *MyCompany* and take a look at international trade relations. So that it does not get too confusing, I will limit myself to three countries A, B, and C. The foreign trade relations between these countries will be described by the following matrix:

$$\begin{array}{c} \\ \text{A} \\ \text{B} \\ \text{C} \end{array} \begin{array}{c} \begin{array}{ccc} \text{A} & \text{B} & \text{C} \end{array} \\ \begin{pmatrix} 0 & 12 & 8 \\ 6 & 0 & 4 \\ 10 & 2 & 0 \end{pmatrix} \end{array},$$

where the rows indicate the respective export and the columns the import. Settlement is made in billions of euros, so that country A exports goods worth 8 billion Euros to C, but in the opposite direction goods worth 10 billion Euros are exported from C to A. If the settlement is to be made in US dollars, each component of the matrix must be multiplied by 1.4, assuming that 1 Euro is equal to 1 dollar and 40 centsas it was in good old times. This gives the new matrix:

$$\begin{pmatrix} 0 & 16.8 & 11,2 \\ 8.4 & 0 & 5.6 \\ 14 & 2.8 & 0 \end{pmatrix}.$$

So I multiplied the old matrix by the number 1.4 by multiplying every single entry of the matrix by 1.4. To make it obvious that this operation has nothing to do with A, B, or C, I left out the column and row labels; the multiplication does not care.

Thus you already know the principle of multiplication number · matrix, which I will now formulate in general.

Multiplication of a Matrix by a Number
For a matrix

$$A = \begin{pmatrix} a_{11} & a_{12} & \cdots & a_{1n} \\ a_{21} & a_{22} & \cdots & a_{2n} \\ \vdots & \vdots & & \vdots \\ a_{m1} & a_{m2} & \cdots & a_{mn} \end{pmatrix}$$

with m rows and n columns and a real number λ one sets:

$$\lambda \cdot A = \begin{pmatrix} \lambda \cdot a_{11} & \lambda \cdot a_{12} & \cdots & \lambda \cdot a_{1n} \\ \lambda \cdot a_{21} & \lambda \cdot a_{22} & \cdots & \lambda \cdot a_{2n} \\ \vdots & \vdots & & \vdots \\ \lambda \cdot a_{m1} & \lambda \cdot a_{m2} & \cdots & \lambda \cdot a_{mn} \end{pmatrix}.$$

It is best to do some quick math yourself:

Exercise 5.4 Calculate:

(a)

$$2 \cdot \begin{pmatrix} 1 & 0 & -2 \\ -3 & 5 & 1 \\ 4 & 2 & -1 \end{pmatrix} + 3 \cdot \begin{pmatrix} -6 & 3 & 4 \\ 1 & 2 & -2 \\ -3 & -4 & 0 \end{pmatrix}$$

(b)

$$-3 \cdot \begin{pmatrix} -1 & 5 & 2 \\ -2 & 1 & 0 \\ 3 & 1 & -2 \end{pmatrix} + \begin{pmatrix} 2 & -5 & -2 \\ 2 & -1 & 0 \\ -3 & -1 & 3 \end{pmatrix} \quad \blacksquare$$

To put it in words again: You multiply a matrix by a number by multiplying each entry in the matrix by that number. Again, this is not so difficult as to cause you sleepless nights, but after all, matrices are quite complex entities, and so it cannot hurt to briefly write down the calculation rules for the operations I have shown you so far. You would not find them surprising, though, because it turns out that with regard to the addition of matrices and to the multiplication of matrices by scalars, everything is quite the same as you have already learned in ordinary number crunching. So, for example, you can swap the order of addition, you can multiply out parentheses, and you can pre-combine numbers that occur as common factors.

Rules of Calculation for Matrices

If A and B are matrices with m rows and n columns and λ and μ are real numbers, then the following rules of arithmetic apply:

(a) $A + B = B + A$;
(b) $\lambda \cdot (A + B) = \lambda \cdot A + \lambda \cdot B$;
(c) $(\lambda + \mu) \cdot A = \lambda \cdot A + \mu \cdot A$.

As advertised, none of this is surprising, and the rules basically just say that everything works the same as it always has. Still, an example or two is never amiss:

Example 5.16 I calculate the expression

$$3 \cdot \begin{pmatrix} 2 & 5 \\ -1 & 7 \end{pmatrix} - 2 \cdot \left[\begin{pmatrix} 1 & 3 \\ 4 & 0 \end{pmatrix} + 5 \cdot \begin{pmatrix} 2 & 1 \\ 0 & -1 \end{pmatrix} \right].$$

First, I tackle the inner bracket and determine:

$$\begin{pmatrix} 1 & 3 \\ 4 & 0 \end{pmatrix} + 5 \cdot \begin{pmatrix} 2 & 1 \\ 0 & -1 \end{pmatrix} = \begin{pmatrix} 1 & 3 \\ 4 & 0 \end{pmatrix} + \begin{pmatrix} 10 & 5 \\ 0 & -5 \end{pmatrix} = \begin{pmatrix} 11 & 8 \\ 4 & -5 \end{pmatrix}.$$

So

$$\begin{aligned} 3 \cdot \begin{pmatrix} 2 & 5 \\ -1 & 7 \end{pmatrix} - 2 \cdot \left[\begin{pmatrix} 1 & 3 \\ 4 & 0 \end{pmatrix} + 5 \cdot \begin{pmatrix} 2 & 1 \\ 0 & -1 \end{pmatrix} \right] &= \begin{pmatrix} 6 & 15 \\ -3 & 21 \end{pmatrix} - 2 \cdot \begin{pmatrix} 11 & 8 \\ 4 & -5 \end{pmatrix} \\ &= \begin{pmatrix} 6 & 15 \\ -3 & 21 \end{pmatrix} - \begin{pmatrix} 22 & 16 \\ 8 & -10 \end{pmatrix} = \begin{pmatrix} -16 & -1 \\ -11 & 31 \end{pmatrix}. \end{aligned}$$

But with the same right you could have multiplied the factor 2 into the square bracket first, then multiplied the single prefactors into the matrices and finally added everything, which I highly recommend as an exercise. The result must be the same in each case.□

It seems it is time once again for a practice assignment.

Exercise 5.5 Calculate

$$2 \cdot \begin{pmatrix} 1 & 4 \\ -2 & 6 \end{pmatrix} - 3 \cdot \left[\begin{pmatrix} 0 & 2 \\ 3 & -1 \end{pmatrix} + 4 \cdot \begin{pmatrix} 1 & 0 \\ -1 & -2 \end{pmatrix} \right]. \quad \blacksquare$$

Probably you will admit that so far everything has been quite harmless. Even matrix multiplication, which I will introduce next, does not stand out for being too complicated, but at least it is a little more complicated than the other operations – though not by much. The best thing will be, we will look at an example again.

Example 5.17 We return to your good old company *MyCompany* and assume that your company operates a two-stage production. From the raw materials R_1 and R_2 the intermediate products Z_1, Z_2 and Z_3 are produced in the first production stage, while in the second production stage the final products E_1 and E_2 are produced from these intermediate products. The material consumption in each production stage can then be described with the help of the following two matrices:

$$\begin{array}{c} \\ R_1 \\ R_2 \end{array} \begin{array}{c} \begin{array}{ccc} Z_1 & Z_2 & Z_3 \end{array} \\ \begin{pmatrix} 1 & 4 & 5 \\ 2 & 3 & 6 \end{pmatrix} \end{array}$$

and

$$\begin{array}{c} \\ Z_1 \\ Z_2 \\ Z_2 \end{array} \begin{array}{c} \begin{array}{cc} E_1 & E_2 \end{array} \\ \begin{pmatrix} 1 & 3 \\ 5 & 3 \\ 4 & 1 \end{pmatrix} \end{array}$$

Of course, before calculating with such matrices, it is necessary to agree on their interpretation. The first matrix describes how many units of the two raw materials are needed in the production of one unit each of the intermediate products. Thus, to produce one unit of Z_1, you need exactly one unit of the raw material R_1 and two units of the raw material R_2. On the other hand, to produce one unit of the intermediate Z_2, you need four units of R_1 and three units of R_2. You can read the second matrix in the same way: The intermediate products are now used to produce the final products, and for one unit of the final product E_1 you will have to record the consumption of one unit of Z_1, five units of Z_2 and four units of Z_3.

So much for the initial situation. However, since I need to know at the end how many raw materials I should procure for further production, I am primarily interested in the respective raw material consumption per unit of the end product and therefore want to set up a total material consumption matrix that gives me exactly this raw material consumption. Once I have this matrix, I can then answer questions such as "What is the raw material requirement if the plant is to produce 1000 units of E_1 and 1200 units of E_2?" .

From the second matrix you can read that for one unit E_1 you need exactly 1 Z_1, 5 Z_2 and 4 Z_3. For one unit Z_1 you need 1 R_1 and 2 R_2, for one unit Z_2 you need 4 R_1 and 3 R_2, for one unit Z_3 you need 5 R_1 and 6 R_2. Consequently, for one unit E_1 one needs the following quantity R_1:

$$1 \cdot 1 + 4 \cdot 5 + 5 \cdot 4 = 41$$

and the following quantity R_2:

$$2 \cdot 1 + 3 \cdot 5 + 6 \cdot 4 = 41.$$

In this simple way you have already calculated the consumption of raw materials in the production of one unit of the final product E_1, and perhaps the method of calculation reminds you of a calculation method I showed you in the section on vectors. We had already seen this combination of multiplying and adding in the case of the scalar product, and in fact the scalar products are here

$$\begin{pmatrix} 1 \\ 4 \\ 5 \end{pmatrix} \cdot \begin{pmatrix} 1 \\ 5 \\ 4 \end{pmatrix} \quad \text{and} \quad \begin{pmatrix} 2 \\ 3 \\ 6 \end{pmatrix} \cdot \begin{pmatrix} 1 \\ 5 \\ 4 \end{pmatrix}$$

You see that the vector $\begin{pmatrix} 1 \\ 4 \\ 5 \end{pmatrix}$ corresponds exactly to the first row of the first matrix, just written as a column, while the vector $\begin{pmatrix} 1 \\ 5 \\ 4 \end{pmatrix}$ is nothing else than the first column of the second matrix. In the second calculation you have the same situation: The vector $\begin{pmatrix} 2 \\ 3 \\ 6 \end{pmatrix}$ corresponds to the second row of the first matrix, just written as a column, while the vector $\begin{pmatrix} 1 \\ 5 \\ 4 \end{pmatrix}$ still represents the first column of the second matrix.

Thus, one obtains the total raw material consumption for a unit E_1, by calculating the scalar product of the first or second *row vector* of the first matrix with the first *column vector* of the second matrix, considering the row vector as column vector for a moment, so that one can form the scalar product at all. Analogously one finds for a unit E_2:

$$1 \cdot 3 + 4 \cdot 3 + 5 \cdot 1 = 20$$

and

$$2 \cdot 3 + 3 \cdot 3 + 6 \cdot 1 = 21,$$

from which you can see that both 20 units of R_1 and 21 units of R_2 are required to produce one unit of your final product E_2.

So the total material consumption matrix is:

$$\begin{matrix} & E_1 & E_2 \\ R_1 & \multicolumn{2}{c}{\multirow{2}{*}{}} \end{matrix} \quad \begin{array}{c} \\ R_1 \\ R_2 \end{array} \begin{array}{c} E_1 \quad E_2 \\ \begin{pmatrix} 41 & 20 \\ 41 & 21 \end{pmatrix} \end{array}$$

And now we are ready to write down a matrix product. One calls the just carried out connection of the two matrices their product and writes for it:

$$\begin{pmatrix} 1 & 4 & 5 \\ 2 & 3 & 6 \end{pmatrix} \cdot \begin{pmatrix} 1 & 3 \\ 5 & 3 \\ 4 & 1 \end{pmatrix} = \begin{pmatrix} 41 & 20 \\ 41 & 21 \end{pmatrix}.$$

You have just seen how to calculate this product: You flip the columns of the second matrix in order onto the rows of the first matrix and form the scalar products. This works not only for these two matrices, but also for all sorts of others, under one small condition, which I will come to in a moment. Before we do that, let us briefly explore the question, "What is the raw material requirement if the company is to produce 1000 units of E_1 and 1200 units of E_2?" With the help of the calculated consumption matrix, it is now easy to answer, because for 1000 units of E_1 and 1200 units of E_2, one needs R_1 units:

$$41 \cdot 1000 + 20 \cdot 1200 = 65,000$$

and to R_2 units:

$$41 \cdot 1000 + 21 \cdot 1200 = 66,200.$$

You write for it:

$$\begin{pmatrix} 41 & 20 \\ 41 & 21 \end{pmatrix} \cdot \begin{pmatrix} 1000 \\ 1200 \end{pmatrix} = \begin{pmatrix} 65,000 \\ 66,200 \end{pmatrix},$$

and this also works the same way as before, because you first have to form the scalar product of the first row of the matrix and the given vector, and then in the same way you have to form the scalar product of the second row of the matrix and the vector.

You define general matrix products the same way I showed you in Example 5.17. As soon as you put this into a general formula, you get somewhat unwieldy expressions that are difficult to understand at first, but sometimes this cannot be avoided.

Matrix Multiplication

Let A and B be matrices, the rows of A being of the same length as the columns of B. The product $C = A \cdot B$ of the two matrices then has as many rows as A and as many columns as B, and is calculated as follows. If

$$A = \begin{pmatrix} a_{11} & a_{12} & \cdots & a_{1n} \\ a_{21} & a_{22} & \cdots & a_{2n} \\ \vdots & \vdots & & \vdots \\ a_{m1} & a_{m2} & \cdots & a_{mn} \end{pmatrix}$$

is a matrix with m rows and n columns and

$$B = \begin{pmatrix} b_{11} & b_{12} & \cdots & b_{1k} \\ b_{21} & b_{22} & \cdots & b_{2k} \\ \vdots & \vdots & & \vdots \\ b_{n1} & b_{n2} & \cdots & b_{nk} \end{pmatrix}$$

is a matrix with n rows and k columns, then the rows of A are as long as the columns of B. The matrix $C = A \cdot B$ then has the form

(continued)

$$C = \begin{pmatrix} c_{11} & c_{12} & \cdots & c_{1k} \\ c_{21} & c_{22} & \cdots & c_{2k} \\ \vdots & \vdots & & \vdots \\ c_{m1} & c_{m2} & \cdots & c_{mk} \end{pmatrix}.$$

The individual entries c_{ij} in C are calculated as the scalar product of the *i-th* row of A and the *j-th* column of B, that is:

$$c_{ij} = \begin{pmatrix} a_{i1} \\ a_{i2} \\ \vdots \\ a_{in} \end{pmatrix} \cdot \begin{pmatrix} b_{1j} \\ b_{2j} \\ \vdots \\ b_{nj} \end{pmatrix} = a_{i1}b_{1j} + a_{i2}b_{2j} + \cdots + a_{in}b_{nj}.$$

If $\boldsymbol{x}$ is a vector of the same length as the rows of A, then the product $A \cdot \boldsymbol{x}$ is a vector of the same length as the columns of A and is calculated by writing in turn the scalar product of the first row of A with $\boldsymbol{x}$, the scalar product of the second row of A with $\boldsymbol{x}$, ..., the scalar product of the last row of A with $\boldsymbol{x}$ into a vector.

Maybe this sounds a bit abstract and unfamiliar now, but this will soon become clear after the following examples.

Example 5.18

(a) I consider the two matrices

$$A = \begin{pmatrix} 1 & 1 \\ 1 & -1 \end{pmatrix}, \quad B = \begin{pmatrix} 2 & -1 \\ 0 & 1 \end{pmatrix}.$$

Both matrices have two rows and two columns, so surely the rows of A are as long as the columns of B and I can get on with multiplying. The product matrix $C = A \cdot B$ must also consist of two rows and two columns, and using $C = \begin{pmatrix} c_{11} & c_{12} \\ c_{21} & c_{22} \end{pmatrix}$ provides the calculation rule:

$$c_{11} = 1 \cdot 2 + 1 \cdot 0 = 2, \quad c_{12} = 1 \cdot (-1) + 1 \cdot 1 = 0$$

and

$$c_{21} = 1 \cdot 2 + (-1) \cdot 0 = 2, \quad c_{22} = 1 \cdot (-1) + (-1) \cdot 1 = -2.$$

Consequently

$$A \cdot B = \begin{pmatrix} 2 & 0 \\ 2 & -2 \end{pmatrix}.$$

(b) Now it is about the matrices

$$A = \begin{pmatrix} 1 & 4 & 2 \\ 4 & 0 & -3 \end{pmatrix} \quad \text{and} \quad B = \begin{pmatrix} 1 & 1 & 0 \\ -2 & 3 & 5 \\ 0 & 1 & 4 \end{pmatrix}.$$

The row length of A is three, and the columns of B also have length three, so that nothing stands in the way of multiplication. The product matrix $C = A \cdot B$ then has as many rows as A, i.e. two, and as many columns as B, i.e. three. Now all I have to do is write the right scalar products in the right place, strictly according to the rules, and I get:

$$A \cdot B = \begin{pmatrix} 1 \cdot 1 + 4 \cdot (-2) + 2 \cdot 0 & 1 \cdot 1 + 4 \cdot 3 + 2 \cdot 1 & 1 \cdot 0 + 4 \cdot 5 + 2 \cdot 4 \\ 4 \cdot 1 + 0 \cdot (-2) + (-3) \cdot 0 & 4 \cdot 1 + 0 \cdot 3 + (-3) \cdot 1 & 4 \cdot 0 + 0 \cdot 5 + (-3) \cdot 4 \end{pmatrix}$$
$$= \begin{pmatrix} -7 & 15 & 28 \\ 4 & 1 & -12 \end{pmatrix}.$$

In contrast, you cannot compute $B \cdot A$ at all, because now the matrix that is in front (i.e., B) has a row length of three, while the matrix that is behind (and that is A) is plagued with a column length of two – this does not match, and so this matrix product does not exist.

I have mentioned several times that you cannot multiply all the matrices together: For $A \cdot B$ to be a meaningful expression, the rows of A must be as long as the columns of B. But now the row length of A is just the number of columns of A, and on the other hand the column length of B is equal to the number of its rows. So you can work out $A \cdot B$ if and only if the number of columns of A is equal to the number of rows of B. The reason I wanted to point out this formulation is that it can be easily translated into a formula. Indeed, we abbreviated the set of all matrices with m rows and n columns at the beginning of the section as $\mathbb{R}^{m \times n}$, and if A is in the set $\mathbb{R}^{m \times n}$, then B must be in the set $\mathbb{R}^{n \times k}$, that is, have n rows and who knows how many columns. In this case, A has exactly n columns and B has n rows, which is why everything fits together. And you already know how many rows and columns $A \cdot B$ has, because the product has as many rows as A and as many columns as B. So $A \cdot B$ has exactly m columns and k rows.

Condition for Matrix Multiplication
Exactly then one can form the matrix product $A \cdot B$ if $A \in \mathbb{R}^{m \times n}$ and $B \in \mathbb{R}^{n \times k}$ holds. In this case, the product $A \cdot B$ has m rows and k columns, so $A \cdot B \in \mathbb{R}^{m \times k}$.

It could be that you still find it a little difficult to handle the matrix product; on the one hand, this could be due to the novelty of the calculation method, but on the other hand, it could also be due to the fact that we do not yet have a really clear means at hand with which matrix multiplications can be carried out. But this need not remain so, for there is a simple and easily comprehensible scheme which makes the calculation of matrix products a mere routine: the so-called Falk's scheme, which I will show you in the next example.

Example 5.19 I want to calculate the product from

$$A = \begin{pmatrix} 1 & 0 \\ -2 & 1 \\ 0 & 3 \end{pmatrix} \quad \text{and} \quad B = \begin{pmatrix} 2 & 1 & 9 \\ 0 & -1 & -2 \end{pmatrix}$$

Falk's scheme then looks like this:

$$\begin{array}{cc|ccc} & & 2 & 1 & 9 \\ & & 0 & -1 & -2 \\ \hline 1 & 0 & 2 & 1 & 9 \\ -2 & 1 & -4 & -3 & -20 \\ 0 & 3 & 0 & -3 & -6 \end{array}$$

You write down the two matrices A and B in this arrangement: A is at the bottom left, B is at the top right. Exactly one 3×3 matrix fits into the space enclosed by A and B, because since A is a 3×2 matrix and B is a 2×3 matrix, the product $A \cdot B$ must be a 3×3 matrix. We can then easily determine the entries of the product matrix. Each position in the result matrix is obtained by continuing one A row to the right and one B column down, because these two lines must intersect exactly within the space for the result matrix. For the first row of A and the first column of B, the continued lines intersect exactly where the number 2 is entered. This number 2 is created by multiplying the A row *vector* (1,0) with the B *column vector* $\begin{pmatrix} 2 \\ 0 \end{pmatrix}$ component by component and then adding the results, i.e. calculating the scalar product required at this point. The same principle is used again. The next entry in the first row of the product matrix is created by combining the first *A row* (1,0) with the second *B column* $\begin{pmatrix} 1 \\ -1 \end{pmatrix}$ in the specified way, and so on until all entries are filled.

Thus, to apply Falk's scheme to calculate $A \cdot B$, all you have to do is write A down to the left and B up to the right, and then fill the resulting empty space with the appropriate scalar products. You will have the opportunity to do this in the following exercises.

Exercise 5.6 Calculate:

(a)

$$\begin{pmatrix} 1 & 0 & -2 \\ -3 & 5 & 1 \\ 4 & 2 & -1 \end{pmatrix} \cdot \begin{pmatrix} -6 & 3 & 4 \\ 1 & 2 & -2 \\ -3 & -4 & 0 \end{pmatrix}.$$

(b)

$$\begin{pmatrix} -1 & 5 & 2 \\ -2 & 1 & 0 \end{pmatrix} \cdot \begin{pmatrix} 2 & -5 & -2 & 1 \\ 2 & -1 & 0 & 2 \\ -3 & -1 & 3 & 0 \end{pmatrix}.$$ ■

Once again, we have reached the point where a new operation is available and I need to report back to you on how it stacks up against the other operations. In short, it is time for some calculation rules. They offer little that is surprising, but there is at least one point that is quite out of the ordinary. First of all, the rules:

Calculation Rules for the Matrix Product

Given matrices A, B and C for which the following operations are to be feasible. Then the following rules apply:

(a)

$$A \cdot (B \cdot C) = (A \cdot B) \cdot C.$$

(b)

$$A \cdot (B + C) = A \cdot B + A \cdot C.$$

(c)

(continued)

$$(A+B)\cdot C = A\cdot C + B\cdot C.$$

(d) In general, $A \cdot B \neq B \cdot A$, if one can compute both products at all.

Let us go through this for a moment. The opening sentence of these rules is important, because in the rules themselves I have refrained from always writing down exactly what type each matrix should be, so that you can add or even multiply them. So I assume throughout that the numbers of rows and columns are such that all operations go well. The first rule then only states, that you can do without parentheses, just as you do in number crunching, as long as you are only dealing with multiplications, because the same thing will come out of every way of replacing parentheses anyway. The second rule formulates what is known as the distributive law, which is just a fancy name for the simple fact that you are allowed to multiply out and pre-combine at any time, even with respect to the matrix product: So matrix multiplication is compatible with matrix addition. But why is multiplication and pre-combining done again in the third rule? In c) is actually the same as in b)! It looks like it, but it is not. A closer look shows that in b) the single matrix is on the left and the sum on the right, while in c) it is the other way round. And that actually makes a difference, because according to rule d) you cannot just swap the order when multiplying, so I had to write down the rule about multiplying out in two variations.

So the really surprising thing is rule d): $A \cdot B$ and $B \cdot A$ are – if both exist at all – two different matrices. It is easy to see that often enough one cannot form both products at all. For example, if A has two rows and three columns, while B has three rows and four columns, then A and B fit together beautifully, because A has as many columns as B has rows. Conversely, however, nothing works at all because B has four columns while A has only two rows, so the product $B \cdot A$ is not possible. But even if you can calculate both products, the result is usually not the same.

Example 5.20

(a) For the matrices

$$A=\begin{pmatrix} 1 & 2 \\ -2 & 1 \end{pmatrix} \quad \text{and} \quad B=\begin{pmatrix} 0 & 3 \\ 1 & -1 \end{pmatrix}$$

I want to calculate the products $A \cdot B$ and $B \cdot A$. Using Falk's scheme, you can see that

$$A\cdot B=\begin{pmatrix} 2 & 1 \\ 1 & -7 \end{pmatrix} \text{ and } B\cdot A=\begin{pmatrix} -6 & 3 \\ 3 & 1 \end{pmatrix}$$

is valid. Consequently, you can form both products, but you get two different results.

(b) Let us take a look at how the distributive law plays out. For this purpose I calculate the product

$$\begin{pmatrix} 1 & 2 \\ -2 & 1 \end{pmatrix} \cdot \left[\begin{pmatrix} 1 & 1 \\ 0 & -1 \end{pmatrix} + \begin{pmatrix} -1 & 2 \\ 1 & 0 \end{pmatrix} \right]$$

in two ways. One way is to first add the two matrices in the square brackets and then do the multiplication. The addition results in:

$$\begin{pmatrix} 1 & 1 \\ 0 & -1 \end{pmatrix} + \begin{pmatrix} -1 & 2 \\ 1 & 0 \end{pmatrix} = \begin{pmatrix} 0 & 3 \\ 1 & -1 \end{pmatrix}$$

and thus the whole thing reduces to the multiplication

$$\begin{pmatrix} 1 & 2 \\ -2 & 1 \end{pmatrix} \cdot \begin{pmatrix} 0 & 3 \\ 1 & -1 \end{pmatrix} = \begin{pmatrix} 2 & 1 \\ 1 & -7 \end{pmatrix},$$

because I had just calculated this in example (a). Now the distributive law from rule (b) tells me that I can also proceed differently and multiply out the bracket. This then results in:

$$\begin{pmatrix} 1 & 2 \\ -2 & 1 \end{pmatrix} \cdot \begin{pmatrix} 1 & 1 \\ 0 & -1 \end{pmatrix} + \begin{pmatrix} 1 & 2 \\ -2 & 1 \end{pmatrix} \cdot \begin{pmatrix} -1 & 2 \\ 1 & 0 \end{pmatrix} = \begin{pmatrix} 1 & -1 \\ -2 & -3 \end{pmatrix}$$
$$+ \begin{pmatrix} 1 & 2 \\ 3 & -4 \end{pmatrix} = \begin{pmatrix} 2 & 1 \\ 1 & -7 \end{pmatrix}. \quad \blacksquare$$

A little of your own calculating cannot hurt either:

Exercise 5.7 Calculate in two ways

$$\begin{pmatrix} 2 & 3 & -1 \\ -2 & 4 & 1 \\ 0 & 5 & -2 \end{pmatrix} \cdot \left[2 \cdot \begin{pmatrix} 1 & 2 & -2 \\ -1 & 3 & 1 \\ 4 & 0 & 3 \end{pmatrix} + \begin{pmatrix} 0 & 4 & -1 \\ -2 & 3 & 1 \\ 1 & 5 & -2 \end{pmatrix} \right]. \quad \blacksquare$$

You now know how to multiply matrices, but that immediately raises another question: If you can already multiply matrices together and the result is a matrix again, could not there also be something like a matrix division? This sounds a bit theoretical at first, but it has a very concrete practical background. Recall for a moment the two-step production process that led us to the matrix product. A similar example also shows why the inversion can be useful. So suppose that in your company, *MyCompany,* there is a two-stage production process with two raw materials R_1, R_2, two intermediate products Z_1, Z_2, and two final products E_1, E_2. Through measurements or useful bookkeeping, one knows the consumption matrix B at the transition from Z to E and the total consumption matrix C, which describes how many units of raw materials R_1, R_2 are consumed to produce one unit of E_1 and E_2, respectively. Both matrices are, of course, matrices with two rows and two columns, since you are dealing with two raw materials as well as two intermediate products and two final products. The problem is: How can you determine the consumption matrix A from this when moving from raw materials to intermediate products?

In Example 5.17, we had established that $A \cdot B = C$ holds. Now I know both B and C, so unfortunately I cannot calculate by simple multiplication. If we were dealing with numbers, then things would be simple: the equation $a \cdot b = c$, provided $b \neq 0$, I would easily divide by b and get $a = c/b$ or even $a = c \cdot 1/b$. I would need something like that for matrices too, and my problem would be solved. So I have to think of some kind of "reciprocal matrix", which plays the role of 1/b. Since these are matrices, you then do not write 1/B, but rather somewhat more prettily B^{-1}, but basically mean exactly the same thing, because in the end the eq. $A = C \cdot B^{-1}$ should hold. In this way, you can then calculate the matrix A you are looking for.

All well and good, but how do I get to the inverse matrix? It is supposed to play the role of 1/b, and therefore I first have to find out what represents a one in the realm of matrices. Among numbers, one is the only number you can multiply by without fear of consequences, because $1 \cdot a = a \cdot 1 = a$, no matter which $a \in \mathbb{R}$ you might choose. So I need a matrix that has no effect when multiplied, and it is easy to find. For example

$$\begin{pmatrix} 1 & 0 \\ 0 & 1 \end{pmatrix} \cdot \begin{pmatrix} a & b \\ c & d \end{pmatrix} = \begin{pmatrix} a & b \\ c & d \end{pmatrix},$$

as you can easily calculate, and for matrices with three rows and three columns the matrix $\begin{pmatrix} 1 & 0 & 0 \\ 0 & 1 & 0 \\ 0 & 0 & 1 \end{pmatrix}$ has the same miraculous effect. This is not a coincidence, but a hint how the "one matrix" must look like.

Unit Matrix

If A is a matrix with n rows and n columns, then with

$$I_n = \begin{pmatrix} 1 & 0 & \cdots & 0 \\ 0 & 1 & \cdots & 0 \\ \vdots & \ddots & \ddots & \vdots \\ 0 & \cdots & 0 & 1 \end{pmatrix} \in \mathbb{R}^{n \times n}$$

always:

$A \cdot I_n = I_n \cdot A = A$.

Therefore, the matrix I_n *is* called **unit matrix**.

The unit matrix has n rows and n columns and is characterized by the fact that its product with any matrix from $\mathbb{R}^{n \times n}$ results in the matrix itself. Its structure is simple enough: In the diagonal running from top left to bottom right – the so-called main diagonal – there are ones, everywhere else there are zeros.

Now I have the matrix one in hand, and thus nothing stands in the way of a definition of the inverse matrix. Remember our starting point: Here I am concerned with finding an inverse matrix for a given matrix, which is to play the role of the reciprocal. Now, however, for $a \neq 0$, the reciprocal $1/a$ is characterized by the fact that $a \cdot 1/a = 1$. Nothing overly new, but very helpful, because you can now apply this directly to the matrices. The number a becomes the matrix A, the reciprocal 1/a becomes the inverse matrix A^{-1} and the number 1 becomes the unit matrix I_n.

Inverse Matrix

A matrix $A \in \mathbb{R}^{n \times n}$ is called *invertible*, or *regular*, if there exists a matrix $A^{-1} \in \mathbb{R}^{n \times n}$ with

$$A \cdot A^{-1} = A^{-1} \cdot A = I_n.$$

In this case, A^{-1} is called the inverse matrix of A.

Note that here I am still only considering so-called square matrices, i.e. matrices that are as long as they are wide; only in this situation can we really speak of invertibility. The inverse matrix of A is thus the matrix which, when multiplied by A, yields the unit matrix; here (and only here!) it does not matter whether you calculate $A \cdot A^{-1}$ or $A^{-1} \cdot A$ – the unit matrix must always come out.

Let us go back to my initial example, where I wanted to reconstruct the matrix A from B and C. To get to A, I would have to determine B^{-1}. From $A \cdot B = C$ then follows

$$A \cdot B \cdot B^{-1} = C \cdot B^{-1},$$

because of course I can multiply on both sides with the same matrix. So is

$$A \cdot I_n = C \cdot B^{-1},$$

because, after all, $B \cdot B^{-1} = I_n$. And thus

$$A = C \cdot B^{-1},$$

because multiplying by the unit matrix does not change the matrix A. So this way you can actually calculate the matrix A, as long as the inverse matrix of B is available.

Nothing is given to you in life, not even inverse matrices. Firstly, I have not yet said a word about how to calculate such an inverse matrix, and secondly, there are – in contrast to the familiar numbers – unfortunately matrices that do not consist only of zeros and still do not have an inverse matrix, to which you cannot form an "inverse matrix".

Example 5.21

(a) I am trying to calculate the inverse matrix of $A = \begin{pmatrix} 1 & 1 \\ 2 & 2 \end{pmatrix}$. For this purpose, I will just assume that A actually is invertible, that is, it has an inverse matrix. Then there exists an inverse matrix $A^{-1} = \begin{pmatrix} a & b \\ c & d \end{pmatrix}$, with the property $A \cdot A^{-1} = I_2$, because that is exactly how the inverse matrix was defined. However, this means:

$$\begin{aligned} \begin{pmatrix} 1 & 0 \\ 0 & 1 \end{pmatrix} &= I_2 \\ &= A \cdot A^{-1} \\ &= \begin{pmatrix} 1 & 1 \\ 2 & 2 \end{pmatrix} \cdot \begin{pmatrix} a & b \\ c & d \end{pmatrix} \\ &= \begin{pmatrix} a+c & b+d \\ 2a+2c & 2b+2d \end{pmatrix}, \end{aligned}$$

where I ended up doing nothing more than multiplying the two matrices together in the usual way. So all in all the equation

$$\begin{pmatrix} 1 & 0 \\ 0 & 1 \end{pmatrix} = \begin{pmatrix} a+c & b+d \\ 2a+2c & 2b+2d \end{pmatrix}$$

and thus these two matrices must have the same entries. It follows, in particular, that $1 = a + c$ and $0 = 2a + 2c$. But this is completely impossible, since $2a + 2c$ is exactly twice $a + c$. Now if $a + c = 1$, then $2a + 2c$ has no choice but to take the value 2, and certainly $2 \neq 0$. So if the matrix A had an inverse matrix, then that would result in a number being able to be 0 and 2 at the same time, which is rather rare after all. So my matrix A cannot have possessed an inverse matrix, and it follows that A is not invertible.

(b) Now I look at the slightly larger matrix

$$A = \begin{pmatrix} 1 & 0 & -1 \\ -8 & 4 & 1 \\ -2 & 1 & 0 \end{pmatrix} \in \mathbb{R}^{3\times 3}.$$

How can you calculate that

$$A^{-1} = \begin{pmatrix} 1 & 1 & -4 \\ 2 & 2 & -7 \\ 0 & 1 & -4 \end{pmatrix}$$

holds? Quite simply, by calculating $A \cdot A^{-1}$ and finding that the unit matrix I_3 comes out. That is easy, but how did I come up with this matrix A^{-1}? That is a lot harder and will be covered in the next section.

Another small example to practice:

Exercise 5.8 Prove that the matrix $A = \begin{pmatrix} 1 & 2 \\ 2 & 4 \end{pmatrix}$ has no inverse.?

You have seen in Example 5.21a) that inverting matrices probably has something to do with linear systems of equations, because while calculating I came across two equations with two unknowns, which together formed a linear system of equations. Already in the context of linear dependence of vectors, these systems of equations have come to the fore, and now again! It is obviously about time that I bring them a little closer to you.

5.3 Linear Systems of Equations

That linear systems of equations are needed to solve one or the other of the problems with which I have maltreated you in the first two sections has become clear by now. But not only for the self-made problems you have to use them, they also appear in the most different application contexts, be they more technical or more economic oriented. I will show you what it is all about with a very simple application case: the calculation of breakfast costs.

Example 5.22 A student gets two rolls and a litre of milk for breakfast. He pays 1 Euro and 10 cents for them. If we use B to denote the price of bread rolls and M to denote the price of milk, we can translate this situation into the equation $2B + M = 1.1$. Unfortunately, it does not have a unique solution; for example, $B = 0.5$ and $M = 0.1$ are solutions, but so are $B = 0.25$ and $M = 0.6$. Of course, my goal is to determine food prices, so I need more information about my student's eating habits. I get that the next day, because now he has a visitor for breakfast and buys five rolls, two liters of milk, and six eggs for the price of 3 Euros and 35 cents. I will not dwell on why the two breakfasters need such heavy fortification early in the morning, but rather state the new equation that results. It is:

$$5B + 2M + 6E = 3.35,$$

where E stands for the price of an egg. Both equations together are still not uniquely solvable; for example, $B = 0.25$, $M = 0.6$, $E = 0.15$, but also $B = 0.3$, $M = 0.625$, $E = 0.1$ are solutions. So I am still not able to assign unique prices to each food item. The following day, however, this will change, as the student under observation again changes his breakfast purchases, this time resulting in the eq. $4B + 2M + 4E = 2.8$.

After three days, I have now put together three equations with three unknowns, and this is what is called a system of linear equations. So it is:

$$\begin{array}{ccccccc} 2B & + & M & & & = 1.1 \\ 5B & + & 2M & + & 6E & = 3.35 \\ 4B & + & 2M & + & 4E & = 2.8. \end{array}$$

What I am looking for, of course, are B, M and E. To make the first and the third equation a little more similar, I multiply the first equation by two and get $4B + 2M = 2.2$. Now I can combine this well with the third equation, because it starts the same way as the doubled second one. For this purpose I write both under each other

$$\begin{array}{ccccccc} 4B & + & 2M & & & = 2.2 \\ 4B & + & 2M & + & 4E & = 2.8 \end{array}$$

and note that it might be helpful to subtract the first equation from the second, that is: the left side of the first equation from the left side of the second and the right side of the first equation from the right side of the second. Since we are dealing with equations, these operations are allowed, and fortunately the unknowns B and M are dropped in the process. What remains is:

$$4E = 0.6, \quad \text{so } E = 0.15.$$

This is good, because now I know that an egg costs fifteen cents. I can now substitute this value into the first two equations of my original system of equations and get:

$$\begin{aligned} 2B &+ M &= 1.1 \\ 5B &+ 2M &= 2.45, \end{aligned}$$

because the old second equation was $5B + 2M + 6E = 3.35$, so $5B + 2M + 0.9 = 3.35$, and so $5B + 2M = 2.45$. Using the same principle as before, I now try again to get rid of one unknown, and for this purpose I multiply the first equation by two. The result is

$$\begin{aligned} 4B &+ 2M &= 2.2 \\ 5B &+ 2M &= 2.45. \end{aligned}$$

Subtracting the first equation from the second then gives $B = 0.25$, and the price of the roll is calculated. If we insert it into the very first eq. $2B + M = 1.1$, we get $0.5 + M = 1.1$, i.e. $M = 0.6$. In total I have $B = 0.25$; $M = 0.6$; $E = 0.15$.

The procedure I have just demonstrated here will perhaps have sounded familiar to you: it is the classic addition procedure with which you have certainly also been plagued in school. In this section, I do not want to do anything other than systematize this procedure a bit and make a scheme out of it that you can use without having to think too hard. To do this, I first need to write down what exactly is meant by a system of linear equations. But that is not too difficult. A single linear equation has some unknowns, but you do not necessarily know how many, so you say it has the unknowns $x_1, x_2, \ldots, x_n$. So in Example 5. 22, $n = 3$. Since the equation is supposed to be linear, you may only do with the unknowns what I did with them in the example: You may multiply them by constants and then add or subtract them. Now I just have to invent names for the constant prefactors of the unknowns, and I do that just like I did with the entries for matrices. If my linear equation is the first equation of the system, I write

$$a_{11}x_1 + a_{12}x_2 + \cdots + a_{1n}x_n = b_1.$$

The constant a_{11} thus belongs in the *first* equation to the *first* unknown, a_{12} belongs in the *first* equation to the *second* unknown and so on, until finally with b_1 the right side of the first equation appears. So here again we read a_{11} as a one one and not as a eleven. The second equation is then of course called according to the same pattern

$$a_{21}x_1 + a_{22}x_2 + \cdots + a_{2n}x_n = b_2.$$

The summary of some – say m – linear equations with the same n unknowns is then a linear system of equations.

Linear System of Equations

The system consisting of m linear equations with n unknowns $x_1, \ldots, x_n$

$$\begin{array}{ccccccccc} a_{11}x_1 & + & a_{12}x_2 & + & \cdots & + & a_{1n}x_n & = b_1 \\ a_{21}x_1 & + & a_{22}x_2 & + & \cdots & + & a_{2n}x_n & = b_2 \\ \vdots & & \vdots & & & & \vdots & \vdots \\ a_{m1}x_1 & + & a_{m2}x_2 & + & \cdots & + & a_{mn}x_n & = b_m, \end{array}$$

where the numbers a_{ij} and b_i are known, is called a linear system of equations. With the **coefficient matrix**

$$A = \begin{pmatrix} a_{11} & \cdots & a_{1n} \\ a_{21} & \cdots & a_{2n} \\ \vdots & & \vdots \\ a_{m1} & \cdots & a_{mn} \end{pmatrix},$$

the right side $\mathbf{b} = \begin{pmatrix} b_1 \\ \vdots \\ b_m \end{pmatrix}$ and the vector $\mathbf{x} = \begin{pmatrix} x_1 \\ \vdots \\ x_n \end{pmatrix}$ one writes for it also often

$$A \cdot \mathbf{x} = \mathbf{b}.$$

You can see that I have written down the system of equations in two different ways. The first method is the conventional one, which I am sure you are familiar with from school. The second way is a matrix notation, which is based on what you learned in the section about matrices. Namely, since every equation is based on the same scheme and, most importantly, always has the same unknowns, you save yourself the trouble of writing down the same unknowns over and over again, and just collect the coefficients in a large matrix A and the right-hand side in a vector $\boldsymbol{b}$. If you then work out the product $A \cdot \boldsymbol{x}$ as we discussed in the second section, you will find that exactly the gathering of the left-hand sides of my system of equations comes out. So the somewhat terse notation $A \cdot \boldsymbol{x} = \boldsymbol{b}$ is nothing more than a truncating notation for a system of linear equations.

Well, you can now write down such a linear system of equations, but how can you solve it systematically? I will show you in a moment in the next example. Before that, however, I need to briefly go through something else with you. In Example 5.22, I manipulated the given equations again and again without doing any harm, because these manipulations did not change anything in the solutions of the system of equations. It is best to note down briefly which manipulations can be carried out without any risk:

Permitted Manipulations of Systems of Equations

The solution set of a linear system of equations is not changed by the following operations:

(a) Multiplication of an equation by a non-zero number.
(b) Adding a multiple of one equation to another.
(c) Swapping two equations.

There is probably not much to add to this, except perhaps the fact that with b) you can of course subtract equations from other equations, since you are also allowed to multiply with negative numbers. In the next example, I will show you how to take advantage of the license you have just been given to manipulate and write it down succinctly, too. The trick will be to write a system of equations in matrix form and then change the rows of this matrix so that you can read off the solution directly at the end.

Example 5.23 I consider the linear system of equations

$$\begin{aligned} -x + 2y + z &= -2 \\ 3x - 8y - 2z &= 4 \\ x \qquad\quad 4z &= -2. \end{aligned}$$

with the unknowns x, y and z. The coefficient matrix then is

$$A = \begin{pmatrix} -1 & 2 & 1 \\ 3 & -8 & -2 \\ 1 & 0 & 4 \end{pmatrix}$$

and on the right side is the vector

$$\mathbf{b} = \begin{pmatrix} -2 \\ 4 \\ -2 \end{pmatrix}.$$

However, a system of equations always includes complete equations including the right-hand side, which is why I add the right-hand side of the matrix A that I just wrote down. This then results in the matrix that is one column wider

$$(A, b) = \begin{pmatrix} -1 & 2 & 1 & -2 \\ 3 & -8 & -2 & 4 \\ 1 & 0 & 4 & -2 \end{pmatrix}.$$

Each row of this matrix now corresponds to an equation of the given system of equations: the first three numbers give the coefficients of the unknowns, while the last number describes the right-hand side. Now I want to make sure that one unknown vanishes, and in matrix notation this means that, for example, the coefficients of the unknown x become zero, because $0 \cdot x$ is nothing. I therefore add three times the first row to the second as well as the first row to the third. By doing this, I have actually done nothing but add the triple of the first equation to the second, and the first equation itself to the third, and such operations are always allowed. The new matrix is:

$$\begin{pmatrix} -1 & 2 & 1 & -2 \\ 0 & -2 & 1 & -2 \\ 0 & 2 & 5 & -4 \end{pmatrix}.$$

The second and third lines now each have zeros at the beginning. So the coefficient of x has become zero, in other words x has been eliminated from these two equations and the second and third lines together correspond to two equations with two unknowns, namely $-2y + z = -2$ and $2y + 5z = -4$. From the last equation y should now also disappear, so that only one equation with one unknown remains. To do this, I simply add the second line to the third and get

$$\begin{pmatrix} -1 & 2 & 1 & -2 \\ 0 & -2 & 1 & -2 \\ 0 & 0 & 6 & -6 \end{pmatrix}.$$

The last line now corresponds to the eq. $6z = -6$, and that already makes it clear what I have to do now. Dividing the last line by 6 gives:

$$\begin{pmatrix} -1 & 2 & 1 & -2 \\ 0 & -2 & 1 & -2 \\ 0 & 0 & 1 & -1 \end{pmatrix}.$$

Now, of course, I can read directly from the third line that $z = -1$ holds. It would be nice to be able to produce the same situation for the other unknowns, but that is not difficult now. In order to calculate only with y in the second line, the entry 1 must disappear from the third place, because in this case both the prefactor of x and that of z are zero. To this end, I subtract the third line from the second and find:

$$\begin{pmatrix} -1 & 2 & 1 & -2 \\ 0 & -2 & 0 & -1 \\ 0 & 0 & 1 & -1 \end{pmatrix}.$$

The second line now corresponds to the equation $-2y = -1$, which of course should be divided by -2. Therefore, I divide the second line by -2 with the result:

$$\begin{pmatrix} -1 & 2 & 1 & -2 \\ 0 & 1 & 0 & \frac{1}{2} \\ 0 & 0 & 1 & -1 \end{pmatrix}.$$

You probably already know what has to happen now. Except for the -1 at the very front, all coefficients must be eliminated from the first row, which I can easily do by subtracting the duplicated second row and the unchanged third row from the first:

$$\begin{pmatrix} -1 & 0 & 0 & -2 \\ 0 & 1 & 0 & \frac{1}{2} \\ 0 & 0 & 1 & -1 \end{pmatrix}.$$

Multiplying the first line by -1 finally gives:

$$\begin{pmatrix} 1 & 0 & 0 & 2 \\ 0 & 1 & 0 & \frac{1}{2} \\ 0 & 0 & 1 & -1 \end{pmatrix}.$$

Finally it is done. If you now translate the individual lines into equations again, then only x appears in the first equation, only y in the second and only z in the third, and you can see at a glance:

$$x = 2, \quad y = \frac{1}{2}, \quad z = -1. \quad \blacksquare$$

Did you see what this all boils down to? I turned three equations with three unknowns into two equations with two unknowns by letting the coefficients of x become zero in the second and third rows of my matrix. Adding or subtracting rows here is exactly the same as adding or subtracting equations, and that was explicitly allowed. I then extracted an equation with only one unknown from the two resulting equations by condemning the coefficient of y to insignificance in the third row of the matrix. At this state of affairs, I had achieved a matrix that consisted of only zeros on the left below the main diagonal, and that is always the first goal when solving systems of equations. Then I laboriously worked my way up again from the bottom to the top to have only zeros on the right side above the main diagonal (with the exception of the right side, of course) and to be able to read off the solutions directly at the end.

This method is called the **Gaussian algorithm**. It is based on converting the matrix (A, b) with simple row transformations to the form

$$\begin{pmatrix} 1 & 0 & \cdots & 0 & x_1 \\ 0 & 1 & \cdots & 0 & x_2 \\ \vdots & \ddots & \ddots & \vdots & \vdots \\ 0 & \cdots & 0 & 1 & x_n \end{pmatrix}$$

where $x_1, \ldots, x_n$ are the solutions of the system of equations. The Gaussian algorithm exists in several variants, but they all actually amount to the same thing. First, we look at it for the case where the system of equations leads to exactly one solution for each of the unknowns:

Gauss Algorithm

Let $A\mathbf{x} = \mathbf{b}$ be a linear system of equations with n equations and n unknowns that has a unique solution. We start with the matrix

$$(A \mid b) = \begin{pmatrix} a_{11} & \cdots & a_{1n} & b_1 \\ a_{21} & \cdots & a_{2n} & b_2 \\ \vdots & & \vdots & \vdots \\ a_{n1} & \cdots & a_{nn} & b_n \end{pmatrix},$$

in which the coefficients of the system of equations and its right side are listed. Then you need to perform the following steps:

(continued)

(a) If $a_{11} = 0$, find a line whose first element is different from zero, swap this line with the first line and rename it. In any case, $a_{11} \neq 0$ is then valid.
(b) Subtract a suitable multiple of the first row from the second, third, . . ., last row, so that these rows each start with zero. The new matrix then has only zeros in the first column below a_{11}.
(c) Subtract a suitable multiple of the second row from the third, . . ., last row, so that these rows each start with zero and also have a zero in the second place. The new matrix has then in the first column below a_{11} and in the second column below a_{22} only zeros.
(d) Repeat this procedure for the following columns of the matrix until there are only zeros left below the main diagonal elements.
(e) Divide the last row by the last non-zero coefficient so that there is a one at the end of the main diagonal. Then proceed with the appropriate row transformations from bottom to top, so that the matrix has the following form at the end:

$$\begin{pmatrix} 1 & 0 & \cdots & 0 & x_1 \\ 0 & 1 & \cdots & 0 & x_2 \\ \vdots & \ddots & \ddots & \vdots & \vdots \\ 0 & \cdots & 0 & 1 & x_n \end{pmatrix}.$$

The last column contains the solution of the linear system of equations.

If this procedure does not lead to the goal, the linear system of equations does not have a unique solution.

Pretty abstract at first glance, I know. But you have no choice but to get a little abstract if you want to write down a reasonably elaborate procedure in a way that is somehow generally valid. Another example will show you that there is actually not that much behind it.

Example 5.24 I want to solve the linear system of equations

$$\begin{array}{ccccccccc} x & & & + & z & + & u & = 0 \\ x & + & y & + & 2z & + & u & = 1 \\ & - & y & & & + & u & = 0 \\ x & & & & & + & 2u & = 0 \end{array}$$

The matrix representation of this system of equations is:

$$(A, b) = \begin{pmatrix} 1 & 0 & 1 & 1 & 0 \\ 1 & 1 & 2 & 1 & 1 \\ 0 & -1 & 0 & 1 & 0 \\ 1 & 0 & 0 & 2 & 0 \end{pmatrix}.$$

Now I need to get zeros only in the first column below the top one. Subtracting the first row from the second and fourth gives the desired matrix:

$$\begin{pmatrix} 1 & 0 & 1 & 1 & 0 \\ 0 & 1 & 1 & 0 & 1 \\ 0 & -1 & 0 & 1 & 0 \\ 0 & 0 & -1 & 1 & 0 \end{pmatrix}.$$

Adding the second to the third line returns:

$$\begin{pmatrix} 1 & 0 & 1 & 1 & 0 \\ 0 & 1 & 1 & 0 & 1 \\ 0 & 0 & 1 & 1 & 1 \\ 0 & 0 & -1 & 1 & 0 \end{pmatrix},$$

and already I have only zeros in the second column below the diagonal element and can devote myself to the third column. Adding the third to the fourth row yields:

$$\begin{pmatrix} 1 & 0 & 1 & 1 & 0 \\ 0 & 1 & 1 & 0 & 1 \\ 0 & 0 & 1 & 1 & 1 \\ 0 & 0 & 0 & 2 & 1 \end{pmatrix}.$$

You can see that there are now only zeros below the main diagonal, but the last element of the main diagonal still has the unfavorable value two. Therefore I divide the fourth row by two and get:

$$\begin{pmatrix} 1 & 0 & 1 & 1 & 0 \\ 0 & 1 & 1 & 0 & 1 \\ 0 & 0 & 1 & 1 & 1 \\ 0 & 0 & 0 & 1 & \frac{1}{2} \end{pmatrix}.$$

The first part of the procedure is done with this, I worked my way from top to bottom and created zeros below the main diagonal. Now I change the direction of view and fight my way from the bottom to the top in order to also zero everything on the right above the diagonal. I do the fourth column in one step by subtracting the fourth row from the third and the first with the result:

$$\begin{pmatrix} 1 & 0 & 1 & 0 & -\frac{1}{2} \\ 0 & 1 & 1 & 0 & 1 \\ 0 & 0 & 1 & 0 & \frac{1}{2} \\ 0 & 0 & 0 & 1 & \frac{1}{2} \end{pmatrix}.$$

Then I subtract the third row from the first and the second, which in the end leads me to the matrix

$$\begin{pmatrix} 1 & 0 & 0 & 0 & -1 \\ 0 & 1 & 0 & 0 & \frac{1}{2} \\ 0 & 0 & 1 & 0 & \frac{1}{2} \\ 0 & 0 & 0 & 1 & \frac{1}{2} \end{pmatrix}$$

from which I can directly read off the results, because now

$$x = -1, \quad y = z = u = \frac{1}{2}. \quad \blacksquare$$

So basically it is not a big deal. First you work your way from the top left to the bottom right and make sure that all the columns below the main diagonal in the sequence have only zeros, starting with the first column. Once that is done, you reverse the direction of work and make sure that all the columns above the main diagonal are zeros in order, starting at the back and ending at the front.

To do this, once again, exercises.

Exercise 5.9 Solve the following systems of linear equations:

$$\begin{array}{rcrcrcl} x & + & y & - & 5z & = & -10 \\ 3x & - & 4y & + & z & = & 9 \\ 6x & + & 2y & + & 4z & = & 12 \end{array}$$

and

$$\begin{array}{rcrcrcrcl} x & + & 3y & + & 2z & - & 6u & = & 1 \\ x & + & 2y & - & 2z & - & 5u & = & -1 \\ 2x & + & 4y & - & 2z & - & 9u & = & 0 \\ 2x & + & 4y & - & 6z & - & 9u & = & 2 \end{array}$$

■

This is already excellent, but unfortunately it is not quite enough, because not all linear systems of equations have a unique solution.

Example 5.25 I am looking for the solutions of the linear system of equations

$$\begin{array}{rcrcrcrcl} x & + & 2y & + & z & + & 3u & = & 3 \\ 2x & + & 4y & + & 2z & + & 6u & = & 5 \\ 3x & + & 6y & + & 2z & + & 10u & = & -3 \end{array}$$

with three equations and the four unknowns x, y, z, and u. By now, writing down the matrix form is just routine for you; it reads:

$$(A, b) = \begin{pmatrix} 1 & 2 & 1 & 3 & 3 \\ 2 & 4 & 2 & 6 & 5 \\ 3 & 6 & 2 & 10 & -3 \end{pmatrix}.$$

According to my Gaussian scheme, I have to subtract twice the first line from the second and three times the first line from the third. This gives:

$$\begin{pmatrix} 1 & 2 & 1 & 3 & 3 \\ 0 & 0 & 0 & 0 & -1 \\ 0 & 0 & -1 & 1 & -12 \end{pmatrix}.$$

The second row makes a somewhat strange impression, since it contains only zeros in the area of the coefficients and we are actually used to rows with so many zeros only at the end of the matrix. So I swap the second row with the third row, which is just like swapping two equations and is therefore allowed. This gives me the matrix

$$\begin{pmatrix} 1 & 2 & 1 & 3 & 3 \\ 0 & 0 & -1 & 1 & -12 \\ 0 & 0 & 0 & 0 & -1 \end{pmatrix}.$$

And now turn your attention once to the third line and translate it back into an equation. Since I only have zeros in the coefficients here, the equation is:

$$0x + 0y + 0z + 0u = -1, \quad \text{so } 0 = -1.$$

Does something like this happen? But rather rarely, I got here a nonsensical equation that can never be satisfied. Whatever values you may put in for the four unknowns, you will never manage to get the eq. $0 = -1$ to be satisfied. But if I already know that I have an equation here that I can never satisfy, then I might as well save myself the trouble of searching further and lie down on the couch or in the bathtub: It would lead to nothing anyway. No combination of values in the world can save this impossible eq. $0 = -1$, and that just means there is no solution! This system of linear equations is simply unsolvable. You then either say this very phrase "The system of equations is unsolvable" or state the solution set as the empty set $\emptyset$, that is, the somewhat desolate set in which you cannot find a single element.□

So there are also linear systems of equations that do not allow a solution and whose solution set is therefore empty. The nice thing is that you do not have to grow gray hairs over the question of how to recognize such unreliable systems of equations, because the example has shown it quite clearly: You start the Gauss algorithm exactly as you have learned it, and if it leads to a row where there is no zero only in the last column, then the system of equations has no solution.

Unsolvable Linear Systems of Equations

If $A \cdot \boldsymbol{x} = \boldsymbol{b}$ is a system of linear equations where the Gaussian algorithm leads to a row of the form $\begin{pmatrix} 0 & 0 & \cdots & 0 & b \end{pmatrix}$ with $b \neq 0$ in which all entries except the last one are zero, then the system of linear equations has no solution.

It does not matter where this unsuitable line is; as soon as it appears, you can safely refrain from any further calculations and be content with the fact that the system of equations is unsolvable.

Exercise 5.10 Determine the solution to the system of equations

$$\begin{array}{ccccccl} x & + & y & + & 2z & = 1 \\ x & + & 5y & + & 2z & = 3 \\ 2x & - & 2y & + & 4z & = 5. \end{array} \quad ■$$

So there are systems of equations where you can find exactly one solution for each unknown, and there are systems of equations that have no solution at all. Is that it? Unfortunately no, there is also a third possibility, namely an infinite number of solutions. You can see that already in the clear example of the system of equations, which consists only of the one equation $x + y = 0$, because obviously I can use $x = y = 0$ or $x = 1, y = -1$ or $x = 2, y = -2$, etc. as solutions. Using a somewhat larger example, I will now show you how to deal with such a situation.

Example 5.26 I examine the system of equations

$$\begin{aligned} x + \quad y + z &= 0 \\ 2x + \quad y - z &= 3 \\ 3x + 2y \qquad &= 3. \end{aligned}$$

In matrix form it says

$$\begin{pmatrix} 1 & 1 & 1 & 0 \\ 2 & 1 & -1 & 3 \\ 3 & 2 & 0 & 3 \end{pmatrix}.$$

This still looks normal, so I will run the Gaussian algorithm as usual. To produce the necessary zeros in the first column, I subtract the double first row from the second and the triple first from the third. This gives:

$$\begin{pmatrix} 1 & 1 & 1 & 0 \\ 0 & -1 & -3 & 3 \\ 0 & -1 & -3 & 3 \end{pmatrix}.$$

Still no problems occur, but they will soon if I continue the scheme and subtract the second row from the third to eliminate the interfering -1 in the third row. The matrix then reads:

$$\begin{pmatrix} 1 & 1 & 1 & 0 \\ 0 & -1 & -3 & 3 \\ 0 & 0 & 0 & 0 \end{pmatrix}.$$

This is now a new situation, because only zeros appear in the third line, which I usually only know from sessions. Translated into an equation, the third line is just $0 = 0$, and that is not very helpful information. In other words, by applying the Gaussian algorithm, it turned out that the system of equations actually only had the information from two equations with three unknowns packed into it, not three eqs. A closer look at the equations confirms this diagnosis, because the third equation is just the sum of the first two equations and therefore could not offer anything new.

The zero-line is not usable as an informant. It is also no longer possible to produce all ones along the main diagonal, because in the last line there are only zeros. So I cannot reach the original goal of the Gauss algorithm anymore. But that does not matter. So far, in the course of the Gauss algorithm, we have tried to produce ones in the main diagonal and zeros otherwise, and that only means that a unit matrix should be produced. And I can still get a unit matrix on the row; it would not be quite as big, but better a small unit matrix than none at all. As things stand, with the system of equations at hand, I cannot get a unit matrix with three rows and three columns, but only one with two rows and two columns. To do this, I multiply the second row by -1 and then subtract it from the first. This results in:

$$\begin{pmatrix} 1 & 0 & -2 & 3 \\ 0 & 1 & 3 & -3 \\ 0 & 0 & 0 & 0 \end{pmatrix}.$$

As you can see, I now have a slightly smaller unit matrix of two rows and two columns at the top left, and that is all man can ask for. In fact, if you translate the remaining two rows into equations, they are $x - 2z = 3$ and $y + 3z = -3$. That is all the information there is. Since I had already made a small unit matrix, I can now easily solve for the two unknowns x and y, because both already have the prefactor 1. So it holds:

$$x = 3 + 2z \quad \text{und} \quad y = -3 - 3z.$$

This system of equations does not give any further information, however long you may search. Obviously, based on these two equations with three unknowns, it is not possible to calculate unique solutions for x, y and z, I have too little information for that. But there are solutions anyway: For example, if $z = 0$, it immediately follows from the two calculated equations that $x = 3$ and $y = -3$. If, on the other hand, $z = 1$, you will find $x = 5$ and $y = -6$. Whatever you put in for z, you can immediately *calculate* x and y from the two calculated equations without any problems. Therefore, this system of linear equations has an infinite number of solutions, because you can substitute whatever you want for z and calculate the values for x and y from it. So the solutions have the form

$$x = 3 + 2z, \quad y = -3 - 3z, \quad z \in \mathbb{R} \text{ any}.$$

In this case, since you can use whatever you want for z, z is also called a free parameter.

This is how it always works. If zero rows occur in the course of the Gaussian algorithm, then you can safely ignore them and build up as large a unit matrix as possible. If this unit matrix has, for example, two rows and two columns and there are more than two unknowns, then you can insert any values for the excess unknowns (in Example 5.26 this was z) and calculate the values for the other unknowns directly from the equations resulting from the last calculated matrix.

Gauss Algorithm

To solve the linear system of eqs. $A \cdot \boldsymbol{x} = \boldsymbol{b}$, bring the matrix (A, b) by the known operations to a form in which there is a unit matrix at the top left that is as large as possible, with only pure zero rows or rows of the form $\begin{pmatrix} 0 & 0 & \cdots & 0 & b \end{pmatrix}$ occurring below this unit matrix.

If there are rows of the form $\begin{pmatrix} 0 & 0 & \cdots & 0 & b \end{pmatrix}$ with $b \neq 0$ below the unit matrix, the system of equations is unsolvable. If there are pure zero rows, they can be ignored. If the unit matrix does not quite reach the last column (i.e. the right-hand side), the "excess" unknowns can be freely chosen and the other unknowns calculated from the new matrix.

Again, this is a fairly abstract formulation, so another example would not hurt:

Example 5.27 Given the linear system of equations

$$\begin{array}{rcrcrcrl} x & + & 5y & + & 2z & + & 3u & = 4 \\ 4x & + & 18y & + & 2z & + & 8u & = 12 \\ 3x & + & 11y & - & 6z & + & u & = 4 \\ & & 2y & + & 6z & + & 4u & = 4. \end{array}$$

In matrix form the system is

$$(A, b) = \begin{pmatrix} 1 & 5 & 2 & 3 & 4 \\ 4 & 18 & 2 & 8 & 12 \\ 3 & 11 & -6 & 1 & 4 \\ 0 & 2 & 6 & 4 & 4 \end{pmatrix}.$$

Once again, the Gaussian algorithm comes into play, which must provide zeros below the top left one. To do this, I subtract the quadruple first row from the second and the triple first from the third. This gives the matrix:

$$\begin{pmatrix} 1 & 5 & 2 & 3 & 4 \\ 0 & -2 & -6 & -4 & -4 \\ 0 & -4 & -12 & -8 & -8 \\ 0 & 2 & 6 & 4 & 4 \end{pmatrix}.$$

Since I prefer to have ones in the main diagonal, I divide the second row by −2. The result is:

$$\begin{pmatrix} 1 & 5 & 2 & 3 & 4 \\ 0 & 1 & 3 & 2 & 2 \\ 0 & -4 & -12 & -8 & -8 \\ 0 & 2 & 6 & 4 & 4 \end{pmatrix}.$$

Below the newly created one, I now want to create zeros again completely according to plan. I do this by adding the quadruple second row to the third and subtracting the double second row from the fourth. With this I have:

$$\begin{pmatrix} 1 & 5 & 2 & 3 & 4 \\ 0 & 1 & 3 & 2 & 2 \\ 0 & 0 & 0 & 0 & 0 \\ 0 & 0 & 0 & 0 & 0 \end{pmatrix}.$$

Two rows and thus also two equations have proved to be superfluous, so the unit matrix I am looking for can only consist of two rows and two columns. It is also easy to find, because all I have to do is subtract five times the second row from the first. This leads to:

$$\begin{pmatrix} 1 & 0 & -13 & -7 & -6 \\ 0 & 1 & 3 & 2 & 2 \\ 0 & 0 & 0 & 0 & 0 \\ 0 & 0 & 0 & 0 & 0 \end{pmatrix}.$$

Here I have clearly separated the achieved unit matrix from the rest of the world, so that you have it better in view. The last two rows are just zeros, so they do not matter anymore. The small unit matrix of two rows and two columns shows that you can compute the unknowns x and y from z and u. Indeed, the corresponding equations are:

$$x - 13z - 7u = -6 \quad \text{and} \quad y + 3z + 2u = 2.$$

It follows:

$$x = 13z + 7u - 6 \quad \text{sowie} \quad y = -3z - 2u + 2.$$

So you can freely choose the values for z and u – these are the already mentioned free parameters – and thus get the solutions:

$$x = 13z + 7u - 6, \quad y = -3z - 2u + 2, \quad z, u \in \mathbb{R} \text{ any}.$$

So here, too, there are an infinite number of solutions. You can grab the values for the unknowns z and u out of thin air at will, and easily calculate the values for x and y using the calculated equations. For example, if you take $z = 0$, $u = 1$, you get $x = 1$, $y = 0$, while for $z = 1$, $u = 0$ you immediately get $x = 7$, $y = -1$. Thus, there are no limits to the number of solutions.

As you have seen, there are three possibilities for a linear system of equations: it can have no solution at all, or it can have exactly one solution for each unknown, or it can have an infinite number of solutions. But that does not matter, because you also saw that you can test all these possibilities in the same way using the Gauss algorithm and calculate the solutions.

Now we were doing some math again.

Exercise 5.11 Determine the solutions of the following systems of linear equations.

$$\begin{aligned} x - y \quad\; &= 2 \\ x + y - 2z &= 0 \\ 2x - y - z &= 3 \end{aligned} \quad \text{and} \quad \begin{aligned} x - \quad y \quad + \quad\; u &= 0 \\ -2y - z + 2u &= 0 \\ -2x + \quad 5y + 2z - 4u &= 0 \\ 4x - \quad 13y - 6y + 10u &= 0 \end{aligned}$$

You will remember that I had to leave one problem open in each of the sections on vectors and on matrices, because we did not quite know how to handle the systems of linear equations yet. Now nothing stands in the way of the final solution of these problems. Let us start with the linear dependence of vectors. In the first section I told you that the vectors $\boldsymbol{x}_1, \ldots, \boldsymbol{x}_m$ are linearly independent exactly when from

$$c_1 \cdot \mathbf{x}_1 + c_2 \cdot \mathbf{x}_2 + \cdots + c_m \cdot \mathbf{x}_m = \mathbf{0}$$

always follows:

$$c_1 = c_2 = \cdots = c_m = 0,$$

where $\mathbf{0}$ is understood to be the zero vector. If, on the other hand, the vectors can be combined to form the zero vector without each of the prefactors $c_1, c_2, \ldots, c_m$ having to become zero, then the vectors are linearly dependent. If we now want to test this for concrete given vectors, we quickly arrive once again at a linear system of equations.

Example 5.28 I want to determine whether the vectors

$$\begin{pmatrix} 1 \\ 1 \\ 0 \end{pmatrix}, \begin{pmatrix} 1 \\ 0 \\ 1 \end{pmatrix}, \begin{pmatrix} 1 \\ 3 \\ -2 \end{pmatrix}$$

are linearly dependent or linearly independent. For this I set the equation

$$c_1 \cdot \begin{pmatrix} 1 \\ 1 \\ 0 \end{pmatrix} + c_2 \cdot \begin{pmatrix} 1 \\ 0 \\ 1 \end{pmatrix} + c_3 \cdot \begin{pmatrix} 1 \\ 3 \\ -2 \end{pmatrix} = \begin{pmatrix} 0 \\ 0 \\ 0 \end{pmatrix}$$

and have to calculate whether $c_1 = c_2 = c_3 = 0$ or whether there are other possibilities. I translate this vector equation into a linear system of equations by writing down what must be true for each component of the vectors. This then results in the system of equations

$$\begin{array}{ccccccc} c_1 & + & c_2 & + & c_3 & = 0 \\ c_1 & & & + & 3c_3 & = 0 \\ & & c_2 & - & 2c_3 & = 0. \end{array}$$

In matrix form, this means:

$$\begin{pmatrix} 1 & 1 & 1 & 0 \\ 1 & 0 & 3 & 0 \\ 0 & 1 & -2 & 0 \end{pmatrix}.$$

From now on, everything is routine. To have only zeros in the first column below the main diagonal, I subtract the first row from the second. This results in:

$$\begin{pmatrix} 1 & 1 & 1 & 0 \\ 0 & -1 & 2 & 0 \\ 0 & 1 & -2 & 0 \end{pmatrix}.$$

Now I add the second line to the third and get:

$$\begin{pmatrix} 1 & 1 & 1 & 0 \\ 0 & -1 & 2 & 0 \\ 0 & 0 & 0 & 0 \end{pmatrix}.$$

With this I have already found a zero row, which does not have to bother me in the following. However, I would like to have as large a unit matrix as possible and will have to limit myself to two rows and two columns for this purpose. Adding the second row to the first and then multiplying the second row by -1 finally leads to:

$$\begin{pmatrix} 1 & 0 & 3 & 0 \\ 0 & 1 & -2 & 0 \\ 0 & 0 & 0 & 0 \end{pmatrix}.$$

Translating this again into two equations, they read as follows.

$$c_1 + 3c_3 = 0 \quad \text{and} \quad c_2 - 2c_3 = 0.$$

So you can choose the unknown c_3 arbitrarily and get

$$c_1 = -3c_3, \quad c_2 = 2c_3, \quad c_3 \in \mathbb{R} \text{ any}.$$

For example, $c_1 = -3$, $c_2 = 2$, $c_3 = 1$ is a solution that obviously does not consist only of zeros. Therefore, the three vectors are linearly dependent.□

You can test the vectors from Example 5.11(b) for linear dependence yourself in the following exercise.

Exercise 5.12 Determine whether the vectors listed in Example 5.11(b) are linearly dependent or linearly independent.

One problem is done with that. Another problem occurred in the second section, when we had to calculate the inverse of a given matrix, because that also had something to do with systems of linear equations. As is often the case, I will explain it to you with an example. But first I should remind you briefly what it is all about: If you have given a square matrix A, i.e. a matrix that has as many rows as columns, then you are looking for a matrix A^{-1}, which corresponds to the inverse of this matrix A, so to speak. More precisely, let the eq. $A \cdot A^{-1} = I_n$ be satisfied, where by I_n we mean the n-row unit matrix consisting of ones in the main diagonal and zeros otherwise. We had discussed all this in the second section. You will now see how to get the inverse A^{-1} in the course of Example 5.29.

Example 5.29 I am looking for the inverse of the matrix $A = \begin{pmatrix} 1 & 2 \\ 1 & 3 \end{pmatrix}$, hoping that an inverse matrix exists at all. If it does exist, it has the same format as A itself, so it consists of two rows and two columns. Since I still have no idea what might be in the inverse matrix, I write it down as $A^{-1} = \begin{pmatrix} a & b \\ c & d \end{pmatrix}$. But now I know that $A \cdot A^{-1}$ should give exactly the unit matrix, because that is how the inverse was defined. And I also know what the two-row unit matrix looks like, which is $I_2 = \begin{pmatrix} 1 & 0 \\ 0 & 1 \end{pmatrix}$. So it must hold:

$$\begin{pmatrix} 1 & 0 \\ 0 & 1 \end{pmatrix} = A \cdot A^{-1} = \begin{pmatrix} 1 & 2 \\ 1 & 3 \end{pmatrix} \cdot \begin{pmatrix} a & b \\ c & d \end{pmatrix} = \begin{pmatrix} a+2c & b+2d \\ a+3c & b+3d \end{pmatrix}.$$

Now what on earth is that supposed to help me with? Quite a lot, because I have found out that two matrices are supposed to be equal, viz.

$$\begin{pmatrix} 1 & 0 \\ 0 & 1 \end{pmatrix} = \begin{pmatrix} a+2c & b+2d \\ a+3c & b+3d \end{pmatrix}.$$

But if the two matrices are to be equal, then their entries must be equal, and that means:

$$\begin{array}{rcrl} a & + & 2c & = 1 \\ a & + & 3c & = 0 \end{array} \quad \text{and} \quad \begin{array}{rcrl} b & + & 2d & = 0 \\ b & + & 3d & = 1. \end{array}$$

That is handy, because now I have been given two systems of linear equations, one with unknowns a and c, the other with unknowns b and d. You know how to solve something like that, and I am going to solve the first of the two systems now. Its matrix form is:

$$\begin{pmatrix} 1 & 2 & 1 \\ 1 & 3 & 0 \end{pmatrix}.$$

Subtracting the first line from the second provides

$$\begin{pmatrix} 1 & 2 & 1 \\ 0 & 1 & -1 \end{pmatrix}.$$

Now I subtract the duplicate second line from the first. This leads to:

$$\begin{pmatrix} 1 & 0 & 3 \\ 0 & 1 & -1 \end{pmatrix}.$$

Consequently, $a = 3$ and $c = -1$, and I have already calculated two entries of the inverse matrix. In the same way, of course, I get b and d. The matrix form of the system of equations is:

$$\begin{pmatrix} 1 & 2 & 0 \\ 1 & 3 & 1 \end{pmatrix}.$$

Subtracting the first line from the second provides

$$\begin{pmatrix} 1 & 2 & 0 \\ 0 & 1 & 1 \end{pmatrix}.$$

Now I subtract the duplicate second line from the first. This leads to:

$$\begin{pmatrix} 1 & 0 & -2 \\ 0 & 1 & 1 \end{pmatrix}.$$

Consequently, $b = -2$ and $d = 1$. With that all done, I just need to write down the inverse matrix. It reads

$$A^{-1} = \begin{pmatrix} 3 & -2 \\ -1 & 1 \end{pmatrix}.$$

Actually, I am done now, but you should have noticed something during the calculation: No matter whether I calculated a and c or b and d, the coefficient matrix was the same in each case. Also the operations were the same in each case, which you can already see by the fact that I wrote the same intermediate texts for both systems of equations. Only the right-hand sides were different, of course. So if the coefficient matrix is always the same and basically the same operations are to be done, then you can make your life easier and do everything at once by treating both right-hand sides at once and not just one. The starting matrix then looks like this:

$$\begin{pmatrix} 1 & 2 & 1 & 0 \\ 1 & 3 & 0 & 1 \end{pmatrix}.$$

Subtracting the first line from the second provides

$$\begin{pmatrix} 1 & 2 & 1 & 0 \\ 0 & 1 & -1 & 1 \end{pmatrix}.$$

Now I subtract the duplicate second line from the first. This leads to:

$$\begin{pmatrix} 1 & 0 & 3 & -2 \\ 0 & 1 & -1 & 1 \end{pmatrix}.$$

Again, I did the same thing, only with two right sides at once. And what do you notice? The right half of the last matrix contains exactly the inverse matrix! So in one operation, I calculated the inverse matrix that previously took me two operations.

This is how inverting any matrix works, if it has an inverse at all. You write the matrix in the left half of a large matrix, write the unit matrix in the right half and manipulate the whole matrix with the familiar methods of the Gauss algorithm until the unit matrix has appeared on the left. Once you have reached this pleasant state, you will find the inverse matrix you are looking for in the right half. And it gets even better: after all, there are matrices that do not have inverses, so what about them? Simple. If the matrix is not invertible, you will not succeed with the best will in the world to conjure up a complete unit matrix on the left side of the large matrix. So the Gauss algorithm is at the same time a way to determine the invertibility of a matrix: If the calculation does not lead to the end, there can be no inverse.

Inversion of Matrices

Let $A \in \mathbb{R}^{n \times n}$ a matrix whose inverse matrix A^{-1} is to be computed. We combine A and the unit matrix I_n into one matrix by writing the matrix A in the left half and the unit matrix in the right half, and reshape this matrix using the well-known operations from the Gaussian algorithm so that the left half is transformed into the unit matrix I_n. Then the inverse matrix A^{-1} is in the right half of the reshaped large matrix. If it is not possible to produce the unit matrix in the left half, A is not invertible.

I conclude this section with one final example.

Example 5.30 We are looking for the inverse of the matrix

$$A = \begin{pmatrix} 1 & 0 & -1 \\ 3 & 1 & -3 \\ 1 & 2 & -2 \end{pmatrix}.$$

As usual, I write the three-row unit matrix I_3 next to A in a large matrix with three rows and six columns:

$$\begin{pmatrix} 1 & 0 & -1 & 1 & 0 & 0 \\ 3 & 1 & -3 & 0 & 1 & 0 \\ 1 & 2 & -2 & 0 & 0 & 1 \end{pmatrix}.$$

This new matrix must be transformed in such a way that the unit matrix I_3 is in its left half, because as soon as this is done, I have the inverse A^{-1} in front of me in the right half, provided all transformations go well. Now subtract the tripled first row from the second row, and then subtract the first row from the third. The result is the matrix

$$\begin{pmatrix} 1 & 0 & -1 & 1 & 0 & 0 \\ 0 & 1 & 0 & -3 & 1 & 0 \\ 0 & 2 & -1 & -1 & 0 & 1 \end{pmatrix}.$$

The doubled second line is subtracted from the third. This gives:

$$\begin{pmatrix} 1 & 0 & -1 & 1 & 0 & 0 \\ 0 & 1 & 0 & -3 & 1 & 0 \\ 0 & 0 & -1 & 5 & -2 & 1 \end{pmatrix}.$$

The third row is multiplied by -1 and then the new third row is added to the first row. This then gives the matrix:

$$\begin{pmatrix} 1 & 0 & 0 & -4 & 2 & -1 \\ 0 & 1 & 0 & -3 & 1 & 0 \\ 0 & 0 & 1 & -5 & 2 & -1 \end{pmatrix}.$$

And already you see the inverse matrix in front of you, because on the left is now the unit matrix, so on the right is listed, without any question, the inverse matrix of A. So it reads:

$$A^{-1} = \begin{pmatrix} -4 & 2 & -1 \\ -3 & 1 & 0 \\ -5 & 2 & -1 \end{pmatrix}. \quad \blacksquare$$

Exercise 5.13 Calculate the inverse of the matrix

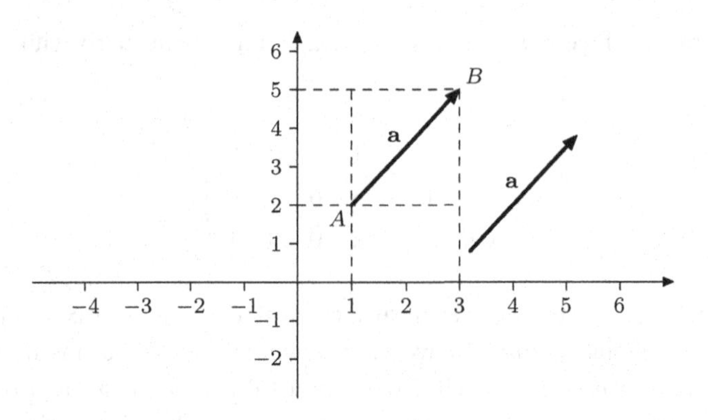

Fig. 5.1 Vector in the plane

$$A = \begin{pmatrix} 1 & 0 & -1 \\ -8 & 4 & 1 \\ -2 & 1 & 0 \end{pmatrix}$$

from Example 5.21.

Now we have firstly discussed everything necessary about linear systems of equations and secondly, in one fell swoop, we have also dealt with the points left open from earlier sections. In the next section I will show you how to apply vectors and linear systems of equations to one or another geometric problem.

5.4 Analytical Geometry

Do not be alarmed by the somewhat lofty title, I am only going to introduce a very small part of analytic geometry here to illustrate a bit of the techniques You have learned so far. While general vector and matrix calculus can be encountered by almost anyone at one time or another, from business economists to engineers, psychologists to physicists, the more geometric interpretation of vector calculus that I am about to talk about is aimed more at the technically and scientifically oriented reader. First, I need to say some things about the representation of vectors in the coordinate system and then I will talk about the equation of a plane in space.

Vectors I have so far considered simply as collections of numbers, completely ignoring the fact that they can also be interpreted geometrically, provided you restrict yourself to two- and three-dimensional vectors. The idea is simple enough, and I will show it to you briefly using two-dimensional vectors as an example. Have a look at Fig. 5.1.

Here I have entered two points A and B into a two-dimensional coordinate system, whereby you already know such coordinate systems from the chapter about functions. If, as usual, we call the horizontal direction the x-direction and the vertical direction the y-direction, then A obviously has the **coordinates** $A = (1,2)$ and B the coordinates $B = (3,5)$. Now, you may be interested in the fastest way to get from A to B, and the shortest connection between the two points is the distance entered in the figure. But I had asked for the way from A to B and not the other way round from B to A, so the route must have a **direction**, and this is symbolized by the small arrow pointing in the direction of B. The route therefore has a direction and, of course, a **length**, because it is not supposed to point beyond B, after all.

What does this have to do with the well-known two-dimensional vectors? It is quite simple: You can describe the directed path just discussed with the help of a vector. To get from A to B, I have to take two steps in the horizontal direction and three steps in the vertical direction. This completely describes the direction and length of the directed path $\boldsymbol{a}$, because if I start at A, take two steps to the right and three steps up, I will have arrived exactly where my directed path should take me. The horizontal and the vertical coordinates therefore describe my directed path, and therefore one sets $\mathbf{a} = \begin{pmatrix} 2 \\ 3 \end{pmatrix}$, which only means that one has to walk two steps to the right and three steps upwards. It is completely irrelevant from where I start this walking. The coordinate notation only indicates in which directions one has to walk, but not where the starting point is. That is why I can move around as I like; as long as I do not change direction and length, i.e. only make parallel shifts, it remains the same vector, which you can also see in Fig. 5.1.

With this we have already clarified how a two-dimensional vector can be interpreted graphically: as a directed distance in the plane, where the first component indicates how far one moves in the horizontal direction, and the second component reveals the number of steps in the vertical direction. The starting point does not matter, only the number of steps is important.

Two Dimensional Vectors

A two-dimensional vector $\begin{pmatrix} a_1 \\ a_2 \end{pmatrix}$ can be drawn into a two-dimensional coordinate system by moving from any starting point by a_1 steps in the horizontal direction and by a_2 steps in the vertical direction. The vector is then a directed distance from the chosen starting point to the reached end point.

Conversely, given two points with coordinates $A = (x_A, y_A)$ and $B = (x_B, y_B)$, the **connecting vector** $\overrightarrow{AB}$ between the two points is calculated by $\overrightarrow{AB} = \begin{pmatrix} x_B - x_A \\ y_B - y_A \end{pmatrix}$.

For example, in Fig. 5.1, $A = (1,2)$ and $B = (3,5)$. Therefore, $\overrightarrow{AB} = \begin{pmatrix} 3-1 \\ 5-2 \end{pmatrix} = \begin{pmatrix} 2 \\ 3 \end{pmatrix}$, just as you had seen earlier, holds. Incidentally, there is of course a very special starting point in the coordinate system that is the easiest to calculate with, and that is the zero point $\mathbf{0} = \begin{pmatrix} 0 \\ 0 \end{pmatrix}$. Namely, if any point $A = (x_A, y_A)$ is given and you are looking for the vector connecting the zero point and A, it is $\overrightarrow{\mathbf{0}A} = \begin{pmatrix} x_A - 0 \\ y_A - 0 \end{pmatrix} = \begin{pmatrix} x_A \\ y_A \end{pmatrix}$. Thus, the vector from the origin to the point A has the same coordinates as the point A itself, and it indicates how one has to walk from the origin of the coordinate system to get to the desired location A. That is why it is also called the **location vector** or **position vector** of A.

For the moment that was enough about two-dimensional vectors, I have to take care of the three-dimensional ones. But that is not a problem now, because everything works the same way as with the two-dimensional ones, only with one dimension more. If you have a three-dimensional coordinate system, the points have three coordinates and not only two, and to find out how to get from point $A = (1, 2, 3)$ to point $B = (-1, 3, 1)$, you only have to calculate the differences. In the first component I have to take two steps in the negative direction, in the second one step in the positive direction, and in the third two steps in the negative direction: it is the same game as just now with the only difference that a spatial component is added. I will therefore also summarize everything that is necessary.

Three Dimensional Vectors

A three-dimensional vector $\begin{pmatrix} a_1 \\ a_2 \\ a_3 \end{pmatrix}$ can be drawn into a three-dimensional coordinate system by moving from any starting point by a_1 steps in the first direction, by a_2 steps in the second direction and by a_3 steps in the third. The vector is then a directed distance from the chosen starting point to the reached end point.

Conversely, given two points with coordinates $A = (x_A, y_A, z_A)$ and $B = (x_B, y_B, z_B)$, the **connecting vector** $\overrightarrow{AB}$ between the two points is calculated by $\overrightarrow{AB} = \begin{pmatrix} x_B - x_A \\ y_B - y_A \\ z_B - z_A \end{pmatrix}$.

So, except for the number of components, nothing has changed. Briefly summarized, we can say that a two-dimensional vector is a directed distance in the plane, while a three-dimensional vector is a directed distance in space. The location vector of a two- or three-dimensional point is then the vector pointing from the origin to the chosen point.

In addition again the one or other exercise:

Exercise 5.14

(a) Calculate the vectors between the points

$A = (2, 3, 1)$ and $B = (-1, 0, 3)$ and between $C = (8, 17)$ and $D = (9, 16)$.

(b) What are the position vectors of the points $A = (3, 6, -1)$ and $B = (4, 7, 9)$?

Now I have not only played around with the vectors, but I have calculated very concretely, and I should show you how these arithmetic operations affect the representation in the coordinate system. Let us start by calculating the vector $-a$ from a vector $\boldsymbol{a}$. For example, while $\boldsymbol{a}$ connects the points A and B, $-a$ should point exactly in the opposite direction and connect the point B with the point A, and this is the case: You get $-\boldsymbol{a}$ from $\boldsymbol{a}$ by **reversing** the direction of a, so to speak, simply attaching the vector arrow to the other side of the vector.

Negative Vector

If $\boldsymbol{a}$ is a two- or three-dimensional vector, then $-\mathbf{a}$ is the vector of the same length, pointing exactly in the opposite direction. So if $\boldsymbol{a}$ is the vector between points A and B, then $-a$ is the vector between B and A.

But I had not only calculated negative vectors, but any vectors multiplied by any real numbers. This is also no problem in the coordinate system. For example, take the vector $\mathbf{a} = \begin{pmatrix} 1 \\ 2 \end{pmatrix}$, which points from the origin to the point (1, 2). As you learned in the first section, then $2 \cdot \mathbf{a} = \begin{pmatrix} 2 \\ 4 \end{pmatrix}$, and that is the vector pointing from the origin to the point (2, 4). So instead of taking one step to the right and two steps up, I am now going to take two steps to the right and four steps up: Everything has doubled, which is no wonder when you multiply by two. And since everything has doubled, I also just need to double the vector drawn in the coordinate system by simply letting it point in the same direction, but stretching it by a factor of two.

Multiplication of a Vector by a Number

If $\boldsymbol{a}$ is a two- or three-dimensional vector and λ is a positive real number, then $\lambda \cdot \boldsymbol{a}$ is the vector pointing in the same direction as $\boldsymbol{a}$ but stretched by the factor λ. If λ is a negative real number, then $\lambda \cdot \boldsymbol{a}$ is the vector pointing in the opposite direction as $\boldsymbol{a}$, stretched by the factor $|\lambda|$.

So if the factor is positive, it is stretched in the given direction, if it is negative, it is stretched in the opposite direction, and that is all there is to it. The next operation, the addition of two vectors, is also nothing really exciting, as you can immediately see from Fig. 5.2.

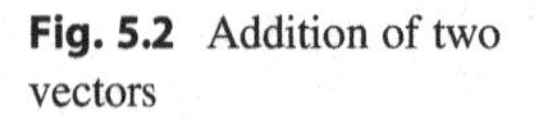

Fig. 5.2 Addition of two vectors

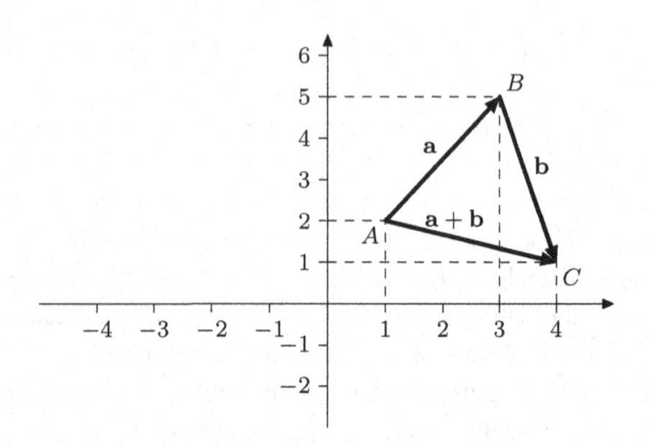

The vector $\boldsymbol{a}$ points here from $A = (1, 2)$ to $B = (3, 5)$, therefore has the coordinates $\mathbf{a} = \begin{pmatrix} 2 \\ 3 \end{pmatrix}$. In contrast, $\boldsymbol{b}$ is the vector pointing from $B = (3, 5)$ to $C = (4, 1)$ and therefore can be described by the coordinates $\mathbf{b} = \begin{pmatrix} 1 \\ -4 \end{pmatrix}$. Then, following the rules from Section 1, $\mathbf{a} + \mathbf{b} = \begin{pmatrix} 2 \\ 3 \end{pmatrix} + \begin{pmatrix} 1 \\ -4 \end{pmatrix} = \begin{pmatrix} 3 \\ -1 \end{pmatrix}$. So if I put the vector $\boldsymbol{a} + \boldsymbol{b}$ at the point A, I have to go three steps to the right and one step down to reach its endpoint, and you see where I end up: exactly at the point C.

This is how it always works. As soon as you want to add two two-dimensional or two three-dimensional vectors, you draw the first one into a coordinate system, attach the second one to the first one, and the sum of the two is then the vector pointing from the starting point of the first vector to the end point of the second vector.

Addition of Vectors

Let two two-dimensional or two three-dimensional vectors $\boldsymbol{a}$ and $\boldsymbol{b}$ be given. If $\boldsymbol{a}$ and $\boldsymbol{b}$ are drawn into a coordinate system in such a way that the starting point of $\boldsymbol{b}$ coincides with the end point of $\boldsymbol{a}$, then $\boldsymbol{a} + \boldsymbol{b}$ is the vector pointing from the starting point of the vector $\boldsymbol{a}$ to the end point of the vector $\boldsymbol{b}$.

You can see an example of this in Fig. 5.2. Remember, by the way, that with vectors it does not matter where you start, only their direction and length matters. Therefore, I can append the vector $\boldsymbol{b}$ to the end of $\boldsymbol{a}$ without any problems.

Here is something else to practice:

Exercise 5.15 Draw the vectors $\mathbf{a} = \begin{pmatrix} -1 \\ 2 \end{pmatrix}$, $\mathbf{b} = \begin{pmatrix} 2 \\ 1 \end{pmatrix}$, $\mathbf{c} = \begin{pmatrix} 0 \\ 1 \end{pmatrix}$ in a coordinate system and determine the vectors $\boldsymbol{a} + 2\,\boldsymbol{b} - \boldsymbol{c}$ and $3\,\boldsymbol{a} + \boldsymbol{b} + \boldsymbol{c}$.

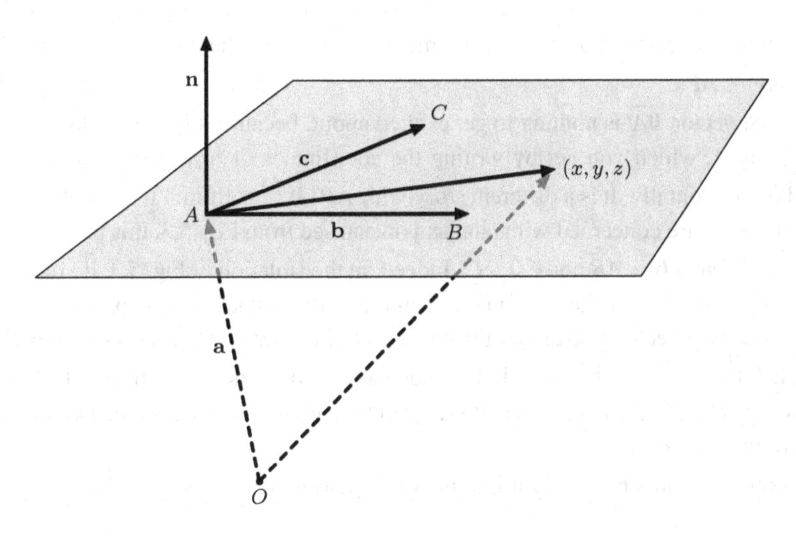

Fig. 5.3 Plane in space

Now I have gathered enough material to tackle the real topic of this section: the equation of a plane in space. I assume that you can imagine a plane in space, you just have to think of a sheet of paper floating around motionless in space, extended to all sides as you like. My goal is to describe such a plane with a simple equation. Now, in the chapter about functions, you have not dealt with planes yet, but you have dealt with straight lines, which are usually described by equations of the form $y = ax + b$. These straight lines were all within a two-dimensional coordinate system, i.e. within a plane. Now I go one step further and consider a space, i.e. a three-dimensional coordinate system, into which I want to place a plane – so the situation is quite similar, and maybe I will get lucky and find a similar equation as well. In fact, that is exactly what will happen. If we denote the three spatial coordinates by x, y, and z, plane equations always take the form $4x - 3y + 2z = 17$ or $-2x + 4y + z = 9$, so in general $ax + by + cz = d$. That is what I am getting at; you will see that it does not get wild at all.

The starting point of the whole thing is the well-known observation that a table with three legs never wobbles, provided the legs are stable. In other words, you can always put exactly one plane through three points in space, as you can also easily determine by holding up three fingers and putting a sheet of paper on it. So I assume that I am supposed to lay a plane through three arbitrary points A, B and C, which are located somewhere in a three-dimensional coordinate system. Figure 5.3 shows roughly what this looks like. For now, all I am interested in about this figure is that the plane is somewhere in space and passes through the three points A, B, and C. But this is not enough to determine an equation, so I added an arbitrary point (x, y, z) and the zero point. How do you get from the zero point to the point (x, y, z)? Quite simple: First you go to the point A, because it was given. And then you have already reached the plane and can walk comfortably from A to (x, y, z). If I denote my point (x, y, z) with P for the moment, then I get the location vector of P by $\overrightarrow{\mathbf{0}P} = \overrightarrow{\mathbf{0}A} + \overrightarrow{AP}$.

The first vector $\overrightarrow{\mathbf{0}A}$ is nothing to get excited about, because it is just the location vector of the point A, which you get by writing the coordinates of A as a vector, i.e. vertically instead of horizontally. It is a different story with $\overrightarrow{AP}$. It runs from A to P, so it lies entirely in the plane we are concerned with, and as you can see from Fig. 5.3, this plane is spanned by the two vectors $\mathbf{b} = \overrightarrow{AB}$ and $\mathbf{c} = \overrightarrow{AC}$. Indeed, in the situation of Fig. 5.3, if you compress both $\boldsymbol{b}$ and $\boldsymbol{c}$ a tiny bit by the appropriate factor, and then attach the compressed $\boldsymbol{b}$ vector to the compressed $\boldsymbol{c}$ vector, you also get from An to P. In other words, there are real numbers λ and μ such that $\overrightarrow{AP} = \lambda \cdot \mathbf{b} + \mu \cdot \mathbf{c}$ holds. You can combine the vector from A to P from the two given vectors $\boldsymbol{b}$ and $\boldsymbol{c}$, because all these vectors are in the same plane, and this plane is spanned by $\boldsymbol{b}$ and $\boldsymbol{c}$.

If I now insert this new insight into the old equation $\overrightarrow{\mathbf{0}P} = \overrightarrow{\mathbf{0}A} + \overrightarrow{AP}$, we get:

$$\overrightarrow{\mathbf{0}P} = \overrightarrow{\mathbf{0}A} + \lambda \cdot \mathbf{b} + \mu \cdot \mathbf{c} = \mathbf{a} + \lambda \cdot \mathbf{b} + \mu \cdot \mathbf{c},$$

because I had denoted the location vector of A with $\boldsymbol{a}$. And since the point P has just the coordinates (x, y, z) and $\overrightarrow{\mathbf{0}P}$ is the location vector of P, we have

$$\begin{pmatrix} x \\ y \\ z \end{pmatrix} = \overrightarrow{\mathbf{0}A} + \lambda \cdot \mathbf{b} + \mu \cdot \mathbf{c} = \mathbf{a} + \lambda \cdot \mathbf{b} + \mu \cdot \mathbf{c}$$

for each point (x, y, z) on the plane. The real numbers λ and μ are called **parameters**, and therefore this form of the plane equation is called the **parameterized form of** the plane equation.

Let us look at an example of this.

Example 5.31 It is about the plane through the points $A = (1, 2, 3)$, $B = (0, 1, 2)$ and $C = (-1, 2, 1)$. Then, of course, $\mathbf{a} = \begin{pmatrix} 1 \\ 2 \\ 3 \end{pmatrix}$, and for the vectors $\boldsymbol{b}$ and $\boldsymbol{c}$ is true because they connect the point A with the points B and C, respectively:

$$\mathbf{b} = \begin{pmatrix} 0-1 \\ 1-2 \\ 2-3 \end{pmatrix} = \begin{pmatrix} -1 \\ -1 \\ -1 \end{pmatrix} \quad \text{and} \quad \mathbf{c} = \begin{pmatrix} -1-1 \\ 2-2 \\ 1-3 \end{pmatrix} = \begin{pmatrix} -2 \\ 0 \\ -2 \end{pmatrix}.$$

So the equation of the plane is:

Fig. 5.4 Vertical vectors

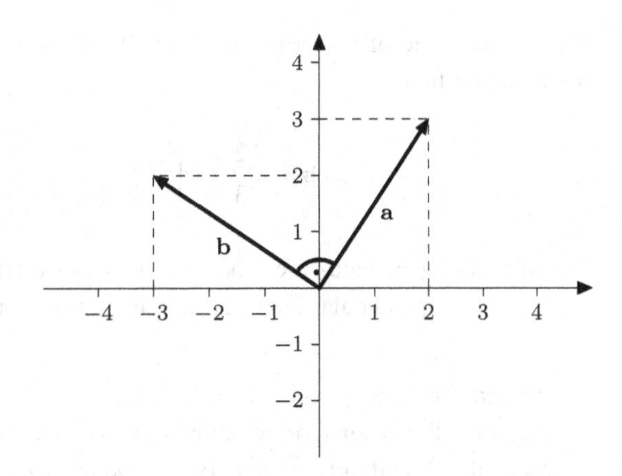

$$\begin{pmatrix} x \\ y \\ z \end{pmatrix} = \begin{pmatrix} 1 \\ 2 \\ 3 \end{pmatrix} + \lambda \cdot \begin{pmatrix} -1 \\ -1 \\ -1 \end{pmatrix} + \mu \cdot \begin{pmatrix} -2 \\ 0 \\ -2 \end{pmatrix} \quad \text{with } \lambda, \mu \in \mathbb{R}.$$

For example, if you take $\lambda = \mu = 0$, you get the plane point (1, 2, 3), which is not very surprising. However, you can use any combination of numbers and you will always end up with a point on the plane; for $\lambda = -1$ and $\mu = 1$, for example, you will find the location vector $\begin{pmatrix} 1 \\ 2 \\ 3 \end{pmatrix} - \begin{pmatrix} -1 \\ -1 \\ -1 \end{pmatrix} + \begin{pmatrix} -2 \\ 0 \\ -2 \end{pmatrix} = \begin{pmatrix} 0 \\ 3 \\ 2 \end{pmatrix}$, from which it immediately follows that the point (0, 3, 2) is also on the plane.

As a small loosening up again an exercise task.

Exercise 5.16 Calculate the parametrized plane equation of the plane through the three points $A = (2, 1, 3)$, $B = (-1, 0, 5)$, and $C = (1, 1, 1)$.

That is already quite good, but not enough, because actually I wanted to achieve a different form of the plane equation. We will have that in a moment. A decisive role will be played by the vector $\boldsymbol{n}$, which I had drawn in Fig. 5.3 as a precaution and had completely ignored up to now. It has the pleasant property of standing perpendicular to the plane, which is why I will now show you how to calculate something like "standing perpendicular".

Example 5.32 Take a quick look at Fig. 5.4. By simply drawing and measuring, you can see that the vectors $\mathbf{a} = \begin{pmatrix} 2 \\ 3 \end{pmatrix}$ and $\mathbf{b} = \begin{pmatrix} -3 \\ 2 \end{pmatrix}$ are perpendicular to each other, so that there is an angle of 90° between them. If you now calculate the scalar product of both vectors, you find:

$$\mathbf{a} \cdot \mathbf{b} = \begin{pmatrix} 2 \\ 3 \end{pmatrix} \cdot \begin{pmatrix} -3 \\ 2 \end{pmatrix} = 2 \cdot (-3) + 3 \cdot 2 = 0.$$

Coincidence? No. Fate? Yes, because that is exactly how it always is: whenever two vectors are perpendicular to each other, their scalar product will be zero, and vice versa.

Vertical Vectors
Two two-dimensional or two three-dimensional vectors $\boldsymbol{a}$ and $\boldsymbol{b}$ ***are*** perpendicular to each other if and only if their scalar product is zero, that is, if $\boldsymbol{a} \cdot \boldsymbol{b} = 0$.

Now what does this have to do with my plane? The vector $\boldsymbol{n}$ in Fig. 5.3 is perpendicular to the two vectors $\boldsymbol{a}$ and $\boldsymbol{b}$, and since these two vectors span the whole plane, it is perpendicular to the whole plane at the same time. This will come in handy in a moment, but for now I need to briefly explore the question of how to calculate such a vector $\boldsymbol{n}$:

Example 5.33 I again use the situation from Example 5.31; the vectors on which $\boldsymbol{n}$ is to be perpendicular are thus called $\mathbf{b} = \begin{pmatrix} -1 \\ -1 \\ -1 \end{pmatrix}$ and $\mathbf{c} = \begin{pmatrix} -2 \\ 0 \\ -2 \end{pmatrix}$. If I denote the ominous $\boldsymbol{n}$ by $\boldsymbol{n} = \begin{pmatrix} n_1 \\ n_2 \\ n_3 \end{pmatrix}$, then according to the principle just established, the scalar product of $\boldsymbol{n}$ with $\boldsymbol{b}$ and also the scalar product of $\boldsymbol{n}$ with $\boldsymbol{c}$ must become zero. So it must hold:

$$0 = \mathbf{n} \cdot \mathbf{b} = \begin{pmatrix} n_1 \\ n_2 \\ n_3 \end{pmatrix} \cdot \begin{pmatrix} -1 \\ -1 \\ -1 \end{pmatrix} = -n_1 - n_2 - n_3$$

and

$$0 = \mathbf{n} \cdot \mathbf{c} = \begin{pmatrix} n_1 \\ n_2 \\ n_3 \end{pmatrix} \cdot \begin{pmatrix} -2 \\ 0 \\ -2 \end{pmatrix} = -2n_1 - 2n_3.$$

I have thus obtained two equations, namely $-n_1 - n_2 - n_3 = 0$ and $-2\,n_1 - 2\,n_3 = 0$. What is this called? That is right, this is a linear system of equations made up of two equations with the three unknowns n_1, n_2, and n_3, and I will now solve it in the manner discussed in the third section of this chapter. For simplicity, if we immediately multiply both equations by -1, the matrix form of the system of equations is

$$\begin{pmatrix} 1 & 1 & 1 & 0 \\ 2 & 0 & 2 & 0 \end{pmatrix}.$$

I subtract the doubled first row from the second, then divide the second by -2, and then subtract the second from the first. This gives the matrix sequence:

$$\begin{pmatrix} 1 & 1 & 1 & 0 \\ 0 & -2 & 0 & 0 \end{pmatrix} \rightarrow \begin{pmatrix} 1 & 1 & 1 & 0 \\ 0 & 1 & 0 & 0 \end{pmatrix} \rightarrow \begin{pmatrix} 1 & 0 & 1 & 0 \\ 0 & 1 & 0 & 0 \end{pmatrix}.$$

Translating this back into equations, we have $n_1 + n_3 = 0$ and $n_2 = 0$. The small unit matrix achieved relates to the unknowns n_1 and n_2, which is why the unknown n_3 is free to be chosen and the other two can be calculated – in this case even very clearly, because finally n_2 is even zero. So I have the solutions:

$$n_1 = -n_3, \quad n_2 = 0, \quad n_3 \in \mathbb{R} \text{ is free to choose.}$$

But the goal of my calculation was to find a vector that is perpendicular to the two vectors $\boldsymbol{b}$ and $\boldsymbol{c}$. And there are actually quite a few of them, because as I just calculated, every vector of the form $\begin{pmatrix} -n_3 \\ 0 \\ n_3 \end{pmatrix}$ does that, so for example $\begin{pmatrix} -1 \\ 0 \\ 1 \end{pmatrix}$ or also $\begin{pmatrix} 2 \\ 0 \\ -2 \end{pmatrix}$. ◻

Just as in Example 5.33, you can always calculate a vector that is perpendicular to two given vectors and thus also to the plane spanned by them by solving the correct system of linear equations. But you can also leave it alone, because in principle it is always the same calculation, and that is why someone earlier came up with a general formula into which you only have to substitute, so that you do not have to solve the basically same system over and over again. The result is called the **vector product** or the **cross product** of two vectors.

Vector Product

If $\boldsymbol{b} = \begin{pmatrix} b_1 \\ b_2 \\ b_3 \end{pmatrix}$ and $\boldsymbol{c} = \begin{pmatrix} c_1 \\ c_2 \\ c_3 \end{pmatrix}$ are two three-dimensional vectors, the vector

(continued)

$$\mathbf{b} \times \mathbf{c} = \begin{pmatrix} b_2c_3 - b_3c_2 \\ b_3c_1 - b_1c_3 \\ b_1c_2 - b_2c_1 \end{pmatrix}$$

is called the vector product or cross product of $\boldsymbol{b}$ and $\boldsymbol{c}$. The vector $\boldsymbol{b} \times \boldsymbol{c}$ is perpendicular to $\boldsymbol{b}$ and to $\boldsymbol{c}$.

This is not pretty, but it is practical. You can save yourself the trouble of solving the equation system from Example 5.33 and calculate it right away:

$$\begin{pmatrix} -1 \\ -1 \\ -1 \end{pmatrix} \times \begin{pmatrix} -2 \\ 0 \\ -2 \end{pmatrix} = \begin{pmatrix} (-1) \cdot (-2) - (-1) \cdot 0 \\ (-1) \cdot (-2) - (-1) \cdot (-2) \\ (-1) \cdot 0 - (-1) \cdot (-2) \end{pmatrix} = \begin{pmatrix} 2 \\ 0 \\ -2 \end{pmatrix},$$

and this is actually one of the solutions to the system of eqs. I had calculated above.

You can also practice this new product a little bit now:

Exercise 5.17 Calculate that the vector product $\begin{pmatrix} 2 \\ -1 \\ 4 \end{pmatrix} \times \begin{pmatrix} -3 \\ 1 \\ 2 \end{pmatrix}$ is perpendicular to the vectors $\begin{pmatrix} 2 \\ -1 \\ 4 \end{pmatrix}$ and $\begin{pmatrix} -3 \\ 1 \\ 2 \end{pmatrix}$.

Now I have everything together to briefly and painlessly calculate the sought plane equation. Just one more term, and then I will get started: A vector that is perpendicular to a plane is also called a **normal vector** of the plane, and that explains why I will denote the perpendicular vector by $\boldsymbol{n}$. Using my old example, I will now show you how all the details I discussed fit together:

Example 5.34 As in Example 5.31, we are concerned with the plane through the points $A = (1, 2, 3)$, $B = (0, 1, 2)$, and $C = (-1, 2, 1)$; you might want to take another look at Fig. 5.3 for orientation. I had already calculated the vectors $\boldsymbol{b}$ and $\boldsymbol{c}$ in Example 5.31 as

$$\mathbf{b} = \begin{pmatrix} 0-1 \\ 1-2 \\ 2-3 \end{pmatrix} = \begin{pmatrix} -1 \\ -1 \\ -1 \end{pmatrix} \quad \text{and} \quad \mathbf{c} = \begin{pmatrix} -1-1 \\ 2-2 \\ 1-3 \end{pmatrix} = \begin{pmatrix} -2 \\ 0 \\ -2 \end{pmatrix}.$$

As a vector $\boldsymbol{n}$, which is perpendicular to $\boldsymbol{b}$ and $\boldsymbol{c}$ and thus perpendicular to the whole plane, I can use the normal vector just determined

$$\mathbf{n} = \mathbf{b} \times \mathbf{c} = \begin{pmatrix} 2 \\ 0 \\ -2 \end{pmatrix}$$

But since this vector $\boldsymbol{n}$ is perpendicular to the plane, it will certainly be perpendicular to any vector that lies entirely in the plane under consideration, for example, to $\boldsymbol{b}$ and $\boldsymbol{c}$, but also to the vector that runs from A to the plane point (x, y, z). Now I can easily calculate this vector, because A has the coordinates $(1, 2, 3)$, while (x, y, z) of course has the coordinates (x, y, z), and therefore the connecting vector is $\begin{pmatrix} x-1 \\ y-2 \\ z-3 \end{pmatrix}$. Now on this vector, $\boldsymbol{n}$ is supposed to be perpendicular, and as you learned, that just means that the scalar product of the two vectors must be zero. So I know:

$$0 = \mathbf{n} \cdot \begin{pmatrix} x-1 \\ y-2 \\ z-3 \end{pmatrix} = \begin{pmatrix} 2 \\ 0 \\ -2 \end{pmatrix} \cdot \begin{pmatrix} x-1 \\ y-2 \\ z-3 \end{pmatrix}.$$

But you also know how to calculate such a scalar product. It is valid namely:

$$\begin{pmatrix} 2 \\ 0 \\ -2 \end{pmatrix} \cdot \begin{pmatrix} x-1 \\ y-2 \\ z-3 \end{pmatrix} = 2 \cdot (x-1) + 0 \cdot (y-2) - 2 \cdot (z-3) = 2x - 2 - 2z + 6 = 2x - 2z + 4.$$

Since the scalar product is zero, it follows:

$$2x - 2z + 4 = 0, \quad \text{so } 2x - 2z = -4 \quad \text{and so } x - z = -2,$$

where in the last step I simply divided the equation by two.

As you can see, I am done. The eq. I am looking for is $x - z = -2$, and that means that all points (x, y, z) on my plane must satisfy this equation. Whichever plane point you choose: The difference between the first and third coordinates will definitely be equal to -2.

This is how it always works, and so that you do not have to laboriously gather the scheme from the examples, I will write it down again now:

Calculation of the Parameter-Free Plane Equation

If $A = (a_1, a_2, a_3)$, $B = (b_1, b_2, b_3)$, and $C = (c_1, c_2, c_3)$ are three points in space that do not all lie on a straight line, then compute the parameter-free equation of the plane that passes through all three points using the following scheme:

(a) One determines

$$\mathbf{a} = \begin{pmatrix} a_1 \\ a_2 \\ a_3 \end{pmatrix}, \quad \mathbf{b} = \begin{pmatrix} b_1 - a_1 \\ b_2 - a_2 \\ b_3 - a_3 \end{pmatrix}, \quad \mathbf{c} = \begin{pmatrix} c_1 - a_1 \\ c_2 - a_2 \\ c_3 - a_3 \end{pmatrix}.$$

(b) Calculate the normal vector with the cross product

$$\mathbf{n} = \mathbf{b} \times \mathbf{c} = \begin{pmatrix} b_2 c_3 - b_3 c_2 \\ b_3 c_1 - b_1 c_3 \\ b_1 c_2 - b_2 c_1 \end{pmatrix}.$$

(c) For the plane points (x, y, z) one makes the approach

$$\mathbf{n} \cdot \begin{pmatrix} x - a_1 \\ y - a_2 \\ z - a_3 \end{pmatrix} = 0.$$

(d) Calculate the scalar product of c) and simplify the resulting equation until an equation of the form $ax + by + cz = d$ is obtained. This equation is then the parameter-free equation of the plane.

By the way, the equation is called parameter-free because – in contrast to the parameterized equation – there are no more parameters λ and μ, but only the coordinates of the points (x, y, z) themselves. And why did I assume that the points do not lie on a straight line? Quite simply, if you have three points on a straight line, then of course there is no unique plane through those three points, but a myriad of them. But you do not have to test this before you start the calculation; if all three points lie on a straight line, you will find that the vector product $\boldsymbol{b} \times \boldsymbol{c}$ gives exactly the zero vector, and there is not much you can do with that.

You can test your new knowledge with the following task.

Exercise 5.18 Determine the parameter-free equation of the plane passing through the three points $A = (4, 0, 0)$, $B = (5, 1, -1)$, and $C = (-3, 2,1)$.

With that, I will end Linear Algebra; in the next chapter, you will learn some things about Differential Calculus.

Differential and Integral Calculus

Chapter 2 dealt with important mathematical objects: **Functions**. By now, you have certainly internalized the things described there a bit more. That is a good thing, too, because I now want to deal with the **analytic** properties of real-valued functions. By this I mean that I want to clarify together with you how to answer natural questions of the following kind: How can the slope of a function be described mathematically? At what points do the values of a function become largest? How can the graph of a function be drawn? How large is the area that the graph of a function encloses with the *x-axis?*

In particular, you will realize together with me that the mathematical answers to the first and the last question just asked are connected in the closest way, although they may seem to have little to do with each other at first sight. This is stated by the main theorem of differential and integral calculus, which I will talk to you about later – but I am not there yet.

6.1 First Derivative of Functions and Derivative Rules

It seems reasonable to me to start with a hike through a mountainous terrain. The gradients that you have to overcome in the course of such a hike depend, as you know, on the location where you are at each moment. Take a look at my hiking suggestion "Palatinate Forest" in Fig. 6.1, which shows you a cross-section of the mountain profile there. Among mathematicians it is quite normal to imagine such a hike as a function $f\colon [a, b] \mapsto \mathbb{R}$. Each point x in the interval $[a, b]$ is assigned the corresponding height value $f(x)$. Figure 6.1 shows you all the points $P = (x \mid f(x))$, where x traverses the interval $[a, b]$ – that is, the **graph of** f. The walk starts at the lowest point (at the point a with height $f(a)$). At the beginning (at the points in the interval between a and x_1) the slopes are relatively large.

G. Walz et al., *Foundations of Mathematics: A Preparatory Course*, https://doi.org/10.1007/978-3-662-67809-1_6

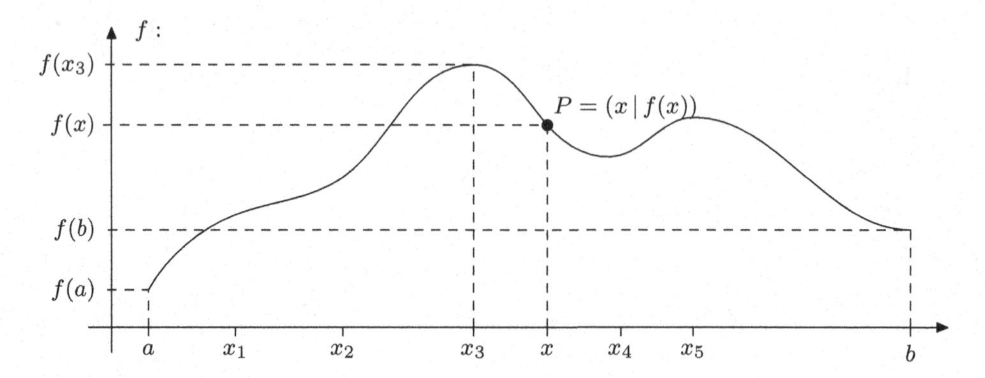

Fig. 6.1 Suggested hike "Palatinate Forest": The gradient of f is not constant, because it depends on the location x at which one is located

After that (between x_1 and x_2), the slopes are somewhat smaller – while the last stretch (between x_2 and x_3) has unnecessarily very large slopes at first, but then – towards the first summit (at the point x_3) – becomes increasingly flatter. Until then, the value of the slopes is always positive – at the summit itself (at the point x_3), fortunately, there is no slope anymore, the value of the slope there is 0. Do not worry – I would not go through all the details of this walk with you now – in the next section, I will come back to them anyway, because there I will chat with you about maxima and minima of functions, among other things. What is important to me right now is that you realize: The slope of this function f is *not constant*. That is the case for a lot of functions.

You have long since noticed what I am concerned with here: My aim is to introduce you to the concept of the **slope of functions** f. The value of the slope of f usually depends on the point x I am looking at. Thus, the varying values of the slope of f are generally described by another function. This is called the **first derivative of** f and is often denoted by f' (pronounced "f dash"). I suggest that I first look at the calculation of the first derivatives of simple functions before giving you the precise and general definition of derivative functions.

The simplest class of functions I can think of are **constant functions**, so $f(x) = c$, where c is a fixed real number. As always, x is a real variable here. The "mudflat walk north of Aurich" (see Fig. 6.2) is describable by such a function, though there I again assume that x merely traverses the interval $[a, b]$. If I assume that the tide is low and the whole hiking takes location 2 meters below the sea surface, I can write down the corresponding function: $f(x) = -2$ for all $x \in [a, b]$. You can easily see that the slope of this function is much simpler here than in the "Palatine Forest": namely, the value of the slope of f is always 0. Thus: $f'(x) = 0$ for all x in the interval $[a, b]$. It is exactly the same if I were to choose a different constant for c, for example $c = 2240$, $c = 4/7$ or even $c = -2/\pi$ – just imagine that the mudflat walk takes location, say, $2/\pi \approx 0.64$ m below the surface of the sea.

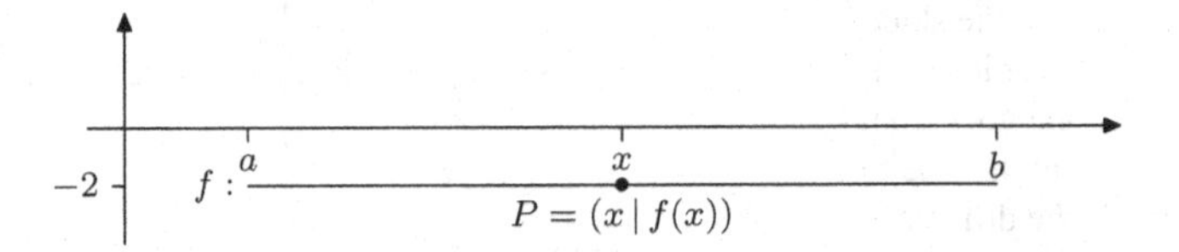

Fig. 6.2 Proposed hike "Mudflat hike north of Aurich": The associated function f is constant -2 and thus 0 is the value of the slope of f at any point x. So for this f the slope disappears everywhere

First Derivative of Constant Functions

The value of the slope of a **constant function** $f(x) = c$ is always 0. The first derivative f' of f thus **vanishes** for all x: $f'(x) = 0$.

I now look at the next most difficult class, namely the **linear functions** $f(x) = c\,x + d$. As always, $x \in \mathbb{R}$ is the variable and d and c are fixed real numbers. Fortunately, the value of the slope of such a function is also very easy to determine. In Chap. 2, you were told that the graph of such a function is always a straight line. However, the *value of the slope* of *a straight line* is known to be *constant*, that is, *independent of* the location of x. An *exceptional situation* – but why is that the case?

For example, take a look at the linear function $f(x) = x$ shown in Fig. 6.3, which I get by substituting $d = 0$ and $c = 1$ above. This function is sometimes called an **identity**. If I move a little bit h ($\neq 0$) to the right from the point 1, that is, to the point $1 + h$, the difference of the corresponding function values, $f(1 + h) = 1 + h$ and $f(1) = 1$, is also h. The slope is obtained by taking the quotient of the difference of the function values with the difference of the *x-values*:

$$\frac{f(1+h)-f(1)}{(1+h)-1} = \frac{(1+h)-1}{h} = \frac{h}{h} = 1.$$

I get the same result if I move a little bit h *to* the right from any other location x, i.e. to the location $x + h$, because then the difference of the corresponding function values $f(x + h) = x + h$ and $f(x) = x$ is also h. To be on the safe side, I write down the quotient of this difference of the function values and the difference of the locations x and $x + h$:

$$\frac{f(x+h)-f(x)}{(x+h)-x} = \frac{(x+h)-x}{h} = \frac{h}{h} = 1.$$

I note two things. First, it does not matter in the result here how h *is* chosen, that is, how far I go away from the fixed point x that I am interested in at the moment. Second, regardless of x, the quotient above always has the value 1. Thus the first derivative f' of f is the function: $f'(x) = 1$.

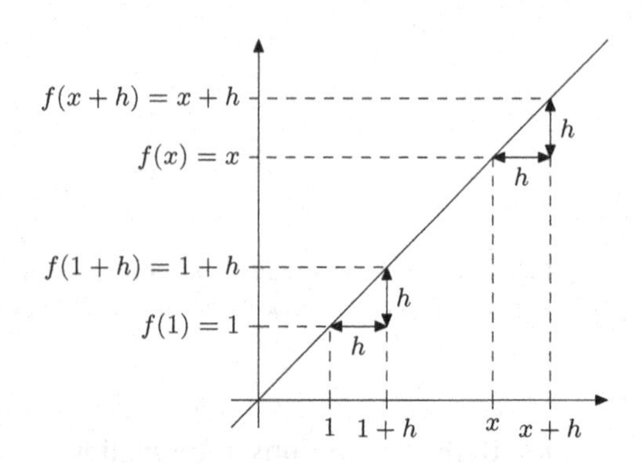

Fig. 6.3 The value of the slope of the identity $f(x) = x$ is equal to 1 for each x: If one goes from any location x by h to the side to the location $x + h$, the difference of the function values $f(x)$ and $f(x + h)$ is also h

If I try the same for the slightly more difficult function $f(x) = 3x$ (i.e. $d = 0$ and $c = 3$), something similar happens. If I move from a fixed (but arbitrary) point x to the side by h – that is, I end up at the point $x + h$ – the difference in the function values is $f'(x + h) - f(x) = 3(x + h) - 3x = 3 \cdot h$. So the quotient is $3 \cdot h/h = 3$. Thus $f'(x) = 3$ is the first derivative of $f(x) = 3x$. In the same way, if I move the graph of this functions f up by – say −2, so consider $f(x) = 3x + 2$ – I then also calculate $f'(x) = 3$ – try for yourself to form the quotient of the differences for this f.

Now I consider a general linear function $f(x) = cx + d$. If I grab an arbitrary fixed x and give myself an h, then as always $(x + h) - x = h$. Moreover, I calculate the difference $f(x + h) - f(x)$ of the function values of f in x and $x + h$ *as follows*:

$$(c(x+h)+d)-(cx+d)=cx+ch+d-cx-d=ch.$$

The quotient of the two differences is thus generally as follows:

$$\frac{f(x+h)-f(x)}{(x+h)-x}=\frac{ch}{h}=c.$$

So the slope has always the value c independent of the location x *and* it does not matter how far I go away from the location x by choosing h ($\neq 0$): I could easily *cancel the h* here.

First Derivative of Linear Functions
The value of the slope of a **linear function** $f(x) = cx +$ dist always c. The first derivative f of f is thus **constant** for all x: $f'(x) = c$.

Constant and linear functions are a big exception, because their first derivative is constant. This is different in general. Already the hiking proposal "Palatinate Forest" showed us that there the values of the gradient depend on the location x and thus do *not* form a *constant* function. It is now time to look at a concrete function, where the values of the slope for different locations x are *actually different*.

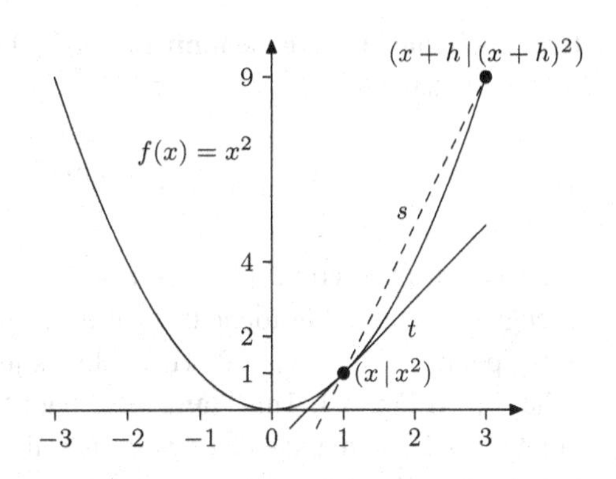

Fig. 6.4 The value of the slope of $f(x) = x^2$ at the point $x = 1$ is 2. This is also the value of the slope of the tangent t to f at $x = 1$, which is the limit of the slopes of the secants $2 + h$ *for* $h \to 0$. For this function f the first derivative f' is not constant, but it holds: $f'(x) = 2\,x$

For $x \in \mathbb{R}$ I consider the function $f(x) = x^2$, whose graph is called a **normal parabola.** You already know this from Chap. 2 – here I have illustrated it in Fig. 6.4 as a solid arc. On a trial basis, I will now proceed as I did above for linear functions. Let us see how far I get with this. For any fixed x and given h ($\neq 0$), the difference of the function values in x and $x + h$ is then $f(x + h) - f(x) = (x + h)^2 - x^2$, while the difference of the digits $(x + h) - x$ gives h as usual. Now, remembering the **binomial formulae** (perhaps look again in Chap. 1), I can further work out the difference in function values: $f(x + h) - f(x) = x^2 + 2\,h\,x + h^2 - x^2 = 2\,h\,x + h^2$. The quotient of the difference in function values with the difference in digits is thus:

$$\frac{f(x+h)-f(x)}{(x+h)-x} = \frac{2hx+h^2}{h} = \frac{h(2x+h)}{h} = 2x+h.$$

Now I see two *crucial differences* compared to what I got out above for linear functions. First: The quotient depends on h – unfortunately h cannot be truncated away completely. Second, here x occurs on the right-hand side – the quotient *depends on* x. To get ahead here, I should now realize what I just calculated there: The calculated quotient $2x + h$ is, according to what I just told you about the slope of linear functions, the slope of the straight line through the points $(x\,|\,f(x)) = (x\,|\,x^2)$ and $(x + h\,|\,f(x + h)) = (x + h\,|\,(x + h)^2)$. In Fig. 6.4 I have illustrated this for $x = 1$ and $h = 2$. The dashed straight line s is the graph of the linear function $s(x) = 4\,x - 3$, which has the same value as f in $x = 1$ and $x = 3$: $f(1) = 1 = s(1)$ and $f(3) = 9 = s(3)$. The slope of this straight line s is 4. This is only *a first approximation* to the actual value of the slope of f in $x = 1$.

So that I can get more accurate approximations to the actual value of the slope of f in $x = 1$, I now imagine that h *is closer and closer to* 0 *without ever becoming* 0 – we say that h **tends towards 0** and write: $h \to 0$. The slope of the straight line through the points $(1\,|\,1)$ and $(1 + h|(1 + h)^2) = (1 + h|1 + 2\,h + h^2)$ is $2 + h$ and represents, for such ever decreasing h, *ever more accurate approximations of the actual value of* the slope of $f(x) = x^2$ in $x = 1$. For example, if I set for h the values 1/10, 1/100, 1/1000, …, I get in turn for this approximation $2 + h$ the results 2.1, 2.01, and 2.001. I think you can see now that the term $2 + h$ **converges to** the **limit** 2 for $h \to 0$. Many mathematicians sometimes like to use the neat notation for this

$$\lim_{h \to 0} (2 + h) = 2$$

and mean that the term $2 + h$ is *arbitrarily close to* the limit (lat.: **limes**) 2, if his just close enough to 0. I have thus found that 2 is the actual value of the slope of the function $f(x) = x^2$ at the point $x = 1$: $f'(1) = 2$. This value is just the slope of the straight line t shown as a solid line in Fig. 6.4. This "**tangents**" (lat.: "touches") the graph of the function f at the point (1|1). I can imagine that this straight line emerges from the straight lines through the two points (1|1) and $(1 + h|(1 + h)^2)$ by the boundary transition $h \to 0$.

I should still make an *important remark* about the meaning of the limit **sign** $\lim\limits_{h \to 0}$, because this occurs even more often in the following. *Actually* this is meant to look if the limit *exists and is equal in each case*, if I consider *all* possibilities to strive with h towards 0. This is of course the case above – in particular, I can look there at *negative hs* with $h \to 0$, so for example $-1/10$, $-1/100$, $-1/1000$, . . ., or perhaps *hs* with $h \to 0$ and **alternating signs**, so for example $-1/2$, $1/4$, $-1/8$, $1/16$, ... You see: In the above term $2 + h$, 2 always *comes* out as the limit.

What is the value of the slope of the function $f(x) = x^2$ now at an arbitrary point x? The quotient $2x + h$ calculated above is merely an approximation to this value, which I can determine as a limit value by forming the limit transition $h \to 0$ in almost exactly the same way as above by (very) casually speaking *substituting* 0 for *h in* $2x + h$:

$$\lim_{h \to 0} (2x + h) = 2x.$$

I see that $2x$ is the *x-dependent* value of the slope of $f(x) = x^2$ at location x: In $x = 1$, 2 is the value of the slope of f: $f'(1) = 2$, while in $x = -3$, I calculate the value of the slope as follows: $f'(-3) = 2(-3) = -6$.

First Derivative of the Quadratic Power Function $f(x) = x^2$

The value of the gradient of the **quadratic power function** $f(x) = x^2$ is $2x$. The first derivative f' of f is thus for all x given by: $f'(x) = 2x$.

A little practice certainly cannot hurt now:

Exercise 6.1 (a) Determine the value of the slope of $f(x) = x^2$ at the points $x = 3$ and $x = -2$. (b) Determine the point x where the first derivative of $f(x) = x^2$ has the value -2.

Exercise 6.2 As above, show that $p_3'(x) = 3x^2$ is the first derivative of $p_3(x) = x^3$. Note: $(x + h)^3 = x^3 + 3x^2h + 3xh^2 + h^3$.

You may have noticed some regularity in the formation of the previous first derivatives. To help you see what I am getting at now, I will tell you that $4x^3$ is the value of the slope of $p_4(x) = x^4$ at any point x. Also, I will tell you that the first derivative of $p_5(x) = x^5$ is the function $p_5'(x) = 5x^4$. So the *rough rule* for finding the first derivative of a general **power function** $p_i(x) = x^i$, where $i \in \mathbb{N}$, is: write the exponent i multiplicatively forward and subtract one from that exponent.

First Derivative of the Power Function $p_i(x) = x^i$
The value of the slope of the **power function** $p_i(x) = x^i$ is $i\,x^{i-1}$. The first derivative p_i' of p_i is thus determined for all x by: $p_i'(x) = ix^{i-1}$.

Example 6.1
So now you can easily calculate the first derivative p_{128}' of $p_{128}(x) = x^{128}$: $p_{128}'(x) = 128\,x^{127}$. Or if someone should approach you on the street tomorrow and ask: "What is the value of the slope of the function $p_7(x) = x^7$ at the point $x = -\sqrt{2}$?", you can confidently answer: The first derivative p_7' of $p_7(x) = x^7$ is $p_7'(x) = 7\,x^6$. Substituting $x = -\sqrt{2} p_7'(-\sqrt{2}) = 7\,(-\sqrt{2})^6 = 7(-1)^6(\sqrt{2})^6 = 7$ $(2^{1/2})^6 = 7\,(2^{(1/2)6}) = 7\,(2^{6/2}) = 7\,(2^3) = 7 \cdot 8 = 56$. Congratulations.

Of course, the above procedure for determining the values of the slopes of a function, or its first derivative, is not limited to power functions $p_i(x) = x^i$. I will now explain how to *generalize* this concept of the first derivative – to do this, I recommend looking again at Fig. 6.4 and ask you to compare the descriptions that follow with the above procedure for $f(x) = x^2$. So now I look at more general functions of a real variable x. If you like, you can think of such funcions as the f of the hike in the "Palatinate Forest" in Fig. 6.1. In what follows, I consider an arbitrary but fixed x and suitable h ($\neq 0$) close to 0. This means that I want to assume that x and $x + h$ both always lie in the **domain of definition** of f (see Chap. 2). ◀

Secant of f Through x and $x + h$
The straight line through the two points $(x \mid f((x)))$ and $(x + h \mid f(x + h))$ is called the **secant of f through x and $x + h$.**

For fixed x the secant of f through x and $x + h$ depends of course also on the choice of h. In Fig. 6.4 I have denoted a secant by s and shown it as a dashed line. The *slope of this secant* is the quotient

$$\frac{f(x+h)-f(x)}{h}$$

from the difference of the function values $f(x + h)$ and $f(x)$ with the difference of the $x + h$ and x. This quotient is readily to as **difference quotient of f in x for h.** Such quotients generally represent only *approximations* to the actual slope of f in x. The transition of these approximate slopes to the value of the *actual slope* of f in x is conceptually described building limits.

Differential Quotient and First Derivative of f in x

If the limit of the difference quotients of f in x for $h \to 0$

$$\lim_{h\to 0}\frac{f(x+h)-f(x)}{h}$$

exists, then this **differential quotient of f in** x is called $f'(x)$ (read: "f dash of x"). If this is the case, the *real number* $f'(x)$ is the value of **the** gradient of f in x, and is called the **first derivative of f at the point** x. It is then said that f is **(once) differentiable in** x.

Some people also talk about f being **differentiable (once) at the point** x – this means exactly the same thing. *If f is differentiable* at the point x, then the value $f'(x)$ of the slope of f in x coincides with the value of the slope of a special straight line – this occurs as the straight line emerging from the secants through the boundary transition $h \to 0$ and "**tangents**" the graph of the function f at the point $(x \mid f(x))$. In the example of Fig. 6.4 I have shown this particular straight line t as a solid line.

Tangent of f in x

The linear function t through the point $(x \mid f(x))$ with slope $f'(x)$ is called the **tangent of f in** x.

I can write down the tangent t of f in x explicitly by substituting into the above general form of linear functions $c = f'(x)$ and $d = -f'(x)\,x + f(x)$. This looks like this

$$t(\overline{x}) = f'(x)\overline{x} - f'(x)x + f(x),$$

whereby I use here for distinction reasoning for the fixed location x as an exception the variable designation $\overline{x}$. If I substitute here x for $\overline{x}$, $f(x)$ comes out, and I also recognize that of course $t'(\overline{x}) = c = f'(x)$ holds.

Example 6.2

In the example of Fig. 6.4, $c = f'(1) = 2$ and $d = -f'(1)\,1 + f(1) = -(2 \cdot 1) + 1 = -1$, and the tangent of $f(x) = x^2$ in $x = 1$ is the linear function $t(\overline{x}) = 2\overline{x} - 1$.

The mathematician experiences special joy if the function f is differentiable at *all* location x. In this case one says that **the function** f *is* **differentiable**, and one then has – just as I wrote at the beginning of this section – *another* real-valued function f' in the variable x, which yields the value of the derivative $f'(x)$ of f for every location x. This is then called the **derivative function of** f, or **the derivative of** f *for* short. By the way, this is what I had in mind when I used the symbol $f'(x)$ above to denote the differential quotient for the fixed location x: Here, one anticipates the connection to the function f', so to speak, because one hopes that the derivative of f exists everywhere.

A little exercise to internalize the things certainly does not hurt: ◄

Exercise 6.3 Calculate the difference quotients of $f(x) = x^2 + x$ in x for h, showing that $f'(x) = 2x + 1$ is the first derivative of f for all x. Hint: See how I proceeded above for $p_2(x) = x^2$.

In the following, I discuss three more examples of differentiable functions. These are standard functions that you know from the previous chapters.

First Derivative of the Exponential Function $\exp(x) = e^x$

The value of the slope of the **exponential function** $exp(x) = e^x$ is e^x. The first derivative $\exp'$ of exp. for all x is given by: $\exp'(x) = e^x$.

You see that the derivative $\exp'$ of the exponential function exp. is the exponential function itself: $\exp' = \exp$. This is quite clear – but how can you now see this again? I am just sketching out a basic idea for you here:

First, I consider an arbitrary x and calculate the difference quotients for $h \neq 0$:

$$\frac{\exp(x+h) - \exp(x)}{h} = \frac{e^{x+h} - e^x}{h} = \frac{e^x e^h - e^x}{h} = e^x \frac{e^h - 1}{h}.$$

In doing so, I applied the power law $e^{x+h} = e^x \cdot e^h$ in the penultimate step and then factored out e^x in the numerator. Now, I have to calculate the differential quotient – i.e. the following limit value:

$$\lim_{h \to 0} \frac{\exp(x+h) - \exp(x)}{h} = \lim_{h \to 0} e^x \frac{e^h - 1}{h} = e^x \lim_{h \to 0} \frac{e^h - 1}{h} = e^x \cdot 1 = e^x.$$

I recognize that for this again e^x comes out – thus: $\exp'(x) = e^x = \exp(x)$. I can write the term e^x here after the second "="- sign, without doing anything wrong, before the limit,

Table 6.1 Values of the term $(e^h - 1)/h$ for choices of h close to 0

$h =$	-0.1	-0.01	-0.001	0.001	0.01	0.1
$(e^h - 1)/h \approx$	0.95163	0.99502	0.99950	1.00050	1.00502	1.05171

because this has nothing to do with the limit for $h \to 0$, because h *does not* occur in this. That is intuitively understandable. But after that, I admittedly did something that actually requires a *more detailed analysis* – namely, I use:

$$\lim_{h \to 0} \frac{e^h - 1}{h} = 1.$$

I do not prove this to you here – but you can trust me completely: This can be deduced from the definition of **Euler's number** e. The above casual procedure of simply substituting 0 for h *in* the difference quotient does not work here, by the way: I would then divide by 0, which I must *never never never* do. But I do not want to leave you all alone with this either, so I have prepared Table 6.1 for you, from which you can read off the values of $(e^h - 1)/h$ for h getting close to 0.

My next two examples of differentiable functions are the **sine function** $\sin(x)$ and the **cosine function** $\cos(x)$. Here one usually proceeds in the same way: First you form the difference quotients and then the differential quotient. You need the **addition theorems** and similar limit considerations as for the exponential function. I do not do this now, because others will probably do this for me later in your studies.

First Derivative of the Sine Function sin(x) and Cosine Function cos(x)

The first derivative $\sin'$ of the **sine function** $\sin(x)$ is defined for all x by: $\sin'(x) = \cos(x)$. The first derivative $\cos'$ of the **cosine function** $\cos(x)$ for all x is given by: $\cos'(x) = -\sin(x)$.

Now I have to tell you something that you might not like so much: Unfortunately, there are also functions that are not *differentiable.* A function f is called **not differentiable** if there *is* (at least) *one location* x for which f is not differentiable. You may have noticed that I was *very careful* above in formulating the notion of differential quotient ("*If the limit … exists, …*"), and there is a good reason for that: namely, it could be that for certain functions f at some locations x something goes wrong with the formation of the limit of the differential quotients for $h \to 0$. By this I mean that – if things go stupidly – it can happen to you that the differential quotient does *not exist.* This means that it is either *not a real number* at all, or that it is not so clear what number this limit should be – that is, it *cannot be uniquely determined.* Before I start confusing you, it is best to show you two classic examples of functions that are not differentiable.

I start with the **(square) primitive** $f(x) = \sqrt{x}$, where $x \geq 0$. I must not use any negative numbers x here, because if I were to take the root of a negative number, this would unfortunately no longer be a real number, and I only treat functions here that are **real-valued:** $f(x) \in \mathbb{R}$ for all x. As a test, I now proceed as usual: first I form the difference quotients for any fixed $x \geq 0$, where h *is* chosen so that $x + h > 0$ holds. The difference quotients are

$$\frac{f(x+h)-f(x)}{h} = \frac{\sqrt{x+h}-\sqrt{x}}{h} = \frac{h}{h\left(\sqrt{x+h}+\sqrt{x}\right)} = \frac{1}{\left(\sqrt{x+h}+\sqrt{x}\right)}.$$

Here I first expanded the quotients after the first "="-sign with $\sqrt{x+h}+\sqrt{x}$, then after the second "="-sign I applied the third **binomial formula** (see Chap. 1) $\left(\sqrt{x+h}-\sqrt{x}\right)\left(\sqrt{x+h}+\sqrt{x}\right) = x+h-x = h$ and finally after the third "="-sign I cancelled the h. Now I try to find out the limit for $h \to 0$ – that is, to calculate the differential quotient:

$$\lim_{h\to 0}\frac{f(x+h)-f(x)}{h} = \lim_{h\to 0}\frac{1}{\left(\sqrt{x+h}+\sqrt{x}\right)} = \frac{1}{\left(\sqrt{x}+\sqrt{x}\right)} = \frac{1}{2\sqrt{x}}.$$

I did not think too much about this and – as above – simply casually inserted 0 for h into the term $1/\left(\sqrt{x+h}+\sqrt{x}\right)$ to determine the limit. It actually looks quite OK, because I get $f'(x) = 1/(2\sqrt{x}) = (1/2)x^{-1/2}$, and you may now be wondering what should have gone wrong here. In fact, everything is fine here as *long as* x is not equal to 0. The primitive is differentiable at all positive locations $x > 0$. However, the remaining location $x = 0$ causes enormous difficulties. If I look at it again for $x = 0$ (h should then be >0), I get a rather annoying problem:

$$\lim_{h\to 0}\frac{f(0+h)-f(0)}{h} = \lim_{\substack{h\to 0\\ h>0}}\frac{1}{\left(\sqrt{0+h}+\sqrt{0}\right)} = \lim_{\substack{h\to 0\\ h>0}}\frac{1}{\sqrt{h}}.$$

If h gets closer and closer to 0, the term in the denominator $\sqrt{h}$ will also approach 0 and the difference quotient $1/\sqrt{h}$ will become larger and larger. I will give this a try: as a test, I will take $h = 1/100 = 10^{-2}$, $1/10{,}000 = 10^{-4}$, $1/1000{,}000 = 10^{-6}$ and obtain the following values in turn by substituting them into the term $1/\sqrt{h}$: $1/\sqrt{10^{-2}} = 1/(10^{-2})^{\frac{1}{2}} = 1/10^{(-2)\frac{1}{2}} = 1/10^{-1} = 10$, $1/\sqrt{10^{-4}} = 100$, $1/\sqrt{10^{-6}} = 1000$. I notice that the term $1/\sqrt{h}$ gets larger and larger as h approaches 0. Thus, unfortunately, the term does not approach a *real number* at all – on the contrary, it grows beyond all positive bounds. As a rough rule, one can imagine that in the limit $h \to 0$ one *would divide*

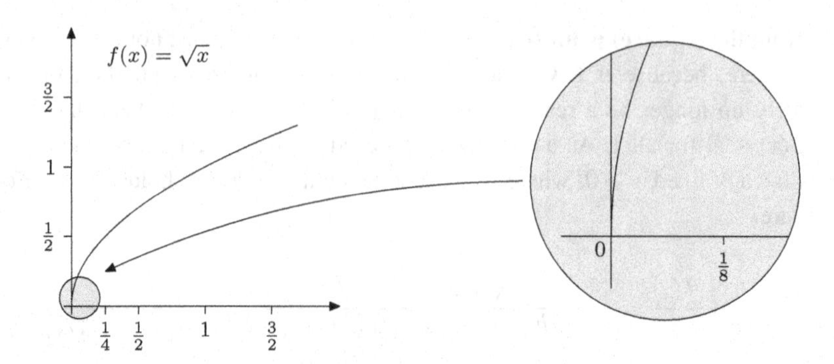

Fig. 6.5 The primitive $f(x) = \sqrt{x}$ is not differentiable at the point $x = 0$: The value of the slope of f becomes infinitely large as $x = 0$ is approached from the right. Zooming in on the location $x = 0$ (*right*) illustrates that the value of the slope grows beyond all positive bounds

by 0, which as we know one must *never never never* do: So here, it is not allowed to substitute $h = 0$ into the term $1/\sqrt{h}$ to find the limit. One now writes this down as follows:

$$\lim_{\substack{h \to 0 \\ h > 0}} \frac{1}{\sqrt{h}} = \infty.$$

The sign ∞ is an abbreviation for **infinitely large**: I cannot find a real number that I could write on the right side next to the "=" sign without cheating. The limit of the difference quotient does not exist for the position $x = 0$. Graphically, this means that the value of the slope of the function becomes larger than any given positive bound if I just get close enough to the location $x = 0$ (see Fig. 6.5). Thus, the primitive is not differentiable at the point $x = 0$. Thus this function is not differentiable – although it worked at all other points, which are larger than 0, with the formation of the limit. By the way, this has nothing to do with the fact that it happens to be the special location $x = 0$: Thus the function $f(x) = \sqrt{x - 2}$, $x \geq 2$, whose graph is obtained by shifting the primitive to the right by two units on the *x-axis, is* not differentiable in $x = 2$.

Fortunately, there is also something positive to report here:

First Derivative of the Root Function $f(x) = \sqrt{x}$, $x \in [a, b]$ with $a > 0$
If $a > 0$, then $1/(2\sqrt{x})$ is the value of the slope of the **Root Function** $f(x) = \sqrt{x}$ for all $x \in [a, b]$. The first derivative f' of f is then for all $x \in [a, b]$ given by: $f'(x) = 1/(2\sqrt{x})$.

Thus, if I consider a sensible interval $[a, b]$ that does not contain the location $x = 0$, the primitive is differentiable there and I can compute the first derivative as I might already suspect according to the above rough rule (for power functions): $(x^{1/2})' = 1/2\,x^{-1/2}$. For

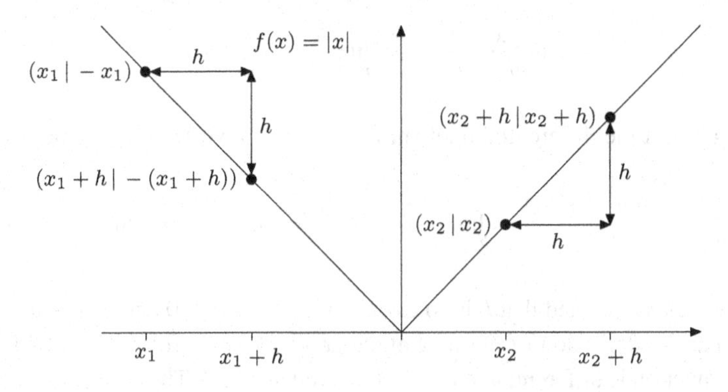

Fig. 6.6 The absolute value function f (x) = |x| is not differentiable, because it is not differentiable at the position $x = 0$: The graph of f has a "kink" there

your information, I briefly whisper to you that the m-th root $f(x) = \sqrt[m]{x} = x^{1/m}$, where $m \in \mathbb{N}$, is compute the derivative ofd in the same way in the admissible domains: $f'(x) = (1/m)x^{(1/m)-1} = (1/m)x^{(1-m)/m} = (1/m)x^{-((m-1)/m)} = (1/m)\left(1/\sqrt[m]{x^{m-1}}\right)$. So you can see that the above rough rule for forming the derivative applies here as well. I should not forget to say that the *m-th primitive* is not differetiable in $x = 0$.

So you can see that in general one has to be careful when forming the limit of the differential quotient for $h \to 0$. It is important that for each x a real number for the differential quotient actually comes out and not ∞. But that is not all: I have to point out another problem to you, which is also related to the formation of the differential quotient. Namely, in exceptional situations, the differential quotient may not exist, *even though* the associated difference quotients do not become infinitely large when h approaches 0. Above I already indicated that this difficulty can occur if it is not clear which value the differential quotient should get – i.e. it *cannot be unambiguously determined*: Then one has *two candidates* for this value, but, similar to many situations in our daily life, unfortunately cannot decide.

To make clear what I mean, let me look at the announced second classic. This is the **magnitude function** defined for $x \in \mathbb{R}$

$$f(x) = \mid x \mid = \begin{cases} x, & \text{falls } x \geq 0, \\ -x, & \text{falls } x < 0, \end{cases}$$

which I have illustrated for you in Fig. 6.6. Note that the magnitude function is *not a* linear function: this function is only **piecewise linear**. If I now grab any $x > 0$ and want to calculate the limit of the difference quotients, this is done in exactly the same way as above for the identity:

$$\lim_{h\to 0}\frac{x+h-x}{h}=\lim_{h\to 0}\frac{h}{h}=\lim_{h\to 0}1=1.$$

The calculation of the differential quotient for any $x < 0$ works in the same way:

$$\lim_{h\to 0}\frac{-(x+h)-(-x)}{h}=\lim_{h\to 0}\frac{-x-h+x}{h}=\lim_{h\to 0}\frac{(-h)}{h}=\lim_{h\to 0}(-1)=-1.$$

Here I have tacitly assumed that h is so close to 0 that $x + h < 0$ and therefore $|x + h| = -(x + h)$ as per the definition of the magnitude above. Wonderful – so I establish that the magnitude function is differentiable for $x > 0$ as well as $x < 0$. The derivative has a value of 1 and -1 respectively. Great, but what about the remaining digit $x = 0$? I will have to look into that a bit more. I put $x = 0$ in each of the two difference quotients above and consider both the **right-hand side limit**(difference quotients are only formed for $h > 0$):

$$\lim_{\substack{h\to 0\\ h>0}}\frac{0+h-0}{h}=\lim_{\substack{h\to 0\\ h>0}}\frac{h}{h}=1$$

as well as the **left-hand limit** (difference quotients formed for $h < 0$):

$$\lim_{\substack{h\to 0\\ h<0}}\frac{-(0+h)-0}{h}=\lim_{\substack{h\to 0\\ h<0}}\frac{(-h)}{h}=-1.$$

Now you see what I meant above by the problem of *two candidates for* the value of the differential quotient. There is obviously a difficulty there: if I am approaching $x = 0$ from the right ($h > 0$), then my candidate for the derivative of $f(x) = |x|$ at $x = 0$ is the value 1, while coming from the left ($h < 0$), there is another candidate for the derivative of f at $x = 0$: -1. True, the candidates for the differential quotient are not infinitely large, as it was the case with the primitive at $x = 0$. Nevertheless, this is not so nice, because I cannot tell exactly which of the two candidates should be the *actual* limit of the differential quotient of fin $x = 0$. This is similar to the casino: Maybe you like the colors "red" and "black" very much and you just cannot decide on which color to bet your chip in roulette. In mathematics, we make it simple: The differential quotient of f in 0 *cannot be uniquely determined* and therefore f has no derivative at $x = 0$. Thus, the magnitude function is not differentiable – figuratively, this means that you exchange your chip for money and go home. You certainly did not do anything wrong with this, and this is also the point of view in mathematics when the unambiguous determination of the differential quotient just does not want to succeed. You can see this clearly in the "kink" that the graph of the absolute value function has at $x = 0$. The functions in Figs. 6.1, 6.2, 6.3, and 6.4 do not all have such a "kink" – the graphs look *smooth*. This is the case for differentiable functions – and above you have seen that these functions are fortunately differentiable at every point x and thus indeed differentiable altogether.

Training is important. That is why I ask you to try your hand at the following tasks. Attention: The (b) part is not so easy.

Exercise 6.4 Investigate whether (a) $f(x) = |x - 2|$ and (b) $f(x) = 1/2\,((x - 2)^3 + |x - 2|^3)$ are differentiable at the point $x = 2$. Are the functions differentiable?▪

I would like to move back in a positive direction now, and in the long run try to make the computation of derivatives a little easier for myself. If I stubbornly cling to the above procedure, I can determine the derivative of a given function by first determining the difference quotients at each point and then trying to calculate the differential quotient. Care must be taken to ensure that this always exists and, in particular, that it is unique. This is generally quite tedious, as you may already have noticed yourself when working through Exercise 6.3.

Example 6.3

Another somewhat tedious example is the function $f(x) = -5x^2 + 4\sqrt{x}$, where I assume x is from an interval $[a, b]$ with $a > 0$ because of the root term. If I use the above procedure, I would first form the difference quotients here:

$$\frac{f(x+h)-f(x)}{h} = \frac{(-5)(x+h)^2 + 4\sqrt{x+h} - (-5x^2 + 4\sqrt{x})}{h}$$
$$= \frac{-5x^2 - 10xh - 5h^2 + 4\sqrt{x+h} + 5x^2 - 4\sqrt{x}}{h}.$$

If I apply the same calculation tricks I showed you above when calculating the difference quotients of the quadratic power function and the primitive again (simultaneously) with a small variation, I get

$$\frac{f(x+h)-f(x)}{h} = \frac{(-10x - 5h)\,h}{h} + \frac{4\sqrt{x+h} - 4\sqrt{x}}{h}$$
$$= (-5)(2x+h) + \frac{4}{\sqrt{x+h} + \sqrt{x}}.$$

Now I form the limit for $h \to 0$, i.e. the differential quotient, by – casually speaking – substituting 0 for h in the difference quotients:

$$\lim_{h \to 0} \frac{f(x+h) - f(x)}{h} = \lim_{h \to 0} \left((-5)(2x+h) + \frac{4}{\sqrt{x+h} + \sqrt{x}} \right)$$
$$= (-5)(2x) + \frac{4}{\sqrt{x} + \sqrt{x}} = -10x + \frac{2}{\sqrt{x}}.$$

I am allowed to do that, because I paid attention that the interval $[a, b]$ does not contain the location $x = 0$. My result here is that f is differentiable, because f is differentiable at all points considered $x \in [a, b]$ and holds true: $f'(x) = -10x + 2/\sqrt{x}$.

As you can see from this example, I have essentially used here the calculation tricks that I had already used above in a similar way for the quadratic power function and the primitive. The legitimate question at this point is whether I could not *have proceeded* much more *simply* by using more directly what I had already learned about the derivative of these two – let us say *elementary* – functions. That is exactly what I intend to do now. So in what follows, I will show you that it is often enough to take a much *simpler route* to determine the derivative of a function. I will now present you a *machinery* of **derivative rules**, which makes it possible to determine the derivative for large classes of *standard functions*, *without explicitly* going the sometimes tedious way over difference and differential quotients, or even to think again deeply about the existence of the latter.

Do you remember my first serious example of determining the derivative of a function? That is right – that was the identity $f(x) = x$ shown in Fig. 6.3, whose derivative is $f'(x) = 1$. Above, I had then told you that I could treat the slightly more difficult function $f(x) = 3x$ similarly, and got: $f'(x) = 3$. Somewhat grungy, but correctly, I can write this down as: $f'(x) = (3x)' = 3(x)' = 3 \cdot 1 = 3$. This shows you that for this function, I can first **factor** *out* the **factor** 3 (second "=" sign) and then merely use what I have already learned about the derivative of the identity. ◀

Example 6.4

I get a more general example if I multiply a power function by a real factor a_i: $f(x) = a_i x^i$. If I remember the derivative of the power function, I can proceed in the same way: $f'(x) = (a_i x^i)' = a_i (x^i)' = a_i i x^{i-1} = i a_i x^{i-1}$.▪. ◀

Example 6.5

Another example is the function $f(x) = 4\sqrt{x}$, where I assume that x is from $[a, b]$ with $a > 0$. I calculate the derivative of this f as follows: $f'(x) = (4\sqrt{x})' = 4(\sqrt{x})' = 4(1/(2\sqrt{x})) = 2/\sqrt{x}$. Here I first factored out the factor 4 (second "=" sign) and then just used what I already learned about the derivative of the primitive.

This little trick with factoring out always works when I multiply a differentiable function f by a real factor λ (read: "lambda") and compute the derivative of the resulting function $\lambda \cdot f$, defined by $(\lambda \cdot f)(x) = \lambda \cdot f(x)$. See why this is so. First, I form the difference quotients

$$\frac{(\lambda \cdot f)(x+h) - (\lambda \cdot f)(x)}{h} = \frac{\lambda f(x+h) - \lambda f(x)}{h} = \lambda \frac{f(x+h) - f(x)}{h}$$

and then the differential quotient

$$\begin{aligned} \lim_{h \to 0} \frac{(\lambda \cdot f)(x+h) - (\lambda \cdot f)(x)}{h} &= \lim_{h \to 0} \lambda \frac{f(x+h) - f(x)}{h} \\ &= \lambda \lim_{h \to 0} \frac{f(x+h) - f(x)}{h} \\ &= \lambda f'(x), \end{aligned}$$

where for the penultimate "=" sign I take advantage of the fact that here I may write the constant factor λ in front of the limit (because this is a fixed number and thus has nothing to do with h), and for the last "=" sign I benefit from the fact that f is *differentiable* in x. ◄

First Derivative of the Factor Product $\lambda \cdot f$ – Factor Rule
If f is a differentiable function and λ is a real constant, then the **factor product** $\lambda \cdot f$ is a differentiable function and the first derivative $(\lambda \cdot f)'$ of $\lambda \cdot f$ is determined for all x by: $(\lambda \cdot f)'(x) = \lambda f'(x)$.

Do you still remember Exercise 6.3? There you had to show that $f'(x) = 2x + 1$ is the first derivative of $f(x) = x^2 + x$ by using the difference quotients. I am sure you have solved the problem confidently by now. As a reward, I will now tell you that you can make this even significantly easier by *directly* using what I told you above about the derivative of the identity and the quadratic power function. These two elementary functions show up as summands of f. If I now compute the derivative of them individually and then sum up the two derivatives, I get f' as well – somewhat grungy, I can write it down as follows:

$$f'(x) = (x^2 + x)' = (x^2)' + (x)' = 2x + 1.$$

With this I have shown you (by the second "="-sign), that the *separation of sums is* allowed when computing derivatives: So the derivative of the sum is here the sum of the derivatives. I can apply the same to the above example $f(x) = -5x^2 + 4\sqrt{x}$, still using the factor rule I just mentioned:

$$f'(x) = (-5x^2)' + (4\sqrt{x})' = (-5)(x^2)' + 4(\sqrt{x})' = (-5)\,2x + \frac{4}{2\sqrt{x}} = -10x + 2x^{-1/2}.$$

I think this is easier than arguing directly about the difference and differential quotients. Why don't you try an example in advance?

Exercise 6.5 As in the examples just shown, determine the first derivative of $f(x) = 3\cos(x) - 2\sin(x)$.

This trick with the difference of the sum always works, if I have two differentiable functions f and g and want to differentiate the sum of the two functions $f + g$, defined by $(f + g)(x) = f(x) + g(x)$. See why this is so: First I form the difference quotients

$$\frac{(f+g)(x+h) - (f+g)(x)}{h} = \frac{f(x+h) + g(x+h) - (f(x) + g(x))}{h} = \frac{f(x+h) - f(x)}{h} + \frac{g(x+h) - g(x)}{h}$$

and then the differential quotient

$$\lim_{h \to 0} \frac{(f+g)(x+h) - (f+g)(x)}{h} = \lim_{h \to 0} \left(\frac{f(x+h) - f(x)}{h} + \frac{g(x+h) - g(x)}{h} \right) = \lim_{h \to 0} \frac{f(x+h) - f(x)}{h} + \lim_{h \to 0} \frac{g(x+h) - g(x)}{h} = f'(x) + g'(x),$$

where at the penultimate "="-sign I simply use, that here I am allowed to pull apart the sums of limits, and at the last "="-sign I profit from the fact, that f and g are *differentiable.*

First Derivative of the Sum of Functions $f + g$ – Sum Rule
If f and g are differentiable functions, then their **sum** $f + g$ is a differentiable function and the first derivative $(f + g)'$ of $f + g$ is defined for all x by: $(f + g)'(x) = f'(x) + g'(x)$.

Sometimes the **difference rule** is formulated: If I write the difference $f - g$ *of* two differentiable functions f and g as the sum of f with the factor product $(-1) \cdot g$, then I get from the two rules above (factor and sum rule):

$$(f - g)' = (f + (-1) \cdot g)' = f' + ((-1) \cdot g)' = f' + (-1)g' = f' - g'.$$

Thus, the derivative of the difference of two differentiable functions is the difference of the derivatives of these functions.

The simple derivative rules discussed so far already allow the derivative to be determined for relatively large classes of functions. For example, looking again at linear functions $p(x) = c\,x + d$ (also called **polynomial functions of degree one**, or **polynomial of degree one** for short), as an alternative to what I showed you pretty much at the beginning of this section, I can work out the derivative as follows: $p'(x) = (cx)' + (d1)' = c(x)' + d\,(1)' = c\,1 + d\,0 = c$. You see, we got the constant c again. But I can now proceed in a similar way if I have to deal with somewhat more difficult functions, which consist of *more than two summands*:

Example 6.6

An example is the following **polynomial of degree five**: $p(x) = 3x^5 - \sqrt{2}x^3 + \frac{8}{7}x^2 - 3$. Multiple application of the factor and sum rule yields the derivative here:

$$\begin{aligned} p'(x) &= (3x^5)' - \left(\sqrt{2}x^3\right)' + \left(\frac{8}{7}x^2\right)' - (3)' = 3\,(x^5)' - \sqrt{2}\,(x^3)' + \frac{8}{7}\,(x^2)' - 3\,(1)' \\ &= 3\,(5x^4) - \sqrt{2}\,(3x^2) + \frac{8}{7}\,(2x) - 3\,(0) = 15x^4 - 3\sqrt{2}x^2 + \frac{16}{7}x, \end{aligned}$$

using what I learned above about derivatives of power functions: $(x^5)' = 5x^4$, $(x^3)' = 3x^2$, $(x^2)' = 2x$ and $(x^0)' = (1)' = 0$. ◀

Example 6.7

Another example is $f(x) = 3\,\exp.(x) - 2\sin(x) + \cos(x)$. Here, the exponential function and the sine and cosine functions, respectively, appear as terms. I now use the knowledge I acquired above about the derivative of these elementary functions and calculate f' using the factor and sum rule:

$$f'(x) = 3\exp'(x) - 2\sin'(x) + \cos'(x) = 3\exp(x) - 2\cos(x) - \sin(x). \quad \blacksquare$$

Polynomial functions are especially close to my heart. I will first show you how I can compute the derivative a general **quadratic polynomial function** (also: **polynomial of degree two**) $p(x) = a_2x^2 + a_1x + a_0$. Here a_0, a_1 and a_2 are fixed real numbers. Using the factor and sum rule, I do this as follows:

$$p'(x) = (a_2 x^2)' + (a_1 x)' + (a_0.1)' = a_2 (x^2)' + a_1 (x)' + a_0 (1)' = 2a_2 x + a_1.$$

Here, as in the example above with the polynomial function of degree five, I have used our knowledge of the derivative of power functions. I thus recognize that the derivative of a quadratic polynomial function is always a linear polynomial function. The general approach succeeds for **polynomial functions of degree** n (in short: **polynomial of degree** n). ◄

First Derivative of Polynomial Functions p

Any **polynomial function** p **of degree** n, $p(x) = a_n x^n + \cdots + a_i x^i + \cdots + a_1 x + a_0$ is differentiable, and the first derivative p' of p is the polynomial of degree $n - 1$, which is fixed for all x as follows:

$$p'(x) = n a_n x^{n-1} + \cdots + i a_i x^{i-1} + \cdots + a_1.$$

You should be able to compute the derivative of polynomials. Therefore, I suggest that you try the following exercise:

Exercise 6.6 Calculate the first derivative of the following polynomial functions: a) $f(x) = -x^3 - 2x^2 + \frac{1}{8}x - \sqrt[3]{2}$ and b) $f_t(x) = t^2 x^4 - 2\,t^4 x^2 + t^2$. Also, determine (c) the derivative of the following polynomials in variable t:

$f_x(t) = t^2 x^4 - 2\,t^4 x^2 + t^2 = -2x^2 t^4 + (x^4 + 1)\,t^2$, which I determined by interchanging the meaning of the **parameter** t and the variable x in (b).

Perhaps the following question now arises: Is there actually something similar to the sum rule for the *product of differentiable functions*? The pleasant answer is: Yes. But beware: Unfortunately – as one might suspect at first glance – the derivative of a product of functions is generally *not* the product of the respective derivatives.

Example 6.8

To do this, look at my example of the cubic power function x^3. I can write this function as the product of the quadratic power function $f(x) = x^2$ and the identity $g(x) = x$: $x^3 = x^2\,x = f(x)\,g(x)$. Above I told you that $f'(x) = 2\,x$ and $g'(x) = 1$ hold. Thus, the product of the derivatives of f and g is $f'(x)\,g'(x) = 2\,x\,1 = 2\,x$. You see: What comes out here does *not at all* always agree with $3\,x^2 = (x^3)'$.▪

Unfortunately, not everything can always go through as smoothly as one might imagine at first sight. To correctly determine the derivative $(f \cdot g)'$ of a product $f \cdot g$ of two differentiable functions f and g, defined by $(f \cdot g)(x) = f(x)\,g(x)$, I have to work a little harder. I now show you that it is possible to express $(f \cdot g)'$ by the functions f, g, f' and g' *which* are assumed to be known. See how I do this in general: first I form the difference quotients

$$\begin{aligned}\frac{(f \cdot g)(x+h) - (f \cdot g)(x)}{h} &= \frac{f(x+h)g(x+h) - f(x)g(x)}{h} \\ &= \frac{f(x+h)g(x+h) - f(x)g(x+h) + f(x)g(x+h) - f(x)g(x)}{h} \\ &= \frac{(f(x+h) - f(x))g(x+h) + f(x)(g(x+h) - g(x))}{h} \\ &= \frac{f(x+h) - f(x)}{h} g(x+h) + f(x)\frac{g(x+h) - g(x)}{h}.\end{aligned}$$

Here I subtracted the term $(f(x)\, g(x+h))/h$ after the second "=" sign and added it right back, which, as you know, does not change the result. With this little trick, I managed to get the difference quotients of the two function f and *g to* appear when forming the difference quotients of the product $f \cdot g$ – see that? Now I calculate the differential quotient:

$$\lim_{h \to 0} \frac{(f \cdot g)(x+h) - (f \cdot g)(x)}{h}$$

$$\begin{aligned}&= \lim_{h \to 0}\left(\frac{f(x+h) - f(x)}{h} g(x+h) + f(x)\frac{g(x+h) - g(x)}{h}\right) \\ &= \left(\lim_{h \to 0}\frac{f(x+h) - f(x)}{h}\right) g(x) + f(x)\lim_{h \to 0}\frac{g(x+h) - g(x)}{h} = f'(x)g(x) \\ &+ f(x)g'(x).\end{aligned}$$

Here I used at the penultimate "="-sign, that here I am allowed to pull apart sums of limits and to write constant factors in front of the limit. Furthermore I profited there also from the fact, that because of the fact that g is differentiable

$$\lim_{h \to 0} g(x+h) = \lim_{h \to 0}(g(x) + g'(x)\,h) = g(x)$$

holds. At the last "="-sign above I also profited from the fact that f and g are differentiable. ◀

First Derivative of the Product of Functions $f \cdot g$ – Product Rule

If f and g are differentiable functions, then their **product** $f \cdot g$ is a differentiable function and the first derivative $(f \cdot g)'$ of $f \cdot g$ is given for all x by: $(f \cdot g)'(x) = f'(x)\, g(x) + g'(x) f(x)$.

I now consider a few examples of the product rule:

Example 6.8 (continued)

First, let me continue our discussion from above and look again at the cubic power function x^3, whose derivative is known to be $3\,x^2$. Now, if I set $x^3 = f(x)\,g(x)$ as above, where $f(x) = x^2$ and $g(x) = x$, the derivative of $f \cdot g$ is calculated by the product rule as follows: $(f \cdot g)'(x) = (x^2)'x + (x)'x^2 = (2x)x + 1(x^2) = 2x^2 + x^2 = 3x^2$. Now we are getting it right. But I should not lose track here – this is *just an example* to illustrate how the product rule works – so I will continue to compute the derivative of the power function x^3 directly as $3x^2$ without using the product rule. ◀

Example 6.9

A serious example is the function defined for $x \in [a, b]$ with $a > 0$

$$(f \cdot g)(x) = \overbrace{(x^2 + x)}^{f(x)}\overbrace{(-5x^2 + 4\sqrt{x})}^{g(x)} .$$

Of course, I could brutally multiply here and then apply the factor and sum rule. However, I would then have the little problem that, for example, the term $4x^2\sqrt{x} = 4x^{5/2}$ would appear and I have not (yet) told you how to compute the derivative of (… at the latest in the next section – I promise). Also, fortunately, I already worked out above the derivatives of the functions $f(x) = x^2 + x$ and $g(x) = -5x^2 + 4\sqrt{x}$: $f'(x) = 2x + 1$ and $g'(x) = -10x + 2/\sqrt{x}$. Therefore, the product rule gives:

$$(f \cdot g)'(x) = \overbrace{(2x + 1)}^{f'(x)}\overbrace{(-5x^2 + 4\sqrt{x})}^{g(x)} + \overbrace{\left(-10x + \frac{2}{\sqrt{x}}\right)}^{g'(x)}\overbrace{(x^2 + x)}^{f(x)} .$$

Now I can multiply and get using the power rules:

$$\begin{aligned}(f \cdot g)'(x) &= -10x^3 + 8x\frac{3}{2} - 5x^2 + 4x^{\frac{1}{2}} - 10x^3 + 2x^{\frac{3}{2}} - 10x^2 + 2x^{\frac{1}{2}} \\ &= -20x^3 - 15x^2 + 10x^{\frac{3}{2}} + 6x^{\frac{1}{2}}. \quad \blacksquare\end{aligned}$$

◀

Example 6.10

The next example is the function $f(x) = \sin^2(x)$. This is the product of the sine function $\sin(x)$ with itself: $f(x) = (\sin \cdot \sin)(x) = \sin(x)\,\sin(x)$. Fortunately, I know from above what the derivative of the sine function is: $\sin'(x) = \cos(x)$. So with that, I can apply the product rule and get:

$$\begin{aligned} f'(x) &= (\sin(x)\sin(x))' = \sin'(x)\sin(x) + \sin'(x)\sin(x) \\ &= 2\sin'(x)\sin(x) = 2\cos(x)\sin(x). \quad \blacksquare \end{aligned}$$

A little training cannot hurt now: ◀

Exercise 6.7 Use the product rule to calculate the derivative of the following functions: a) $f(x) = -x^3(x^2-4x+2)$, b) $f(x) = (x^2+x)\cos(x)$, c) $f(x) = x^2 e^x$, d) $f(x) = \cos(x)e^x$, and e) $f(x) = \cos^2(x)$.

A nice example of applying the product rule is to determine the derivative f' of $f(x) = 1/x = x^{-1}$ for $x \neq 0$. To determine this, I must first recognize that $x\, f(x) = x\,(1/x) = 1$ holds. The derivative of the right-hand side vanishes here: $(1)' = 0$. Now, I compute the derivative of the left-hand side (product of identity and f) using the product rule:

$$(x \cdot f)'(x) = (x)'x^{-1} + (x^{-1})'x = 1x^{-1} + (x^{-1})'x = x^{-1} + (x^{-1})'x.$$

I want to know what $f'(x) = (x^{-1})'$ is. I now already know that $x^{-1} + (x^{-1})'x = 0$ applies. I now reshape the latter by first putting x^{-1} on the right-hand side: $(x^{-1})'x = -x^{-1}$ and then dividing by x: $(x^{-1})' = (-x^{-1})(1/x)$. But $1/x = x^{-1}$ is valid and so I get

$$f'(x) = (x^{-1})' = (-x^{-1})x^{-1} = -x^{-1-1} = -x^{-2} = \frac{-1}{x^2}.$$

Again (for $x \neq 0$) the rough rule for the formation of the first derivative applies, which I have given you for the power function p_i: $(x^{-1})' = (-1)x^{-2}$. It is quite astonishing for how many functions of most different kind the rough rule for finding the derivative is applicable.

It is nice to have the product rule. But what do I do when someone walks in the door and asks the question, "What is the derivative of $\sin^{27}(x)$?" Or, "I'd like to know the derivative of $\exp(\sin(x)) = e^{\sin(x)}$!" All of the previous derivative rules are also of precious little help to us if someone partout forces us to form the derivative of $(3x^2+1)^{71}$ – multiplying out here to then apply the factor and sum rule would simply be far too much work, while using the product rule would also quickly become confusing. I will now tell you how to calculate the derivative of **chained functions** (see Chap. 2), without going into the details of the exact derivative. Something has to be left over for later.

First Derivative of Chained Functions $f \circ g$ – Chain Rule
If f and g are differentiable functions, then the **concatenation** $f \circ g$ **of** f **with** g is a differentiable function and the first derivative $(f \circ g)'$ of $f \circ g$ is determined for all x by: $(f \circ g)'(x) = g'(x) f'(g(x))$.

The best way to understand this is probably by example:

Example 6.11
I first consider $(3x^2 + 1)^{71}$. This example would be much easier if the term in the parenthesis, i.e. $3x^2 + 1$, *were just a variable* $\overline{x}$, because I can easily compute the derivative of the power function $f(\overline{x}) = \overline{x}^{71}: f'(\overline{x}) = 71\overline{x}^{70}$. But unfortunately, this is not so easy here: the term in the parenthesis is *itself a function*, namely $g(x) = 3x^2 + 1$. Fortunately, however, I at least know how to compute the derivative of this function: $g'(x) = 6x$. The concatenation of f with g is $(f \circ g)(x) = (3x^2 + 1)^{71}$. So I get this by *replacing* the variable $\overline{x}$ of the function f *with the function* g: $\overline{x} = g(x)$. Now the chain rule tells me how to find the derivative of $f \circ g$: I need to multiply the derivative of g – that is, $g'(x) = 6x$ – by the derivative of f – that is, $f'(\overline{x}) = 71\overline{x}^{70}$ – *where the variable* $\overline{x}$ *should be relocationd by* $g(x)$. If I do this in this way, I get:

$$(f \circ g)'(x) = \overbrace{6x}^{g'(x)} \overbrace{71 \Bigg(\underbrace{3x^2 + 1}_{\overline{x} = g(x)} \Bigg)^{70}}^{f'(\overline{x}) = f'(g(x))} = 426x(3x^2 + 1)^{70}. \blacksquare$$

◄

Example 6.12
The next example is $\sin^{27}(x)$. This would be easier here if the term in the parenthesis, $\sin(x)$, was just one variable $\overline{x}$, because I can easily compute the derivative of the power function $f(\overline{x}) = \overline{x}^{27}: f'(\overline{x}) = 27\overline{x}^{26}$. But unfortunately this is not the case here: Namely, the term in the parenthesis is the function $g(x) = \sin(x)$, but for which I at least know how to compute the derivative of it: $g'(x) = \cos(x)$. The chain rule shows me how to find the derivative of $(f \circ g)(x) = \sin^{27}(x)$: I have to multiply the derivative g' of g by the derivative f' of f, where the variable $\overline{x}$ should be relocationd by $g(x)$:

$$(f\circ g)'(x) = \overbrace{\cos(x)}^{\sin'(x)} \overbrace{27\left(\underbrace{\sin(x)}_{\overline{x}=\,\sin(x)}\right)^{26}}^{(\overline{x}^{27})'} = 27\cos(x)\sin^{26}(x). \quad \blacksquare$$

Incidentally, I can treat the example $\sin^2(x)$ discussed above in the context of the product rule in the same way: $(f\circ g)(x) = \sin^2(x)$, where $f(\overline{x}) = \overline{x}^2$ and $g(x) = \sin(x)$. Then, using the chain rule, I obtain – $g'(x) = \cos(x)$ and $f'(\overline{x}) = 2\overline{x}$ – $(f\circ g)'(x) = \cos(x)2\,(\sin(x))^1 = =$ $2\cos(x)\sin(x)$, and the result of course agrees with what I calculated above. ◀

Example 6.13

I can also compute the derivative of the function $(f\circ g)(x) = \exp.(\sin(x)) = e^{\sin(x)}$ using the chain rule. I set here $f(\overline{x}) = e^{\overline{x}}$ and $g(x) = \sin(x)$. I know the corresponding derivatives: $f'(\overline{x}) = e^{\overline{x}}$ and $g'(x) = \cos(x)$. Thus, $(f\circ g)'(x) = \cos(x)\,e^{\sin(x)}$.▪

Now it is your turn again. ◀

Exercise 6.8 Calculate the derivative of the following functions: a) $f(x) = (x^2 - 4x + 2)^{16}$, b) $f(x) = \exp.(x^2 - x)$, c) $f(x) = x^{n/m}$, where $x > 0$ and $n, m \in \mathbb{N}$, d) $f(x) = (1 - x^2)(x^2 + 1)^{-2}$.

Perhaps the question has just arisen, what about the derivative of the quotient f/g *of* differentiable functions f and g, defined by $(f/g)(x) = f(x)/g(x)$. *First*, of course, I have to assume here that for all x considered the denominator does not vanish: $g(x) \neq 0$, because otherwise the function f/g would make no sense there, because as we know I must *never never never* divide by 0. Fortunately, I can then determine the derivative of the quotient $f/g = f\cdot(g)^{-1}$ using the functions f, g, f', and g' by *combining* the product and chain rules. Here is how I see it: The product rule first gives

$$(f/g)' = \left(f\cdot(g)^{-1}\right)' = f'\cdot(g)^{-1} + \left((g)^{-1}\right)'\cdot f.$$

Here I already know f', $(g)^{-1} = 1/g$ and f – only how can I write down $((g)^{-1})'$ using g and g'? To do this, I now compute the derivative of $(g)^{-1}$ using the chain rule. To do this, I set $\widetilde{f}(\overline{x}) = \overline{x}^{-1}$ and use that $\widetilde{f}\circ g$ actually gives $(g)^{-1}$: $\left(\widetilde{f}\circ g\right)(x) = (g(x))^{-1}$. I see this by replacing $\overline{x}$ with $g(x)$. Fortunately, I know the derivative of $\widetilde{f}$ from our discussion right after Exercise 6.7: $\widetilde{f}'(\overline{x}) = -\overline{x}^{-2}$. Thus, by the chain rule, I get at all points x with $g(x) \neq 0$:

$$\left((g)^{-1}\right)'(x) = \left(\tilde{f} \circ g\right)'(x) = g'(x)\tilde{f}'(g(x)) = g'(x)\left(-(g(x))^{-2}\right) = -g'(x)(g(x))^{-2}.$$

Now I also know $((g)^{-1})' = -g' \cdot (g)^{-2}$ and may also use this above. The combination of product and chain rule thus results in

$$(f/g)' = f' \cdot (g)^{-1} + \left(-g' \cdot (g)^{-2}\right) \cdot f = f'/g - (g' \cdot f)/g^2 = (f' \cdot g - g' \cdot f)/g^2,$$

where I expanded the first term with g in the last step. That was not quite trivial (math.: easy). As a reward, however, I now know the following:

First Derivative of the Quotient of Functions *f/g* – Quotient Rule
If f and g are differentiable functions, their **quotient** f/g is a differentiable function for all x with $g(x) \neq 0$, and the first derivative $(f/g)'$ of f/g for these x is given by: $(f/g)'(x) = (f'(x)\, g(x) - g'(x)\, f(x))/(g(x))^2$.

Classics for the application of the quotient rule are **rational functions** $f = p/q$, where p and q are polynomial functions.

Example 6.14
An example is the function $f(x) = (x^2 + x)/(x^4 + 1)$, so $p(x) = x^2 + x$ and $q(x) = x^4 + 1$. For all $x \in \mathbb{R}$, $q(x) = x^4 + 1 \geq 1 > 0$ – so fortunately the denominator never vanishes in this example. Moreover, p and q are differentiable and the first derivatives of these polynomial functions are as follows: $p'(x) = 2x + 1$ and $q'(x) = 4x^3$. I use this to calculate

$$\begin{aligned} f'(x). &= \frac{p'(x)q(x) - q'(x)p(x)}{(q(x))^2} = \frac{(2x+1)(x^4+1) - 4x^3(x^2+x)}{(x^4+1)^2} \\ &= \frac{2x^5 + x^4 + 2x + 1 - 4x^5 - 4x^4}{x^8 + 2x^4 + 1} = \frac{-2x^5 - 3x^4 + 2x + 1}{x^8 + 2x^4 + 1}. \end{aligned}$$ ■

◄

Example 6.15
I think it is fun to compute the derivative of the function $k(x) = (x^2 + x - 1)/e^x$. I can do this for all x, because the denominator $g(x) = e^x$ is everywhere differentiable – $g'(x) = e^x$ – and moreover nowhere equal to 0. Moreover, the numerator $f(x) = x^2 + x - 1$ is also everywhere differentiable: $f'(x) = 2x + 1$. I calculate the derivative $k' = (f'\, g - g' \cdot f)/g^2$ of k using the quotient rule:

$$k'(x) = \frac{(2x+1)e^x - e^x(x^2+x-1)}{(e^x)^2} = \frac{(2x+1-x^2-x+1)e^x}{e^{2x}} = \frac{-x^2+x+2}{e^x}.$$

Here I have factored out e^x and then reduced to fraction: $e^x/e^{2x} = 1/e^x$.▪ ◀

Example 6.16

A somewhat more exotic example is formed by the **tangent function** tan(x) = sin(x)/cos (x) for $x \in (-\pi/2, \pi/2)$. As is well known, $\cos(x) \neq 0$ for these x. Moreover, the derivative of the numerator is $\sin'(x) = \cos(x)$, and the derivative of the denominator is $\cos'(x) = -\sin(x)$. I now use this to work out the derivative of tan(x) for $x \in (-\pi/2, \pi/2)$ using the quotient rule:

$$\begin{aligned}\tan'(x) &= \frac{\sin'(x)\cos(x) - \cos'(x)\sin(x)}{\cos^2(x)} = \frac{\cos(x)\cos(x) - (-\sin(x))\sin(x)}{\cos^2(x)} \\ &= \frac{\cos^2(x) + \sin^2(x)}{\cos^2(x)} = \frac{\cos^2(x)}{\cos^2(x)} + \frac{\sin^2(x)}{\cos^2(x)} = 1 + \tan^2(x). \quad \blacksquare\end{aligned}$$

◀

Exercise 6.9 Using the quotient rule, calculate the derivative of the following functions: (a) $f(x) = (x-4)/(2x^2+1)$, (b) $f(x) = (x+4)/(2x^2-1)$, (c) $f(x) = \sin(x)/x$, (d) $f(x) = (1-x^2)/(x^2+1)^2$ and (e) $f(x) = (-x^2+x+2)/e^x$. For which x it is allowed to calculate the derivative? Compare (d) with problem (6.8d) – do you notice anything?

6.2 Applications of Derivatives and Curve Sketching

I would now like to *discuss* special *curves* with you – more precisely, I would like to show you how to determine the graph of given functions in order to illustrate them. In the *curve sketching*, it is particularly important to identify characteristic points of the graph and to describe its behavior between and outside such points. For such an analysis of functions I apply derivatives – they are a powerful tool for this.

As I promised you at the beginning of the last section, I will first return here to the example of the hike in the "Palatinate Forest", which is shown in Fig. 6.1 as the graph of a function f. Looking at this figure, I notice that the value of the slope for this f is positive for all $x \in [a, x_3) \cup (x_4, x_5)$ (that is, $a \leq x < x_3$ and $x_4 < x < x_5$) – there it goes **monotonically** increasing in each case. On the other hand, the value of the slope is negative for all $x \in (x_3, x_4) \cup (x_5, b]$ (that is, $x_3 < x < x_4$ and $x_5 < x \leq b$) – where each is **monotonically** decreasing. Moreover, the value of the slope is 0 at the peak points $(x_3 \mid f(x_3))$ and $(x_5 \mid f(x_5))$, as well as at the trough point $(x_4 \mid f(x_4))$ – there it goes neither up nor down in each case. Moreover, I observe that *globally* the highest point of the walk is reached at the point x_3: $f(x) \leq f(x_3)$ for all $x \in [a, b]$ – the height value is **globally maximal** there. *Globally,* I have the lowest height at the beginning of the walk: $f(x) \geq f(a)$ for all $x \in [a, b]$ – the height value is **globally minimal** there. In particular, *considered globally,* the value $f(x_4)$ at the valley point $(x_4 \mid f(x_4))$ is larger than the value at the point a: $f(x_4) > f(a)$. However, *locally* this value $f(x_4)$ *is the* smallest: Indeed, in a small neighborhood of x_4, say in an interval of the form $(x_4 - c, x_4 + c)$, where $c > 0$ is chosen small enough, $f(x) \geq f(x_4)$ holds for all $x \in (x_4 - c, x_4 + c)$ – this height value is only **locally minimal.**

You know by now that I also call the value of the slope of f at a point x *the* first derivative of f in x, and denote it by $f'(x)$. In what follows, I will show you that for differentiable functions – roughly speaking – I can have the same discussion as I just had. For example, I will tell you how to find the locations of largest and smallest values of differentiable functions in the global and local sense. In addition, I will show you the connection between the type of slope and the sign of the derivative.

I think this is actually quite a good introduction, and so I will start with the connection of the sign of the derivative to the **monotonicity** of functions. Colloquially, you would possibly think here that this is about functions f *that* are *monotonic* – not to say particularly boring. But I will show you in a moment that this does not necessarily apply to such functions f – but rather to their first derivative f' – because their sign does not change for monotone functions f, as you will see in a moment.

In Chap. 2 you were told that a function f is **monotonically increasing** if it always follows from $x_1 < x_2$ that $f(x_1) \leq f(x_2)$. ***Strictly*** **monotonically increasing** is a function if from $x_1 < x_2$ it always follows that $f(x_1) < f(x_2)$. Similarly, the terms **monotonically decreasing** and ***strictly*** **monotonically decreasing** were also **explained** there. In Fig. 6.9 you can find some graphs of (strictly) monotone functions.

Of course, (strictly) monotone functions are not always differentiable. The (square) primitive is an example of a function which, as I showed you in the last section, is not differentiable in $x = 0$ and thus not differentiable at all. On the other hand, however, it is strictly monotonically increasing – to verify this, you can take a close look at its graphs in Figs. 6.5 and 6.9 (middle), or browse through Chap. 2 again. In the following I will concentrate mainly on differentiable functions, because for these I will be able to present you a simple criteria to test for (strict) monotonicity.

To give you a sense of the relationship between the sign of the derivative and monotonicity, consider, for example, a monotonically increasing derivative function f. The difference quotients are then always greater than or equal to 0:

$$\frac{f(x+h)-f(x)}{(x+h)-x} = \frac{f(x+h)-f(x)}{h} \geq 0.$$

Here is how I see it: For given x, $x < x + h$ holds *if* I choose $h > 0$ (small enough). If I set in my mind $x_1 = x$ and $x_2 = x + h$, then because f is monotonically increasing, it follows: f

$(x) \leq f(x+h)$. I can also write this down by subtracting $f(x)$ on both sides of the inequality as follows: $f(x+h) - f(x) \geq 0$. So in this case the numerator of the difference quotients is ≥ 0 and the denominator is $(x + h) - x = h > 0$. Consequently, the difference quotients themselves are ≥ 0. On the other hand, if I consider the remaining case $h < 0$, the inequality signs under consideration turn around, because now $x + h < x$, and I obtain $f(x + h) - f(x) \leq 0$. So then the numerator is ≤ 0 and the denominator $(x + h) - x = h < 0$ – consequently the difference quotients themselves are again ≥ 0. The first derivative $f'(x)$ of f in x now has no other chance: since this occurs as the limit of the difference quotients and these are always greater than or equal to 0, this one is also greater than or equal to 0:

$$f'(x) = \lim_{h \to 0} \overbrace{\left(\frac{f(x+h) - f(x)}{h}\right)}^{\geq 0} \geq 0.$$

I have thus shown you that the first derivative of monotonically increasing, differentiable functions is always greater than or equal to 0. Fortunately, the inverse also holds, as well as an analogy for monotonically decreasing functions.

Monotonicity and Sign of the First Derivative

A differentiable function f is monotonically increasing if and *only if* the derivative f' of f *is* nonnegative for all x: $f'(x) \geq 0$. A differentiable function f is monotonically decreasing if and *only* if the derivative f' of f *is* nonpositive for all x: $f'(x) \leq 0$.

I am looking at examples:

Example 6.17

How about $g(x) = x^2 - 2x$ for $x \geq 1$ (see Fig. 6.7(left))? I call the function g here for two reasons. First, I am showing you that I do not always need to call functions f. Second, I will look at this function g again in a later example, and for this one, as you will see there, it is better that I do not call this function f. I now briskly calculate the first derivative of g: $g'(x) = 2x - 2$. The function g is monotonically increasing *for* $x \geq 1$, because for these x we have: $g'(x) = 2x - 2 \geq 2 \cdot 1 - 2 = 0$. On the other hand, if I now look at the same function prescription for $x \leq 1$, I find that there $g'(x) = 2x - 2 \leq 0$ holds true – the function g is therefore monotonically decreasing *for* $x \leq 1$.

The type of monotonicity thus depends in particular on the points x for which I consider the function g. This is not so unusual – the hike in the "Palatinate Forest" has already shown you that the graph of a function generally consists of different **monotonicity regions.** ◀

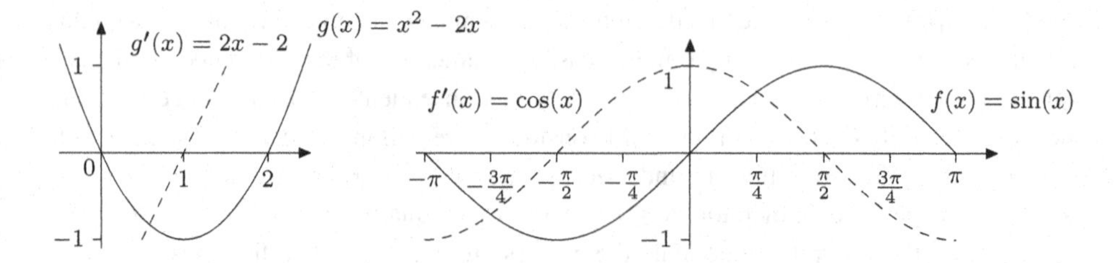

Fig. 6.7 The type of monotonicity often depends on the locations x considered. Left: $g(x) = x^2 - 2x$ is monotonically decreasing for $x \leq 1$ (there: $g'(x) \leq 0$) and monotonically increasing for $x \geq 1$ (there: $g'(x) \geq 0$). Right: the sine function $f(x) = \sin(x)$ is strictly monotonically increasing for $x \in [-\pi/4, \pi/4]$, because there: $\sin'(x) = \cos(x) > 0$. Also, considered on $[-\pi/2, \pi/2]$, $\sin(x)$ is strictly monotonically increasing, while this function is strictly monotonically decreasing on $[-\pi, -\pi/2]$ – thus on $[-\pi, \pi]$ $\sin(x)$ is *not* monotonic

Example 6.18

Another example is $f(x) = -\exp(x^3) = -e^{x^3}$ for $x \in \mathbb{R}$. I calculate the first derivative using the **chain rule**: $f'(x) = -3x^2e^{x^3}$. Because $-3x^2 \leq 0$ and $\exp(\overline{x}) = e^{\overline{x}} > 0$ hold (regardless of what I put in for $\overline{x}$, so also for $\overline{x} = x^3$), it follows $f'(x) \leq 0$ for all $x \in \mathbb{R}$. So this function is monotonically decreasing. Fortunately, here I do not have to *make a* case distinction as for g.

Surely you are wondering at this point whether there is a connection of the *strict* monotonicity with the sign of the first derivative. On this point, without dwelling on the more elaborate arguments, I can tell you the following: ◀

Strict Monotonicity and Sign of the First Derivative

A differentiable function f with $f'(x) > 0$ for all x is strictly monotonically increasing.
A differentiable function f with $f'(x) < 0$ for all x is strictly monotonically decreasing.

Of course, the function $g(x) = x^2 - 2x$ just considered is *strictly* monotonically increasing *for* $x > 1$, because there $g'(x) = 2x - 2 > 0$. A more delicate example of a strictly monotonically increasing function is $f(x) = \sin(x)$, for x from $[-\pi/4, \pi/4]$ (see Fig. 6.7 (right)). To see this, I form the derivative $f'(x) = \cos(x)$ – as shown in the last section – and find that $\cos(x) > 0$ for x from $[-\pi/4, \pi/4]$. See how incomparably easier this is than proving $\sin(x_1) < \sin(x_2)$ for all $-\pi/4 \leq x_1 < x_2 \leq \pi/4$ by possibly other means?

A classical example is the exponential function $f(x) = \exp.(x) = e^x$: This function (see Fig. 6.9 (right)) is strictly monotonically increasing, because $f'(x) = e^x > 0$ for all $x \in \mathbb{R}$.

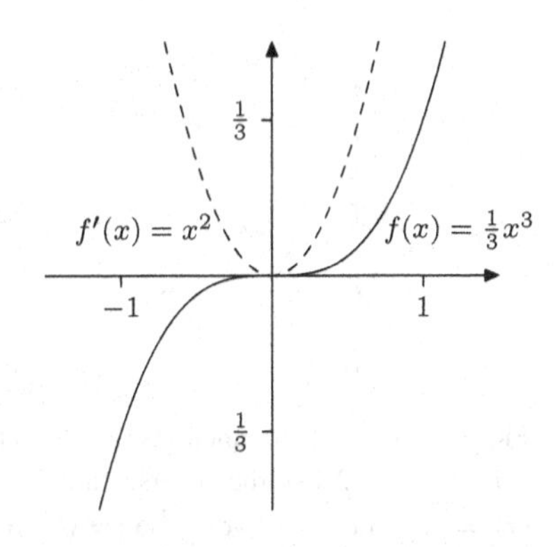

Fig. 6.8 The function $f(x) = 1/3\ x^3$ is strictly monotonic – but its derivative $f'(x) = x^2$ (*dashed*) vanishes at the point 0: $f'(0) = 0$. The point $x = 0$ is nevertheless not a local extremum of f, because f' *does* not change sign there

At this point I must point out that the converse of the last statement is not possible in general: there are *strictly* monotone functions whose derivatives satisfy only $f'(x) \geq 0$ and $f'(x) \leq 0$, respectively.

Example 6.19

The polynomial $f(x) = 1/3\ x^3$, $x \in \mathbb{R}$, is such an example (see Fig. 6.8). If $x_1 < x_2$, it follows that $f(x_1) = 1/3\ x_1^{\ 3} < 1/3\ x_2^{\ 3} = f(x_2)$ – this function f is thus *strictly* monotonically increasing. On the other hand, for its first derivative: $f'(x) = x^2 \geq 0$. Note here that "$\geq$" cannot be relocationd by "$>$", because at the *special location* $x = 0$, I calculate: $f'(0) = 0^2 = 0$. ◄

Since I am just chatting with you so nicely about *strict* monotonicity, it naturally lends itself here to say a few words about **invertible functions**, because in Chap. 2 it was explained to you that *strictly monotonic functions* are *always invertible*. So by now you also know again that a function f^{-1} is called an **inverse function** of f if the **concatenation** $f^{-1} \circ f$ yields the **identity**, i.e. $f^{-1}(f(x)) = x$ *for* all x considered. For simplicity, I now assume that either $f'(x) > 0$ or $f'(x) < 0$ holds for all x considered. According to what is written in the last box, this guarantees me the strict monotonicity of f and thus the existence of the inverse function f^{-1}. Now, assuming for a moment that I already know the derivative of f and hereby want to determine the derivative $(f^{-1})'$ of the inverse function f^{-1} of f, I can do this by applying the **chain rule**. For this I compute the derivative of the left and the right side in the equation $f^{-1}(f(x)) = x$:

$$f'(x)\left(f^{-1}\right)'(f(x)) = (x)'.$$

If in doubt, look again in the previous section – here I have now set $\overline{x} = f(x)$. Considering $f'(x) \neq 0$, I can now divide both sides by $f'(x)$, and then remembering $(x)' = 1$, I get the following statement:

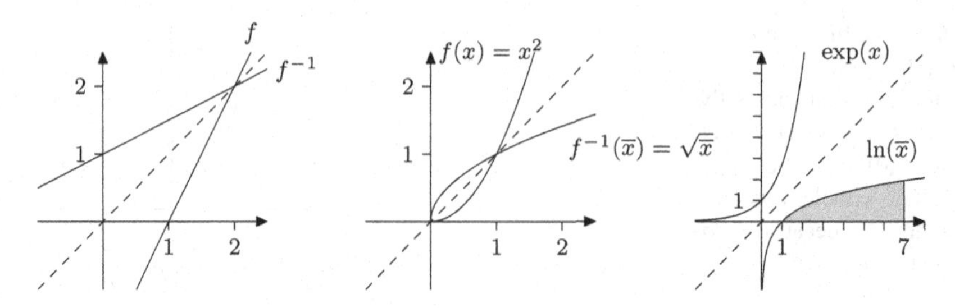

Fig. 6.9 Graphs of functions and the corresponding inverse function. Left: The linear function $f(x) = 2x - 2$ has the inverse function $f^{-1}(\overline{x}) = (1/2)\overline{x} + 1$. Middle: The inverse function of $f(x) = x^2$ is the primitive $f^{-1}(\overline{x}) = \sqrt{\overline{x}}$. Right: the exponential function $f(x) = \exp.(x) = e^x$ has the natural logarithm $f^{-1}(\overline{x}) = \ln \overline{x}$ as its inverse function. All these functions are strictly monotonically increasing

First Derivative of the Inverse Function
If f is a differentiable function with $f'(x) > 0$ or $f'(x) < 0$ for all x, then its inverse function f^{-1} is differentiable and the first derivative $(f^{-1})'$ of f^{-1} is given by: $(f^{-1})'(\overline{x}) = 1/f'(x)$, where and $\overline{x} = f(x)\, f^{-1}(\overline{x}) = x$, respectively.

In particular, I recognize that the sign of the derivative $(f^{-1})'$ of the inverse function f^{-1} coincides with the sign of the derivative f' of f. Thus, for example, if f is strictly monotonically increasing, then f^{-1} is also strictly monotonically increasing. Below I show you three examples of invertible functions and how to calculate the derivative of the corresponding inverse function in each case. You can see the graphs of the functions and their inverse functions in Fig. 6.9: As is well known, these merge into one another by reflection at the **first bisector** (the graph of the identity, dashed line).

Example 6.20
Simple examples of invertible functions are linear functions $f(x) = cx + d$ that are *not* constant: $c \neq 0$. I give you the inverse function directly: $f^{-1}(\overline{x}) = (1/c)\overline{x} - (d/c)$. For good measure, I put $\overline{x} = f(x)$ here and calculate $f^{-1}(f(x)) = (1/c)f(x) - (d/c) = =(1/c)(cx + d) - (d/c) = x$. Now, if I want to work out the derivative of the linear function f^{-1}, I can of course do this directly: $(f^{-1})'(\overline{x}) = 1/c$. Using the above statement, I get the same result – $f'(x) = c$ – as expected: $(f^{-1})'(\overline{x}) = 1/f'(x) = 1/c$. ◀

Example 6.21
Another classic is the quadratic power function $f(x) = x^2$ for $x \geq 0$ (see, for example, Fig. 6.4). From Chap. 2 you know that its inverse function is the (square) primitive $f^{-1}(\overline{x}) = \sqrt{\overline{x}}$: $\sqrt{x^2} = x^{2(1/2)} = x$ for all $x \geq 0$. However, the primitive is only

differentiable for $\overline{x} > 0$. If I want to *calculate* its derivative for $\overline{x} > 0$ using the derivative $f'(x) = 2\,x$ of f, I can do so as follows:

$$\left(\sqrt{\overline{x}}\right)' = (f^{-1})'(\overline{x}) = \frac{1}{f'(x)} = \frac{1}{2x} = \frac{1}{2\sqrt{\overline{x}}},$$

and so I get a result in a new way, which I already know from the previous section. Here I used at the last "="-character that $\overline{x} = x^2$, that is $\sqrt{\overline{x}} = x$. Note that because of $\overline{x} > 0$, $x > 0$ also holds. For $x = 0$, this approach does not work, because I would divide by 0 due to $f'(0) = 0$. It is not too much of a surprise that something has to go wrong here, because the primitive is – as I showed you in the last section – not differentiable in $\overline{x} = 0$.▪ ◀

Example 6.22

Another example is the exponential function $\exp.(x) = e^x$. This is invertible to all $\mathbb{R}$ because of $\exp'(x) = \exp.(x) > 0$. The inverse function of exp. is the **natural logarithm**: $\ln(\exp(x)) = x$ for all x. I now use my knowledge of the derivative of exp. (see previous section) to calculate the derivative of ln for $\overline{x} = \exp(x) > 0$:

$$\ln'(\overline{x}) = \frac{1}{\exp'(x)} = \frac{1}{\exp(x)} = \frac{1}{\exp(\ln(\overline{x}))} = \frac{1}{\overline{x}} = \overline{x}^{-1}.$$

In doing so, I used that $x = \ln(\overline{x})$ holds, and I benefited from the fact that exp. is the inverse function of ln, so $\exp(\ln(\overline{x})) = \overline{x}$ is also valid for all $\overline{x} > 0$.

So much about the connection between strict monotonicity and the determination of the derivative of the inverse function. As announced at the beginning, I now continue by showing you how to find the locations of largest and smallest values of differentiable functions in the global and local sense: ◀

Global Extrema

A location x_0 is called **global maximum of** f if for all x the value of $f(x)$ is less than or equal to the value of f in x_0: $f(x) \leq f(x_0)$. A location x_0 is called a **global minimum of** f if: $f(x) \geq f(x_0)$. If x_0 is a global maximum of f or a global minimum of f, then x_0 is called a **global extremum of** f.

One advantage of (strictly) monotone functions is that one can easily find their global extrema:

Global Extrema of Monotone Functions $f: [a, b] \mapsto \mathbb{R}$

If f is a monotone function given on a closed interval $[a, b]$: $f: [a, b] \mapsto \mathbb{R}$, then a and b are the global extrema of f.

I can easily see this statement. For example, if f is monotonically increasing, then $f(a) \leq f(x) \leq f(b)$ for all $a < x < b$, this means that the left boundary a is a global minimum of f and the right boundary b is a global maximum of f. Such an example is given by the function $f(x) = 1/3\ x^3$, $x \in [-1.1]$, from Fig. 6.8. This is even strictly monotonically increasing and it holds that $f(-1) = -1/3 < f(x) < 1/3 = f(1)$ for all $-1 < x < 1$: Thus $x_0 = -1$ is the global minimum and $x_1 = 1$ is the global maximum of this f. Similarly, I can argue for $f(x) = -\exp((1/3)x^3) = -e^{(1/3)x^3}$, $x \in [0.1]$. Above I had clarified together with you that this function is monotonically decreasing, and thus it follows that 0 is its global maximum and 1 is its global minimum: $f(0) = -1 > f(x) > -e^{1/3} = f(1)$. I have thus seen together with you that finding global extrema of (strictly) monotone functions is quite easy as *long as* they are considered on closed intervals $[a, b]$.

The search for locations with maximum and minimum function values becomes more difficult if the given function f is not monotonic as a whole, but rather consists of different monotonicity ranges. Unfortunately, this is the case with very many functions. The walk in the "Palatinate Forest" (see Fig. 6.1) is of this more difficult kind: there, as I mentioned at the beginning, x_3 inside $[a, b]$ is a global maximum, while the boundary point a is a global minimum. For such functions, where – casually expressed – the kind of monotonicity changes and locations in the interior of $[a, b]$ are candidates for (global) extrema, I generally have to argue more finely and take **local extrema** into account. In the "Palatinate Forest" example these are the locations x_3, x_4 and x_5.

Local Extrema

A point x_0 is called **local maximum of** f if I find *an interval* $(x_0 - c, x_0 + c)$ surrounding x_0 ($c > 0$ is a fixed number) such that for all x the value of $f(x)$ is not greater than the value of f in x_0: $f(x) \leq f(x_0)$, $x \in (x_0 - c, x_0 + c)$. The point $(x_0 \mid f(x_0))$ is then called a **local maximum of the graph of** f. A point x_0 is called a **local minimum of** f if: $f(x) \geq f(x_0)$ for all x from an interval surrounding x_0 $(x_0 - c, x_0 + c)$. The point $(x_0 \mid f(x_0))$ is then called a **local minimum point of the graph of** f. If x_0 is a local maximum of f or a local minimum of f, then x_0 is called a **local extremum of** f. The point $(x_0 \mid f(x_0))$ is then called the **extreme point of the graph of** f.

The magnitude function $|x|$ has a local minimum at $x = 0$. You can see this by studying its graph in Fig. 6.6 thoroughly. However, as I showed you in the previous section, this function is not differentiable in $x = 0$ – so the value of the derivative does not exist there.

The determination of local extrema becomes easier in general, if I consider differentiable functions, because then I can argue **analytically**. Our hike in the "Palatinate Forest" has already shown that the value of the slope at the summit and valley points (which are now called high and local minima) is 0 in each case. I merely made that clear above. *Since the value of the slope of a function f at a point x is just the value of the first derivative f′(x) at this point x*, I can now formulate more generally a criterion for finding local extrema, or the corresponding local maximum and minimum points, of differentiable functions f. For this I argue as follows: If, for example, there is a local maximum of f in x_0, then f is monotonically increasing "to the left of x_0", i.e. in an interval $(x_0 - c, x_0]$, and monotonically decreasing "to the right of x_0", i.e. in an interval $[x_0, x_0 + c)$. Thus, according to what I told you above about the relation between monotonicity and the sign of the first derivative, $f'(x) \geq 0$, $x \in (x_0 - c, x_0]$ and $f'(x) \leq 0$, $x \in [x_0, x_0 + c)$ hold. In particular, $f'(x_0) \geq 0$ and $f'(x_0) \leq 0$, so $f'(x_0) = 0$. I can argue analogously if x_0 is a local minimum (see Fig. 6.11). Thus, if x_0 is a local extremum, the tangent of f in x_0 has slope 0 and is therefore a constant function – this is called a **horizontal tangent**.

Necessary Criterion for Local Extremes

If a differentiable function f has a local extremum in x_0, then the first derivative of f vanishes there, that is, $f'(x_0) = 0$.

Example 6.23

For example, I now look at the **cubic polynomial**

$$f(x) = 1/4x^3 + 3/4x^2 - 3/2x - 2$$

whose graph I have shown as a solid arc in Fig. 6.10, and start looking for local extrema of f. To do this, I first calculate – as was shown more generally for **polynomial functions in** the last section – the first derivative:

$$f'(x) = 3/4x^2 + 3/2x - 3/2 = 3/4\left(x^2 + 2x - 2\right).$$

To find out at which points there might be local extrema, I set the first derivative equal to zero: $3/4\ (x^2 + 2\ x - 2) = 0$. I now remember the procedure for solving **quadratic equations**, in particular the $p - q$ **formula** from Chap. 3. Putting $p = 2$ and $q = -2$ there, I can work out the two solutions of the quadratic equation $f'(x) = 0$:

$$x_1 = -\frac{2}{2} + \sqrt{\frac{2^2}{4} - (-2)} = -1 + \sqrt{\frac{4}{4} + 2} = -1 + \sqrt{3} \quad \text{und} \quad x_2 = -1 - \sqrt{3}.$$

So, as *candidates* for local extrema of f I have identified the two real numbers x_1 and x_2. ◀

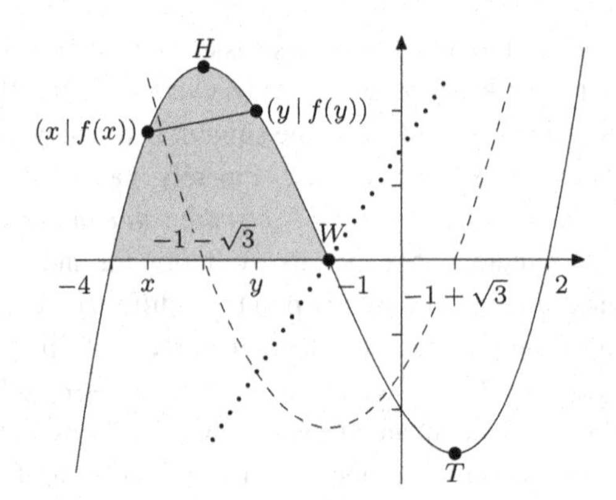

Fig. 6.10 The polynomial $f(x) = 1/4\, x^3 + 3/4\, x^2 - 3/2\, x - 2$ has a local maximum in $x_2 = -1 - \sqrt{3}$ and a local minimum in $x_1 = -1 + \sqrt{3}$. The first derivative $f'(x) = 3/4\, x^2 + 3/2\, x - 3/2$ (*dashed*) vanishes there, $f'(x_1) = f'(x_2) = 0$, *and* changes sign there in each case. If $x \leq -1$, then the graph of this function is concave, because for the second derivative (*dotted straight line*) holds: $f''(x) = 3/2\, x + 3/2 \leq 0$. If $x \geq -1$, then it is convex, because there holds: $f''(x) \geq 0$. In $x_0 = -1$, $f''(-1) = 0$ and $f'''(-1) = 3/2$ hold. Therefore, $x_0 = -1$ is a local minimum of f' and thus an inflection point of f. At the inflection point $W = (-1|0)$ the graph of f passes from concave to convex

But you see by this example, that I have the difficulty to determine quickly, *which kind of* local extremum (maximum or minimum) we have. Somehow I have a bad feeling at this point, because the above criterion only says that the derivative disappears at the local extrema – but here I started the other way round, so to speak, and determined the points with $f'(x) = 0$. But who now guarantees me that there are actually local extrema at these points? I will show you in a moment that for this function f I can quickly check that the locations x_1 and x_2 are indeed local extrema. But first I have to discuss the following example, which shows that the above criterion is *not sufficient* for the existence of a local extremum in x_0: Unfortunately, the condition $f'(x_0) = 0$ does *not* guarantee me in general that there is *indeed* a local extremum in x_0.

Example 6.24

To see this, consider an old acquaintance: the polynomial $f(x) = 1/3x^3$ from Fig. 6.8. Further above I had shown you that its first derivative $f'(x) = x^2$ vanishes (only) at the point $x_0 = 0$, while this function is also strictly monotonically increasing. This means that the point $(x_0|f(x_0)) = (0|f(0)) = (0|0)$ is neither a local maximum nor a local maximum of the graph of f, although $f'(0) = 0$ holds there. Back luck.

The disappearance of the first derivative is just a *necessary* criterion for local extrema: It must be fulfilled at the local extrema of differentiable functions. Conversely,

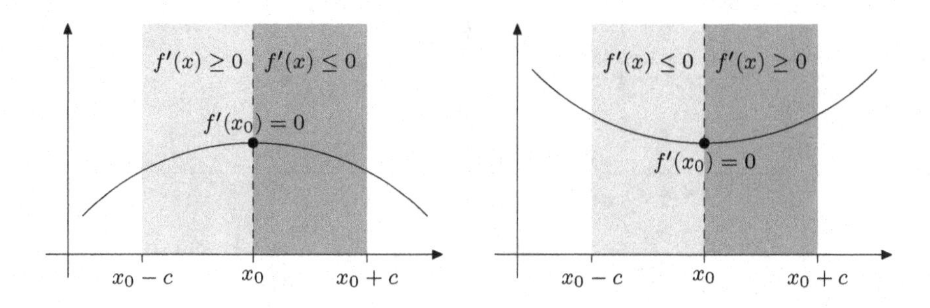

Fig. 6.11 If x_0 is a local extremum, the derivative disappears there: $f'(x_0) = 0$. If the first derivative f' has a sign change from + to − (*left*) there, it is a local maximum of f. If there is a change of sign from − to + for it there (*right*), then it is a local minimum of f

the disappearance of the first derivative at a point (i.e. $f'(x) = 0$) does not guarantee me that this is actually a local extremum.

In order to get a **sufficient criterion for** local extrema, I have to make sure that something like in the example just considered cannot happen. For this I remember again my descriptions of the hike through the "Palatinate Forest" at the beginning of this section. There the first derivative disappears at the locations x_3, x_4 and x_5, and I recognize that these are local extrema. *The reason is* that the value of the slope here **has a sign change in** a neighborhood of these locations. If I consider the location x_3, the first derivative f' changes its **sign from + to -**: in an interval surrounding x_3 $(x_3 - c, x_3 + c)$. Namely, there $f'(x) \geq 0, x \in (x_3 - c, x_3]$ and $f'(x) \leq 0, x \in [x_3, x_3 + c)$ hold. Thus, at the point x_3 there is indeed a local maximum. If I consider the location x_4, the first derivative f' changes its **sign from − to +** in an interval surrounding x_4 $(x_4 - c, x_4 + c)$. Namely, there $f'(x) \leq 0, x \in (x_4 - c, x_4]$ and $f'(x) \geq 0, x \in [x_4, x_4 + c)$ hold. Thus, at the point x_4 there is indeed a local minimum. I have illustrated this **sign change criterion** in Fig. 6.11. Incidentally, in my last example $f(x) = (1/3)x^3$ this is not satisfied at all: The first derivative $f'(x) = x^2$ is always ≥ 0 here – in particular, it does not change sign at the point $x_0 = 0$, which is a candidate for a local extremum. ◀

I now consider again – as promised – the above cubic polynomial $f(x) = 1/4x^3 + 3/4x^2 - 3/2x - 2$ from Example 6.23. As candidates for the local extrema I had calculated – using only the necessary criterion $-x_1 = -1 + \sqrt{3} \approx 0.73$ and $x_2 = -1 - \sqrt{3} \approx -2.73$. Because I want to know whether these are indeed local extrema, I now want to check whether there is a sign change of $f'(x) = 3/4\,(x^2 + 2x - 2)$ at these points. For this I consider the **decomposition into linear factors** (see Chap. 3).

$$f'(x) = 3/4(x - x_1)(x - x_2) = 3/4\left(x + 1 - \sqrt{3}\right)\left(x + 1 + \sqrt{3}\right)$$

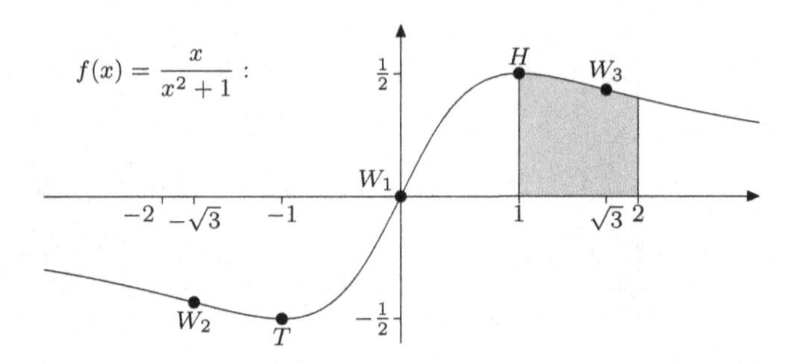

Fig. 6.12 The graph of the function $f(x) = x/(x^2 + 1)$. This has a maximum $H = (1|1/2)$ and a minimum $T = (-1|-1/2)$ and three inflection points $W_1 = (0|0)$, $W_2 = (-\sqrt{3}|-\sqrt{3}/4)$, $W_3 = (\sqrt{3}|\sqrt{3}/4)$

and determine the sign of the derivative as a function of the chosen x. If $x < x_2$, then because $x_2 < x_1$, $x < x_1$ also holds. So in this case $x - x_2 < 0$ as well as $x - x_1 < 0$ and hence $f'(x) > 0$. If $x_2 < x < x_1$, then $x - x_2 > 0$ as well as $x - x_1 < 0$ and hence $f'(x) < 0$. In the remaining case $x_1 < x$ I see in the same way that $f'(x) > 0$ holds. I thus realize: At x_2 the first derivative f' changes sign from + to −, while in x_1 there is a change of sign of f' from − to +. To see this sign behavior of f', you are also welcome to look again at Fig. 6.10: There, as an extra service, I have drawn you the graph of the quadratic polynomial $f'(x)$ in dashed lines. Now, according to the sufficient criterion above, it is clear that the location x_2 is indeed a local maximum and x_1 is indeed a local minimum of f. I have marked the corresponding local maximum and loacal minimum in Fig. 6.10 with *H and T*, respectively.

This was quite laborious, and the legitimate question arises, whether there might not often be a more convenient, sufficient criterion than the direct check for sign changes of the first derivative just shown in the example. Fortunately this is the case – but I need **second derivatives for** this. I will define this a bit more generally in a moment:

Higher Derivatives

A differentiable function f is called **twice differentiable** if the derivative f' of f is in turn differentiable. The derivative $(f')'$ of f' is then called the **second derivative of** f and denoted by f'' (read: "f two dash") or $f^{(2)}$. If $i \in \mathbb{N}$, a function f is called ***i*-times differentiable** if the $(i - 1)$-th derivative $f^{(i-1)}$ of f (exists and) is differentiable. The derivative $(f^{(i-1)})'$ of $f^{(i-1)}$ is then called the ***i*-th derivative of** f and denoted by $f^{(i)}$. If f is **differentiable *i*-times** *for all* $i \in \mathbb{N}$, then f *is* called **infinitely often differentiable**.

For the above cubic polynomial $f(x) = 1/4x^3 + 3/4x^2 - 3/2x - 2$ from Fig. 6.10, I have already calculated the first derivative: $f'(x) = 3/4x^2 + 3/2x - 3/2$. I recognize that f' is a polynomial of degree two and can in turn be differentiated according to our knowledge from the last section. The second derivative f'' of f is thus the linear polynomial

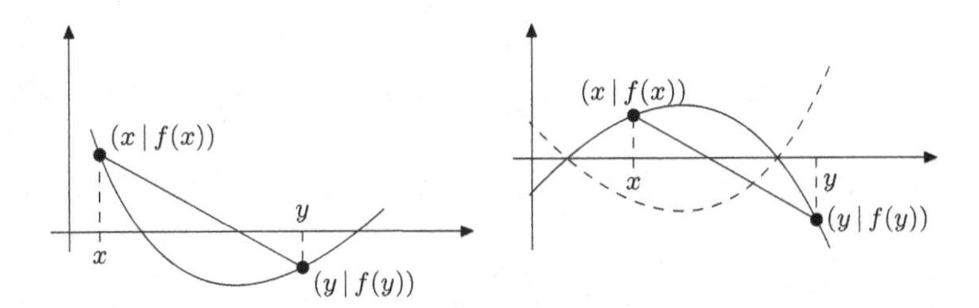

Fig. 6.13 The solid arcs represent convex (*left*) and concave (*right*) regions of the graph of f. If one mirrors at the *x-axis* (substitution of f by $-f$), a concave region becomes convex (*dashed arc*) and vice versa

$f''(x) = (f')'(x) = 3/2x + 3/2$ – whose graph I have illustrated for you as a dotted straight line in Fig. 6.10. Because I enjoy it, I compute the derivative of f again and get the third derivative of f: $f'''(x) = (f'')'(x) = 3/2$. Now I see that the fourth derivative of f **vanishes**: $f''''(x) = (f''')'(x) = 0$. So of course the fifth, sixth, indeed all subsequent derivatives of f vanish: $f^{(i)}(x) = 0$ for all $i \geq 4$. So I see that this polynomial can be differentiated infinitely many times. Something like this holds more generally:

The *i-th* Derivative $p^{(i)}$ of Polynomial Functions p

Every **polynomial function** p **of degree** n can be compute the derivative ofd infinitely often. For $i \in \{1, \ldots, n\}$ the *i-th* derivative $p^{(i)}$ of p is a polynomial of degree $n - i$ and for $i \geq n + 1$ the i-th derivative $p^{(i)}$ vanishes, that is $p^{(i)}(x) = 0$ for all x.

Without commenting further, I will show you the following example, where I compute the derivative ofd a polynomial of degree five seven times:

$$p(x) = x^5 + 3x^3 - 2x, \quad p'(x) = 5x^4 + 9x^2 - 2, \quad p''(x) = 20x^3 + 18x,$$
$$p^{(3)}(x) = 60x^2 + 18, \quad p^{(4)}(x) = 120x, \quad p^{(5)}(x) = 120, \quad p^{(6)}(x) = 0, \quad p^{(7)}(x) = 0.$$

Exercise 6.10 Determine all derivatives of the following polynomial functions: (a) $p(x) = 4x^4 - 6x^2 + e^8$, (b) $f_t(x) = t^3x^3 - tx^2 + tx + t^3$.▪

I can also form higher derivatives for many other functions that are not polynomials.

Example 6.25

An example is $f(x) = e^{x^2 - x}$. Because I did Exercise 6.8b, I know that here the first derivative for all $x \in \mathbb{R}$ is as follows: $f'(x) = (2x - 1)e^{x^2 - x}$. To calculate the second derivative of f I first apply the **product rule** and then use my knowledge about the first derivative of f again:

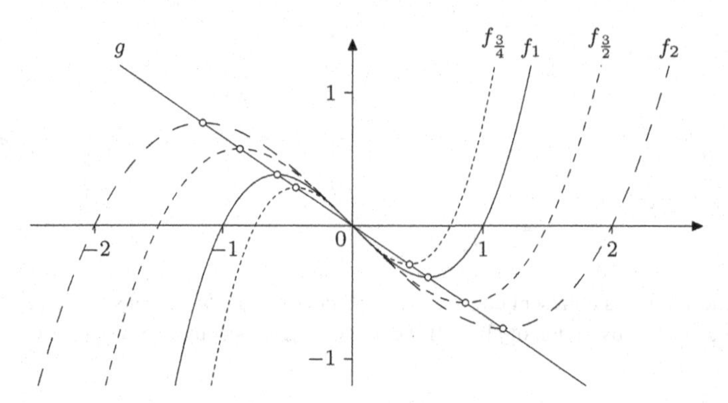

Fig. 6.14 The graph of the cubic polynomial $f_t(x) = 1/t^2 x^3 - x$ for various choices of the parameter t. The extreme points of f_t lie on the straight line g

$$f'(x)$$

$$= (2x-1)'e^{x^2-x} + \left(e^{x^2-x}\right)'(2x-1) = 2e^{x^2-x} + (2x-1)e^{x^2-x}(2x-1)$$

$$= \left(2+(2x-1)^2\right)e^{x^2-x} = (2+4x^2-4x+1)e^{x^2-x} = (4x^2-4x+3)e^{x^2-x}.$$

If I wanted to, I could now compute the next two derivatives of f''. ◄

Exercise 6.11 Determine f'' and f''' from $f(x) = e^{x^2-x}$.▪

If you look at it this way, you might think at first, superficial consideration that you can perhaps always form higher derivatives of (once) differentiable functions. However, this is – you guessed it – not the case. For example, look at the function $f(x) = x^{5/2}$ for $x \geq 0$. This function is differentiable, and I fulfill a promise made in the last section by telling you its derivative: $f'(x) = 5⁄2x^{3/2}$, $x \geq 0$. I also find the second derivative f'' effortlessly by applying the rough rule for forming the derivative of power functions from the last section: $f''(x) = (5/2)(3/2)x^{1/2} = 15/4\sqrt{x}$, $x \geq 0$. But note: I cannot now compute the derivative of f'' again for all $x \geq 0$, because here the primitive appears as a term, and this is known to be not differentiable in $x = 0$. This f is thus a function that is twice, but *not* three times, differentiable. In Exercise 6.4b you will find another such function.

Exercise 6.12 Show that

$$f(x) = \begin{cases} (-1/2)x^2, & \text{falls } x \geq 0, \\ 1/2x^2, & \text{falls } x < 0, \end{cases}$$

is differentiable once, but *not* twice.

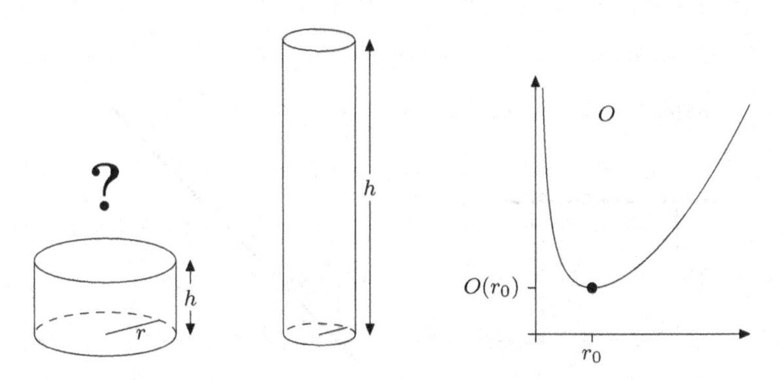

Fig. 6.15 The research manager of a food factory asks: For which choice of radius r is the lateral surface area $O(r)$ of a cylinder with a fixed volume minimal?

Although I have a lot of fun chatting about higher derivatives, I should not forget why I wanted to consider them. Namely, earlier I told you that using the second derivative avoids the somewhat tedious direct check for sign changes of the first derivative in the vicinity of locations x_0, which are candidates for local extrema (i.e.: $f'(x_0) = 0$). I can make this clear to myself as follows: The second derivative f'' of f is the (first) derivative of f' – if f *is* thus *twice differentiable* in x_0, then there exists $f''(x_0)$ – the limit of the difference quotients of f' in x_0 for h:

$$f''(x_0) = \lim_{h \to 0} \frac{f'(x_0 + h) - f'(x_0)}{h} = \lim_{h \to 0} \frac{f'(x_0 + h)}{h}.$$

The last "=" sign is valid here because x_0 is a candidate for a local extreme, so $f'(x_0) = 0$ holds. Now, for example, if $f''(x_0) < 0$, the difference quotients $f'(x_0 + h)/h$ for h close to 0 are less than 0 as well. If I now look at locations "to the left of x_0", i.e. $x_0 + h$, where $h < 0$, there $f'(x_0 + h) > 0$. On the other hand, if I look at locations "to the right of x_0", i.e. $x_0 + h$, where $h > 0$, there $f'(x_0 + h) < 0$. In summary, the first derivative f' in x_0 changes its sign from + to –, because in an interval surrounding x_0 $(x_0 - c, x_0 + c)$, $f'(x) > 0, x \in (x_0 - c, x_0)$ and $f'(x) < 0, x \in (x_0, x_0 + c)$ hold. Thus, by the above sign change criterion, there is a local maximum in x_0. Now asking myself again what I actually used here, I realize that here, in addition to $f'(x_0) = 0$, the central assumption is that the *sign of the second derivative in* x_0 is "negative": $f''(x_0) < 0$.

Sufficient Criterion for Local Extremes

If f is a twice differentiable function and x_0 is such that $f'(x_0) = 0$, then f has a local extremum in x_0 *if, in* addition, $f''(x_0) \neq 0$ holds. In this case, if then $f''(x_0) < 0$, then x_0 is a local maximum. If $f''(x_0) > 0$, then x_0 is a local minimum.

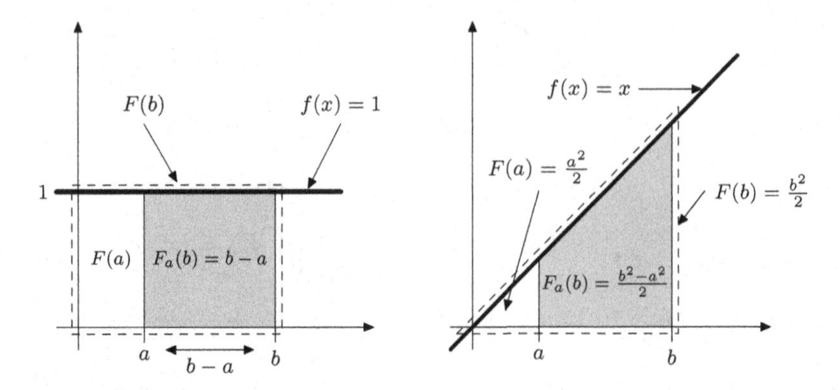

Fig. 6.16 Calculation of certain integrals for $f(x) = 1$ (*left*) and $f(x) = x$ (*right*). The graph of f *encloses* the *grey area* with content $F_a(b)$ when $[a, b]$ passes with the *x-axis*. This is just the difference of the contents of the *outlined* and the *white surface*: $F(b) - F(a)$

I look again at Example 6.23 – that is, the cubic polynomial $f(x) = 1/4\, x^3 + 3/4\, x^2 - 3/2\, x - 2$ from Fig. 6.10. Above I have already calculated the second derivative for this: $f''(x) = 3/2\,(x + 1)$. I now substitute the locations $x_1 = -1 + \sqrt{3}$ and $x_2 = -1 - \sqrt{3}$ into the second derivative and determine its sign:

$$f''(x_1) = 3/2\left(-1 + \sqrt{3} + 1\right) = \left(3\sqrt{3}\right)/2 > 0 \quad \text{and} \quad f''(x_2) = \left(-3\sqrt{3}\right)/2 < 0.$$

From the sufficient criterion just mentioned, it follows that x_1 is a local minimum of f and x_2 is a local maximum of f. This confirms in a more convenient way the result I showed you above in a relatively tedious way by directly checking the sign of the first derivative.

Example 6.26

A little more difficult is finding the local extrema of the fractional-rational function $f(x) = x/(x^2 + 1)$, whose graph I sketched for you in Fig. 6.12. To find out which locations are candidates for local extrema, I should form the first derivative. I do this using the **quotient rule** from the last section:

$$f'(x) = \frac{(x)'\,(x^2+1) - (x^2+1)'\,x}{(x^2+1)^2} = \frac{(x^2+1) - 2x \cdot x}{(x^2+1)^2} = \frac{1 - x^2}{(x^2+1)^2}.$$

Now I determine the locations where f' vanishes: $(1 - x^2)/(x^2 + 1)^2 = 0$. Because I can multiply both sides by the positive term $(x^2 + 1)^2$, I only have to look at the locations where the numerator vanishes: $1 - x^2 = 0$. I realize with my knowledge from Chap. 3 that this equation has the two solutions $x_0 = 1$ and $\tilde{x}_0 = -1$ – so: $f'(1) = 0$ and $f'(-1) = 0$. Now I want to *use* the second derivative f'' of f to check if these candidates are indeed local

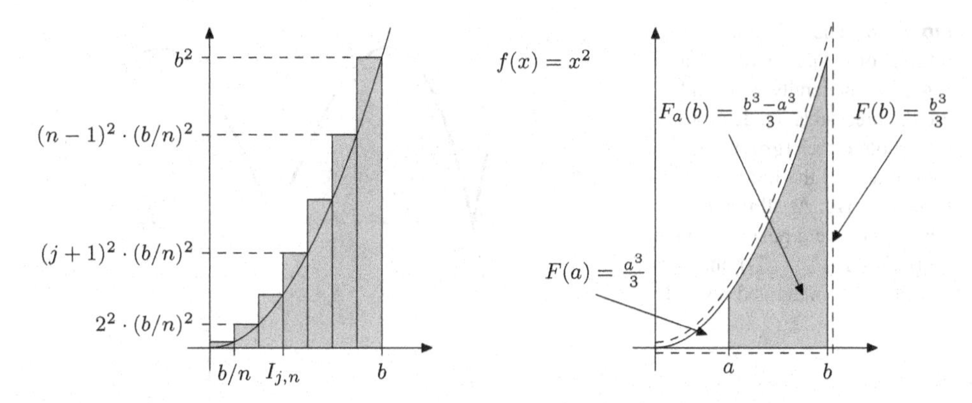

Fig. 6.17 Approximation of the area content $F(b)$ for $f(x) = x^2$ by using a piecewise constant function (*left*). The graph of this f encloses the *grey area* with content $F_a(b) = F(b) - F(a)$ – this is again the difference of the contents of the *outlined* and the *white area*(*right*) when $[a, b]$ passes the *x-axis*

extrema of f. Fortunately, I already know f'', because I am sure you did Exercise 6.8d or 6.9 d for me:

$$f''(x) = (f')'(x) = \left(\frac{1-x^2}{(x^2+1)^2}\right)' = \frac{2x(x^2-3)}{(x^2+1)^3}.$$

I now put $x_0 = 1$ and $\widetilde{x}_0 = -1$ here and get $f''(1) = (2(-2))/2^3 = -4/8 = -1/2 < 0$ as well as $f''(-1) = 1/2 > 0$. The sufficient criterion now says that there is a local maximum of f in $x_0 = 1$ and a local minimum in $\widetilde{x}_0 = -1$. If I wish, I can determine the corresponding maximum H and minimum T of the graph of f, respectively, by substituting $x_0 = 1$ and $\widetilde{x}_0 = -1$ into the function f: $H = (1|f((1))) = (1|1/2)$, $T = (-1|f(-1)) = (-1|-1/2)$. I have shown these as black dots in Fig. 6.12. ◀

Exercise 6.13 Show that $f(x) = e^{x^2 - x}$ has exactly one local extremum and determine the corresponding extreme point of the graph of f. Note that I have already given you f' and f'' above.

What does one actually do if for a candidate x_0 with $f'(x_0) = 0$ the sufficient criterion is not fulfilled – it could be that with a lot of bad luck I compute $f''(x_0) = 0$? Of course, as shown above, I can always try to argue with the changing of the sign of f'. But I will tell you something now that can also sometimes be used with success for higher differentiable, say i *times* differentiable, functions: If i is an *even number* and

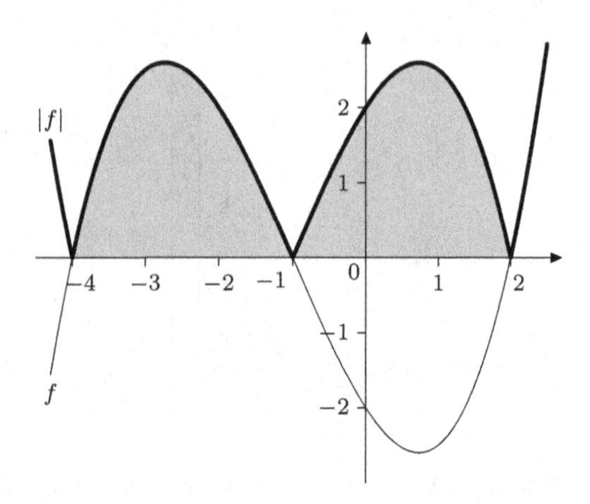

Fig. 6.18 The indefinite integral of f (*thin curve*, see also Fig. 6.10) generally gives only the oriented area. To determine the absolute area (*gray*), one must integrate the function $|f|$ (*broad curve*): At all points where $f(x)$ is not positive (in the figure: for $x \in [-1, 2]$), the function $-f$ is integrated instead of f

$$f'(x_0) = f''(x_0) = \cdots = f^{(i-1)}(x_0) = 0 \quad \text{sowie} \quad f^{(i)}(x_0) \neq 0$$

hold, then f has a local extremum in x_0. If $f^{(i)}(x_0) < 0$ holds, then x_0 is a local maximum of f. Otherwise, x_0 is then a local minimum of f.

Example 6.27

An example for which I can apply this criterion is $f(x) = 1/12x^4$. I calculate the first four derivatives:

$$f'(x) = 1/3x^3, \quad f''(x) = x^2, \quad f'''(x) = 2x, \quad f''''(x) = 2.$$

Now I note that $f'(0) = f''(0) = f'''(0) = 0$ as well as $f''''(0) = 2 > 0$ hold. Fine, $T = (0 \mid 0)$ is local minimum of the graph of this f.

But beware: a comparable criterion *cannot* work for *odd numbers i.* ◀

Example 6.28

I see this by considering the age-old familiar $f(x) = 1/3\, x^3$ from Fig. 6.8. I note for this one without difficulty that $f'(0) = f''(0) = 0$ as well as $f'''(0) = 2 \neq 0$ hold. The sufficient criterion just mentioned fails here, because $i = 3$ is odd. This is not surprising, since I have already shown you above that $x = 0$ is not a local extremum of this f, one can also convince oneself by intensively looking at Fig. 6.8. ▪

Now I think I have covered thoroughly enough the search for local and global extrema of differentiable functions, and now I propose to chat with you a bit more about **convexity** and **concavity of** graphs as well as **inflection points. Convex regions of** the graph of a function f are characterized by the fact that for every choice of two points $x < y$ the curve segment of the graph of f with endpoints $(x|f((x)))$ and $(y|f((y)))$ there always lies below the **secant** (known from the previous section) through these

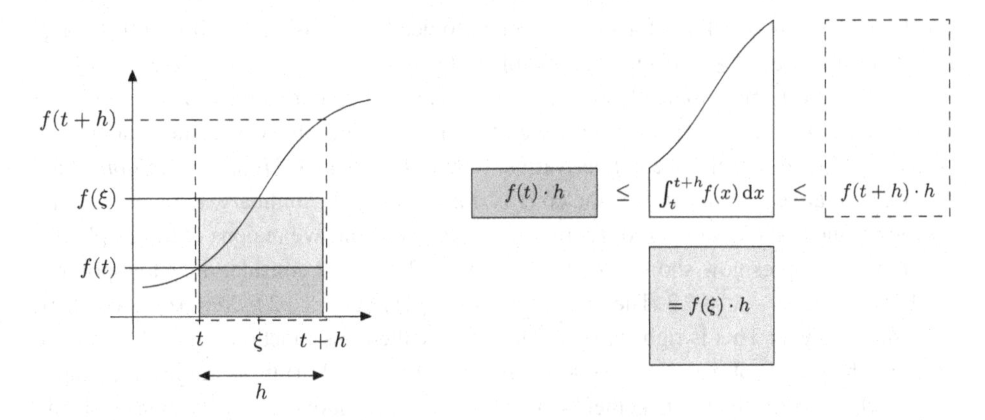

Fig. 6.19 The value of the integral $\int_t^{t+h} f(x)dx$ can be enclosed by the content of the rectangular areas with side length *h*(*grey and outlined*) (*top right*). If *f* is a *continuous* function, then there exists a rectangular surface with side lengths *h* and *f*(ξ) (*bottom right*) such that its content is *exactly* the value of the integral $\int_t^{t+h} f(x)dx$. This is the statement of the **mean value theorem of integral calculus**. The mean value $f((\xi)) \in [f((t)), f(t + h)]$ here is the function value of *f* at some (unspecified) point $\xi \in [t, t + h]$

points – I have illustrated this in Fig. 6.13 (left). Similarly, in the **concave** regions, all such curve segments lie above the corresponding secants (see solid arc in Fig. 6.13 (right)). **Inflection points of** *f* are points x_0, where the graph of *f* changes from a convex to a concave region or vice versa – the corresponding points $(x_0|f(x_0))$ are also called **inflection points**. In Fig. 6.10 I have labeled one such point *W*, while in Fig. 6.12 there are three inflection points – W_1, W_2, and W_3. For *twice* differentiable functions there is a connection between these geometrical terms (convexity and concavity) and the sign of the second derivative. ◀

Convexity and Sign of the Second Derivative

The graph of twice differentiable functions *f* is *convex* exactly if the second derivative f'' becomes nonnegative for all x: $f''(x) \geq 0$.

With the dashed arc in Fig. 6.13, I am trying to make it clear to you that the graph of a function *f* is concave exactly when the graph of –f is convex. This thus implies that the graph of twice-differntiable functions *f* is concave exactly if the second derivative f'' *does* not become positive for all x: $f''(x) \leq 0$. To understand this properly, you can make clear, if necessary, that according to the **factor rule** from the last section, $(-f)f'' = -f''$ always holds.

I can now find convex regions of the graph of a function *f* by noting that in these regions the first derivative f' of *f* is monotone. Our very first statement in this section says that a function – I will now call it g for short – is monotonically increasing if its first derivative g' *is* non-negative: $g'(x) \geq 0$. If I now apply this statement *to the first derivative* $g = f'$ of f, it follows that this is monotonically increasing exactly if its first derivative $(f')' = f''$ – i.e. the *second derivative of* f – is non-negative: $f''(x) \geq 0$. The above statement about the convexity and sign of the second derivative, however, says that *this sign condition of* the second derivative f'' characterizes the convex domains of f. In summary, the *monotonicity regions of* the first derivative f' of f form the convex and concave regions of the graph of f.

I think examples now show more than a thousand words. I consider the cubic polynomial $f(x) = 1/3\,x^3 - x^2 + 2/3$. The first derivative of f is $f'(x) = x^2 - 2\,x$. Maybe f' looks kind of familiar to you. That is right-in Example 6.17, I called this function g: $g = f'$. You can still find the graphs of g and its first derivative g' in Fig. 6.7 (left)-these are just the graphs of f' *and* f'', respectively. Together with you, I clarified above that g is monotonically decreasing for $x \leq 1$ and monotonically increasing for $x \geq 1$ – this means for f that $g'(x) = f''(x) \leq 0$ for $x \leq 1$ as well as $g'(x) = f''(x) \geq 0$ for $x \geq 1$. Thus for *the function* f I establish that the graph is concave for $x \leq 1$ as well as convex for $x \geq 1$.

Another example is the function $F(x) = -\cos(x) + 4$ for $x \in [-\pi/4, \pi/4]$. I calculate its first derivative F' using the knowledge from the last section as follows: $f'(x) = (-1)(\cos(x))' + (4)' = (-1)(-\sin(x)) + 0 = \sin(x)$. The fact that I choose a capital "F" as a designation here has – as I will explain to you later – on the one hand to do with the contents of the next section, and on the other hand, however, I do not want to confuse this with the function $f = F'$ whose graph can be seen in Fig. 6.7 (right), because F and $F' = f$ obviously differ. Further above I had already shown you that $F'' = f'$ *is* always greater than 0 in the interval under consideration $[-\pi/4, \pi/4]$ and therefore f is strictly monotonically increasing there. For F this means that its graph is convex there.

Our interest now turns to the points x_0, where the graph of f changes from a convex to a concave region and vice versa. These also form characteristic points of the graph of f which are different from the maximum and minimum points in general.

Inflection Points

A point x_0 is called an **inflection point of** f if f' has a local extremum in x_0. The point $(x_0|f(x_0))$ is then called the turning **point of the graph of** f.

The cubic polynomial $f(x) = 1/3\,x^3 - x^2 + 2/3$ from just now has the turning point $x_0 = 1$, because $f'(x) = x^2 - 2\,x$ has a local extremum in x_0. To do this, I look at $f''(x) = 2\,x - 2$ and effortlessly check that the necessary criterion $f''(1) = 0$ is satisfied. To see that this is indeed a local extremum of f', I use our convenient sufficient criterion. So I form the third derivative $f'''(x) = 2$ and find that $f'''(1) = 2 > 0$ holds. The point $x_0 = 1$ is thus a local minimum of f'. It follows that the point $W = (1|f((1))) = (1|0)$ is inflection point of f, and I even learn that the graph of f in W transitions from a concave to a convex domain. A similar example is discussed in the signature to Fig. 6.10.

Example 6.26 (continued)

The function $f(x) = x/(1 + x^2)$ illustrated in Fig. 6.12 has three inflection points. To see this, I first recall its second derivative: $f''(x) = 2x(x^2 - 3)/(x^2 + 1)^3$. The necessary condition $f''(x) = 0$ implies that $2x(x^2 - 3) = 0$ must hold. So I make out $x_1 = 0, x_2 = -\sqrt{3}$ and $x_3 = \sqrt{3}$ as candidates for the inflection points. I now need to check whether they are indeed local extrema of f'. I do this first by directly checking f'' with respect to sign changes. Using the **binomial formula,** I write the numerator $z(x) = 2x(x^2 - 3)$ of f'' as follows: $z(x) = 2x(x^2 - 3) = 2x(x + \sqrt{3})(x - \sqrt{3})$. Then I realize that $z(x) < 0$ holds if $x < -\sqrt{3}$ or $0 < x < \sqrt{3}$ and otherwise $z(x) \geq 0$. Moreover, the denominator of f'' is always greater than 0, and it follows that f'' has a sign change from $-$ to $+$ in x_2 and x_3, as well as a sign change from $+$ to $-$ in x_1. Thus x_2 and x_3 are local minima of f' and x_1 is a local maximum of f'. This means that the graph of f *changes* from a concave to a convex region in $x_2 = -\sqrt{3}$ and $x_3 = \sqrt{3}$ respectively, while it changes from convex to concave in $x_1 = 0$. In particular, $W_1 = (0|f((0))) = (0|0)$, $W_2 = (-\sqrt{3}|f(-\sqrt{3})) = (-\sqrt{3}|-\sqrt{3}/4)$ and $W_3 = (\sqrt{3}|f(\sqrt{3})) = (\sqrt{3}|\sqrt{3}/4)$ are inflection points of f. Instead of checking the sign of f'', I can of course use our "convenient" sufficient criterion here. To do this, I need the third derivative f''' of f. Okay – I should work these out. First, I use the **quotient rule**:

$$f'''(x) = \left(\frac{2x(x^2 - 3)}{(x^2 + 1)^3}\right)' = \frac{(2x(x^2 - 3))'(x^2 + 1)^3 - \left((x^2 + 1)^3\right)'(2x(x^2 - 3))}{(x^2 + 1)^6}.$$

So here I need to find the first derivative of $2x(x^2 - 3)$ and $(x^2 + 1)^3$. For the first term, this is easy: $(2x(x^2 - 3))' = (2x^3 - 6x)' = 6x^2 - 6 = 6(x^2 - 1)$. For the second term, I use the **chain rule**: $((x^2 + 1)^3)' = 6x(x^2 + 1)^2$. I now factor out the term $6(x^2 + 1)^2$ in the numerator of f''' and simplify it by multiplying out the remainder terms, using, among other things, the **binomial formula** $(x^2 - 1)(x^2 + 1) = x^4 - 1$:

$$6(x^2 + 1)^2\left((x^2 - 1)(x^2 + 1) - x(2x^3 - 6x)\right) = 6(x^2 + 1)^2(-x^4 + 6x^2 - 1).$$

Now f''' is obtained by cancelling the term $(x^2 + 1)^2$:

$$f''(x) = \frac{6(x^2+1)^2(-x^4+6x^2-1)}{(x^2+1)^6} = \frac{6(-x^4+6x^2-1)}{(x^2+1)^4}.$$

I now insert the inflection points in f'': $f''(0) = -6 < 0, f''(-\sqrt{3}) = f''(\sqrt{3}) = 15/64 > 0$ and thus confirm in an alternative way that these are indeed local extrema of f' – i.e. inflection points of f. Moreover, in this way I also see that the graph of f in $x_2 = -\sqrt{3}$ and $x_3 = \sqrt{3}$ changes from a concave to a convex region in each case, while this changes from convex to concave in $x_1 = 0$. I have also shown you this alternative approach because I want to point out that using the "more convenient" sufficient criterion can sometimes turn out to be more laborious than checking directly for sign changes. This is due to the fact that the determination of the third derivative sometimes requires quite complicated calculations.▪

The examples and my remarks on the last three pages show you that I may now use the necessary and sufficient criteria I formulated above initially for determining local extrema of differentiable functions f in the new context. *Since the inflection points of f are local extrema of the first derivative f' of f*, I can determine them by applying the above criteria to the function f'. I summarize what I have already developed: ◀

Necessary and Sufficient Criterion for Inflection Points

Necessary criterion:	If a twice differentiable function f has an inflection point in x_0, then the second derivative of f vanishes there, that is $f''(x_0) = 0$
Sufficient criterion:	If f is a three times differentiable function and x_0 is a location such that $f''(x_0) = 0$, then f has an inflection point in x_0 *if* moreover $f'''(x_0) \neq 0$ holds

At this point I come back to our ancient acquaintance $f(x) = 1/3x^3$, which is illustrated in Fig. 6.8. Earlier I had shown you that the point $x_0 = 0$ is *not a* local extremum of this f. In particular, I told you that $f'(0) = f''(0) = 0$ as well as $f'''(0) = 2 \neq 0$ hold. *Now* with this I immediately recognize that the point $W = (0|f((0))) = (0|0)$ is an inflection point of f. Such inflection points, where also the first derivative vanishes, are called **saddle points of** the graph of f, guided by the appearance of the **graph** there.

I summarize briefly: If the second derivative disappears at a point, I generally cannot yet assume that there is actually an inflection point there. I can check this quickly by looking whether the third derivative at this point is not equal to 0. Attention: Sometimes you have bad luck here.

Exercise 6.14 (a) Determine whether $f(x) = (1/12)x^4$ has an inflection point. (b) Show that $f(x) = e^{x^2 - x}$ has no inflection points.

I think that by now you have seen how to use derivatives and sketch curves. In particular, you have seen how to use derivatives to find characteristic points of the graph, that is, maxima, minima, and inflection points. I also showed you how to use derivatives to describe the behavior of the graph between and outside of such points. If you only use this knowledge, you can already successfully draw the graph of quite a lot of functions into the **coordinate system**. At this point, however, I would like to point out that there are a number of other ways to obtain additional information about the course of graphs besides using derivatives. In the following, I will describe such sources of additional information in brief, using examples of functions already discussed above as a guide:

Zeros of *f*

These are locations x_0, where the function has the value 0: $f(x_0) = 0$. In Chap. 3, the solution of equations was described to you – these methods often help here. I illustrate this with a couple of examples that occurred in this section. For the function $g(x) = x^2 - 2x$ shown in Fig. 6.7 (left), I find zeros as follows. I first factor out x: $x^2 - 2x = x(x - 2)$. Then I set the function term "equal to 0": $x(x - 2) = 0$. Now I can determine the zeros $x_0 = 0$ and $x_1 = 2$. For the sine function in Fig. 6.7 (right), I still know the (infinitely many) zeros from old school days: $\sin(k\pi) = 0, k \in \mathbb{Z}$. For the polynomial $f(x) = 1/4\,x^3 + 3/4\,x^2 - 3/2\,x - 2$ from Fig. 6.10, I guessed the first zero $x_0 = -1$: $f(-1) = 0$. Then – because I know I can now split off the **linear factor** $(x + 1)$ – I did the following approach:

$$1/4x^3 + 3/4x^2 - 3/2x - 2 = \left(ax^2 + bx + c\right)(x + 1).$$

If I calculate the right side here, I get for this side

$$ax^3 + (a + b)x^2 + (b + c)x + c,$$

and I can immediately read by **comparing coefficients** that $a = 1/4$ and $c = -2$ hold. Moreover, $a + b = 3/4$ and $b + c = -3/2$ must be satisfied and both fortunately work out for the choice $b = 1/2$. The remaining two zeros of

$$f(x) = 1/4x^3 + 3/4x^2 - 3/2x - 2 = \left(1/4x^2 + 1/2x - 2\right)(x + 1) = 1/4\left(x^2 + 2x - 8\right) \times (x + 1)$$

I can now calculate it using the ***p–q* formula** from Chap. 3 by solving $x^2 + 2x - 8 = 0$ (with $p = 2$ and $q = -8$): $f(-4) = f(2) = 0$. I do not have to worry too much about the function

$f(x) = e^{x^2 - x}$: it has no zero, because $exp(\overline{x}) = e^{\overline{x}}$ is always greater than 0, no matter what is used for $\overline{x}$. Finally, I look at the fractional-rational function $f(x) = x/(x^2 + 1)$ from Fig. 6.12. To see where this function vanishes, I set the function term equal to zero: $x/(x^2 + 1) = 0$. Because I can multiply both sides here by the positive term $x^2 + 1$, I only need to look to see where the numerator – x – vanishes. I can now easily see that $x_1 = 0$ is the only zero of this f.

Symmetry of the graph

The graphs of some functions f are **symmetric** to an axis i.e. a straight line $x = a$ parallel to the **y -axis** with distance a. If this is so, then for all x we have:

$$f(a + x) = f(a - x).$$

Graphically, this means that the axis $x = a$ is the mirror axis of the graph of f. An example is the function $f(x) = x^2$ (see Fig. 6.4). The graph of this function is asymmetric to the y-axis $x = 0$. I can recalculate this as follows:

$$f(0 + x) = f(x) = x^2 = (-x)^2 = f(-x) = f(0 - x).$$

Other examples of functions whose graphs are symmetric to the *y-axis* are the cosine function, shown dashed in Fig. 6.7 (right): $\cos(x) = \cos(-x)$, and the magnitude function $|x|$ from Fig. 6.6. Figure 6.7(left) illustrates the graph of the function $g(x) = x^2 – 2x$. This is symmetrical about the axis $x = 1$. I can see this by looking at

$$g(1 + x) = (1 + x)^2 - 2(1 + x) = 1 + 2x + x^2 - 2 - 2x = x^2 - 1$$

and

$$g(1 - x) = (1 - x)^2 - 2(1 - x) = 1 - 2x + x^2 - 2 + 2x = x^2 - 1$$

The result is – as you can see – the same. I can see this clearly by looking closely at Fig. 6.7 (left): $x = 1$ is the mirror axis of the graph of g.

The graphs of some functions f are **point-symmetric** with respect to a point $(a|b)$ in the coordinate system. If this is so, then for all x we have:

$$f(a + x) = -f(a - x) + 2b.$$

This means that the point $(a|b)$ is the mirror point of the graph of f. I leave it here with three examples in which the graphs are **point-symmetric to the origin**, that is, $(a|b) = (0|0)$. The graph of the function $f(x) = 1/3x^3$ shown in Fig. 6.8 has this property:

$$f(x) = 1/3x^3 = -\left(1/3(-x)^3\right) = -f(-x).$$

The sine function shown in Fig. 6.7 (right) also has this property: sin(x) = −sin(−x). For the fractional-rational function $f(x) = x/(x^2 + 1)$ from Fig. 6.12, I trace the point symmetry to the origin as follows. I calculate $f(-x)$ by substituting −x for x everywhere in the function term:

$$f(-x) = \frac{(-x)}{(-x)^2 + 1} = (-1)\frac{x}{x^2 + 1}$$

and now realize that this coincides with $-f(x)$. I see this vividly by intensive observation of Fig. 6.12: (0|0) is mirror point of the graph of f.

Behavior for large *x*

If a function f *is* defined for large x, I can think about how the function values $f(x)$ behave as x gets larger and larger. Thus one can investigate how the values of the function change in the limit transition

$$\lim_{x\to\infty} f(x)$$

for x **towards infinity** (in characters: $x \to \infty$). Before reading on, if necessary, please review what I told you in the last section about limits and their existence. In principle, this is also valid here.

Roughly speaking, two situations can occur again. First: A unique limit exists, that is, there is a real number S such that

$$\lim_{x\to\infty} f(x) = S.$$

This means that the function f for large x *approaches* more and more the **constant function** $a_f(x) = S$ defined by S. One then calls S the **limit value of** f and the graph of a_f **horizontal asymptote of** f. Second, the above bound does not exist – so it is either ∞ or else not uniquely definable. In this case there is no horizontal asymptote.

The function $f(x) = 1/x + 5$ has the horizontal asymptote $a_f(x) = 5$. Indeed, I can easily see that

$$\lim_{x\to\infty}(1/x + 5) = 0 + 5 = 5$$

holds. Reason: The larger x *is* chosen, the closer the term $1/x$ will be to 0, so $1/x$ goes towards 0 for $x \to \infty$. Another example is the fractional-rational function $f(x) = x/(x^2 + 1)$

from Fig. 6.12. To be able to perform the limit calculation here, I first factor out the term with the largest exponent in the numerator and denominator respectively and then truncate:

$$\frac{x}{x^2+1} = \frac{x}{x^2}\frac{1}{1+\frac{1}{x^2}} = \frac{1}{x}\frac{1}{1+\frac{1}{x^2}}.$$

Noting that 1⁄x and 1⁄x^2 go toward 0 for $x \to \infty$, I get from this

$$\lim_{x\to\infty}\frac{x}{x^2+1} = \lim_{x\to\infty}\frac{1}{x}\frac{1}{1+\frac{1}{x^2}} = 0\cdot\frac{1}{1+0} = 0.$$

I now realize that the graph of this f for large x *is* getting closer and closer to the graph of the **zero function** $a_f(x) = 0$. Figure 6.12 shows us this: The function values are closer and closer to the x-axis the larger x is *chosen.* If I now consider the function $f(x) = x^2/(x^2+1)$, I can proceed in the same way and obtain

$$\lim_{x\to\infty}\frac{x^2}{x^2+1} = \lim_{x\to\infty}\frac{1}{1+\frac{1}{x^2}} = \frac{1}{1+0} = 1.$$

You see: If the term with the largest exponent in the numerator and denominator is the same (in the example, this is x^2), then a real number also comes out when the limit is formed – $a_f(x) = 1$ is the horizontal asymptote in this example. But if I now consider the function $f(x) = x^3/(x^2+1)$ and proceed in the same way as before, I get

$$\lim_{x\to\infty}\frac{x^3}{x^2+1} = \lim_{x\to\infty} x\frac{1}{1+\frac{1}{x^2}} = \infty,$$

because x obviously goes to infinity for $x \to \infty$. For this example, there is no horizontal asymptote, because the function grows beyond any prescribed bound for large x. It is the same with the functions whose graphs are shown in Figs. 6.7 (left)–6.10. An example of the delicate kind is the sine function in Fig. 6.7 (right). This does *not* grow over all bounds, because $|\sin(x)| \leq 1$ holds for all x (compare the first section in Chap. 4), yet I cannot uniquely determine the limit of $\sin(x)$ for $x \to \infty$, because the values of the sine function always oscillate back and forth between -1 and 1, even for large x. *I have* an infinite number of sinusoidal functions. I have, so to speak, an infinite number of candidates for the limit (all values from $[-1,1]$) – but cannot decide on any.

To conclude this section, I consider two examples of curve sketching:

Example 6.29

The first example deals with the function set

$$f_t(x) = \frac{1}{t^2}x^3 - x,$$

where $t > 0$ is a fixed parameter of the cubic polynomial f_t with the real variable x. I first compute the zeros of f_t. For this I exclude x from the function term:

$$(1/t^2)x^3 - x = x((1/t^2)x^2 - 1).$$

Now I see that $x_0 = 0$ is a zero of f_t: $f_t(0) = 0$. Now I start looking for possible further zeros. To do this, I consider the eq. $(1/t^2)\, x^2 - 1 = 0$, because *a product of terms is* 0 *exactly when one of these terms vanishes*. If I add the number 1 on both sides here and then multiply both sides by t^2, I see that this equation can also be written as $x^2 = t^2$. Thus, the solutions are $x_1 = t$ and $x_2 = -t$. I can now write f_t as the product of linear factors:

$$f_t(x) = (1/t^2)x(x-t)(x+t).$$

I note in passing that the graph of f_t is point-symmetric with respect to the origin. To convince you, I recalculate this by determining $f_t(-x)$:

$$f_t(-x) = (1/t^2)(-x)^3 - (-x) = (-1/t^2)x^3 + x = -((1/t^2)x^3 - x)$$

and realize that this coincides with $-f(x)$. Now I am interested in local extrema of f_t. For this I form the first derivative (to x):

$$f_t'(x) = (3/t^2)x^2 - 1.$$

To find candidates for local extrema of f_t, I need to find the zeros of f_t': $(3/t^2)\, x^2 - 1 = 0$. Again, I add the number 1 on both sides, then multiply both sides by $t^2/3$, and have the equation in the following form: $x^2 = t^2/3$. Now I see that $x_3 = \sqrt{t^2/3} = (\sqrt{3}/3)\,t$ and $x_4 = -(\sqrt{3}/3)\,t$ are candidates for local extrema of f_t. Now I have to check if they are indeed local extrema. To do this, I form the second derivative of f_t (with respect to x):

$$f_t''(x) = (6/t^2)x.$$

Substituting gives: $f_t''(x_3) = (6/t^2)(\sqrt{3}/3)t = (2\sqrt{3}t)/t^2 = (2\sqrt{3})/t > 0$ and $f_t''(x_4) = (-2\sqrt{3})/t < 0$. Thus $T = (x_3|f_t(x_3)) = ((\sqrt{3}/3)t|(-2\sqrt{3}t)/9)$ is the local minimum of the graph of f_t and $H = (x_4|f_t(x_4)) = (-(\sqrt{3}/3)t|(2\sqrt{3}t)/9)$ is the local maximum of the graph of f_t. By the way, if I express the parameter t here by the *x-coordinates* of the local minima, i.e. $t = (3/\sqrt{3})x_3$, and then substitute this into the *y-coordinates of* the local minima $y_3 = (-2\sqrt{3}t)/9 = (-2\sqrt{3}(3/\sqrt{3})x_3)/9 = (-2/3)x_3$, I get a function $g(\overline{x}) = (-2/3)\overline{x}, \overline{x} > 0$, whose graph describes the location of all local minima – this is sometimes called the **locus of** the local minima. Do you see that $g(\overline{x}) = (-2/3)\overline{x}$ for $\overline{x} < 0$ is also the **locus of** the local maxima? Now I am still interested in the inflection points of f_t. To do this, I set the second derivative f_t'' equal to 0: $(6/t^2)\,x = 0$, and I realize that the point $x_0 = 0$ is a candidate for an inflection point. I now need to check whether it is indeed an inflection point. To do this, I form the third derivative $f_t'''(x) = 6/t^2$. This is obviously a constant function – no big surprise, because f_t is a cubic polynomial of the variable x. I realize that f_t''' is always – so also for the point x_0 – not equal to 0. Thus, the point $W = (0|f((0))) = (0|0)$ is the inflection point of f_t. In Fig. 6.14 I have illustrated the graph of f_t for various choices of parameters $t > 0$ together with the locus of the local maxima and minima. ▪ ◀

Exercise 6.15 Perform as complete a curve sketching as possible for the set of functions $f_t(x) = t^3/x^2 - t$, where $t > 0$ is a fixed parameter.

Example 6.30

The second example is of a more practical nature. Imagine you are the head of the research department of a food factory which, among other things, sells canned ravioli dishes. The filling volume of a cylindrical can should be one liter – that is what your long-standing customers want. For cost reasons, you ask yourself what height h and radius r (see Fig. 6.15 (left)) the cans should have so that sheet metal consumption is minimized. To answer this question you can use the methods developed here. First, to do this, remember the formula for the volume V of a cylinder: $V = \pi r^2 h$. By requiring that one liter should fit in the can ($V = 1$), I can now express the height h by the radius r: $h = 1/(\pi r^2)$. The consumption of sheet metal is determined by the lateral surface area O. As is well known, this can be determined for a cylinder as follows: $O = 2\pi r^2 + 2\pi rh$. If I put $h = 1/(\pi r^2)$ here, the function describes

$$O(r) = 2\left(\pi r^2 + \frac{1}{r}\right), \quad r > 0$$

the area of the lateral surface areaas a function of the radius r under the condition $V = 1$. My goal is to minimize this function. To do this, I first look for local minima. I thus calculate the first derivative of O with respect to the variable r:

$$O'(r) = 2(2\pi r - 1/r^2),$$

and try to figure out at which points it vanishes: $2\pi r - 1/r^2 = 0$. I now expand the first term with r^2 and get $(2\pi r^3-1)/r^2 = 0$, which means $2\pi r^3-1 = 0$. Solving for r gives a candidate local extremum: $r_0 = \sqrt[3]{1/(2\pi)}$. Now I need the second derivative of O:

$$O''(r) = 4(\pi + 1/r^3)$$

and there I use my candidate r_0 to check:

$$O''(r_0) = 4(\pi + 1/r_0^3) = 4(\pi + 2\pi) = 12\pi > 0.$$

Lucky: $r_0 = \sqrt[3]{1/(2\pi)} = (2\pi)^{-1/3} \approx 0,54$ is indeed a local minimum. The corresponding minimum T is $T = (r_0|O(r_0)) = (\sqrt[3]{1/(2\pi)}|\sqrt[3]{2\pi})$, because

$$O(r_0) = 2\left(\pi r_0^2 + \frac{1}{r_0}\right) = 2\pi(2\pi)^{-\frac{2}{3}} + 2(2\pi)^{\frac{1}{3}} = 3(2\pi)^{\frac{1}{3}} \approx 5,54.$$

The height h_0 belonging to r_0 is $h_0 = 1/(\pi r_0^2) = 2^{2/3}\pi^{-1/3} \approx 1.08$. For this function O it is now clear that T is even the only *global minimum*, because otherwise I would have to have encountered a local maximum somewhere – similar to the "Palatinate Forest" example. But to be on the safe side, I will check how the function O behaves for $r \to \infty$ and $r \to 0$ – i.e. towards the edge of its definition range. Since the term r^2 grows for $r \to \infty$ over all bounds – i.e. becomes infinitely large – it follows that

$$\lim_{r\to\infty} O(r) = \lim_{r\to\infty} = 2\left(\pi r^2 + \frac{1}{r}\right) = \infty.$$

Moreover, since the term $1/r$ for $r \to 0$ also grows over all bounds, it follows.

$$\lim_{\substack{r\to 0\\ r>0}} O(r) = \lim_{\substack{r\to 0\\ r>0}} 2\left(\pi r^2 + \frac{1}{r}\right) = \infty.$$

This means in particular that towards the edge of the definition range of O there are certainly no locations with smaller values of O as in r_0. I have shown the approximate course of the graph of O for you in Fig. 6.15 (right). In summary, my recommendation is thus to use cans with a radius of about 5.4 cm and the height of about 10.8 cm. The area of the sheet used per can is then about 55.4 cm^2.▪ ◀

You probably do not work in the research department of a food factory – but maybe you are considering building a house:

Exercise 6.16 For your new villa, you need a rectangular building site with an area of 2 square kilometers. Since the costs of the villa are immense, you start thinking about where you could make savings. After thinking about it for a while, you come up with the idea of minimizing the length of the fence around the plot you want to choose – its circumfere. What lengths would the sides of this rectangular plot ideally have to be?▪

6.3 Integration of Functions

Now that you have learned about the applications of derivatives, I will show you another key mathematical tool for analyzing real-valued functions: **Integration**. This allows you to determine the area under their graphs for many functions. Interestingly, integration and the derivative of functions are, in a sense, inverses of each other. I am not alone in thinking that this was certainly one of the fundamental discoveries (I deliberately do *not* say "invention") in mathematics. We owe this essentially to two mathematicians who lived in the seventeenth century: Gottfried Wilhelm Leibniz and Sir Isaac Newton.

I start with two very straightforward examples. In Fig. 6.16 (left) you see the graph of the **constant** function $f(x) = 1$. If x traverses an interval $[a, b]$, the corresponding part of the graph of this f encloses with the *x-axis* a rectangle area whose content $F_a(b)$ is obviously equal to $b - a$. Since the dashed outlined rectangle has area $F(b) = b$ and the content of the area of the white rectangle with base side $[0, a]$ is obviously $F(a) = a$, we can express $F_a(b)$ as the difference of these area contents: $F_a(b) = F(b) - F(a)$. A little more difficult – but not really difficult – is the computation of such enclosing areas for the **identity** $f(x) = x$. If x traverses the interval $[a, b]$, the corresponding part of the graph of this f encloses with the *x-axis* an area $F_a(b)$ whose content is $1/2 b^2 - 1/2 a^2$. Indeed, in Fig. 6.16 (right) you can see that this area is given by the difference of the content of the area $F(b)$ of the dashed outlined triangle with base side $[0, b]$ with the content of the area $F(a)$ of the white triangle with base side $[0, a]$. Since $F(b)$ is just half the area of a square with side length b, $F(b) = 1/2\, b^2$. Similarly, I see that $F(a) = 1/2\, a^2$ holds.

My choice of the names $F_a(b)$, $F(b)$ and $F(a)$, which may seem a bit strange at first sight, has a good reason, which will become clear to you in the course of the following pages. It will certainly be the same with the use of the **integral sign**, with the help of which I already formulate here our first two, just obtained results for the calculation of **certain integrals:**

$$F_a(b) = \int_a^b dx = F(b) - F(a) = b - a \text{ and } F_a(b) = \int_a^b dx = F(b) - F(a) = \frac{b^2}{2} - \frac{a^2}{2}.$$

More interesting – but also more difficult – is the determination of the area for functions whose graph is a curved curve. The first example that comes to mind is the **quadratic power function** $f(x) = x^2$. Figure 6.17 (right) shows that the content of the enclosing area $F_a(b)$ of the graph of this f on $[a, b]$ with the *x-axis* (gray) is given by the difference of the contents of the enclosing area $F(b)$ of this graph with respect to $[0, b]$ (dashed outlined) and the enclosing area $F(a)$ of this graph with respect to $[0, a]$ (outlined white area). Since the graph of f is curved, I cannot calculate these areas $F(b)$ and $F(a)$ directly as easily as in the examples above. I therefore first try to calculate one of these two surface areas – say $F(b)$ – *approximately*. I do this by approximating f *by* **piecewise constant functions** (see Fig. 6.17 (left)), because I can easily calculate their area contents as above. To do this, I divide the interval $[0, b]$ into n subintervals $I_{j,n}$, where $j \in \{0, \ldots, n-1\}$. Testwise, I choose $I_{j,n} = [j\,(b/n), (j+1)\,(b/n)]$. The length of $I_{j,n}$ is $(j+1)\,(b/n) - j\,(b/n) = b/n$. So the larger n is, the closer this length is to 0. Now in each $I_{j,n}$ I choose a location and look at the function value of f there. As a test, I choose the value at the right edge of the interval of $I_{j,n}$: $f((j+1)\,b/n)$. The content of the area under the graph of f with respect to $I_{j,n}$ is there *approximately* the content of the area of the graph of the constant function with value $f((j+1)\,b/n) = (j+1)^2\,(b/n)^2$. The content of the associated rectangular area is obviously $(j+1)^2\,(b/n)^3$, because the side lengths of this rectangle are b/n *and* $(j+1)^2\,(b/n)^2$. The sum of the areas $\mathcal{F}_n$ of *all* these rectangles gives an approximation to the area $F(b)$ we are looking for. I can write down $\mathcal{F}_n$ as follows:

$$\mathcal{F}_n = 1^2\left(\frac{b}{n}\right)^3 + 2^2\left(\frac{b}{n}\right)^3 + \cdots + (j+1)^2\left(\frac{b}{n}\right)^3 + \cdots + n^2\left(\frac{b}{n}\right)^3$$

$$= \left(1 + 4 + \cdots + (j+1)^2 + \cdots + n^2\right)\left(\frac{b}{n}\right)^3 = \sum_{j=0}^{n-1}(j+1)^2\left(\frac{b}{n}\right)^3$$

For small n as in Fig. 6.17 (left) – there I show the case $n = 6$ – this approximation to $F(b)$ is still very rough. The hope, however, is that I can thus *approximate* the content $F(b)$ *arbitrarily precisely* by choosing very large n. You should now believe me that the sum of the first n squares $1 + 4 + \cdots + (j+1)^2 + \cdots + n^2$ can be written simplified by the following formula: $(n(n+1)(2n+1))/6$. To prove this, I could use **complete induction.** I will leave that for today, because you will probably have to deal with this proof technique later. Using the simplifying formula, we now get

$$\mathcal{F}_n = \frac{n(n+1)(2n+1)}{6}\left(\frac{b}{n}\right)^3 = \frac{b^3}{6}\left(1+\frac{1}{n}\right)\left(2+\frac{1}{n}\right),$$

because n can be truncated away once and $(n+1)/n = 1 + 1/n$ as well as $(2n+1)/n = 2 + 1/n$. If I now choose larger and larger n, i.e. finer and finer subdivisions into subintervals, I hope that the approximate contents $\mathcal{F}_n$ will approximate the sought area $F(b)$ better and better. Fortunately, my hope is fulfilled here: The **boundary transition** for $n \to \infty$ leads to a *unique* **boundary value**

$$\lim_{n\to\infty}\mathcal{F}_n = \lim_{n\to\infty}\frac{b^3}{6}\left(1+\frac{1}{n}\right)\left(2+\frac{1}{n}\right) = \frac{b^3}{6}\lim_{n\to\infty}\left(1+\frac{1}{n}\right)\left(2+\frac{1}{n}\right) = \frac{b^3}{6}2 = \frac{b^3}{3},$$

because $1/n$ approaches 0 for $n \to \infty$. $F(b)$ here is indeed equal to $b^3/3$. By mentally *replacing* b by a, I can see in exactly the same way that $F(a)$ has the value $a^3/3$. Thus it follows from the initial reasoning that the content of the enclosing surface $F_a(b)$ of the graph of $f(x) = x^2$ on $[a, b]$ with the *x-axis* has the value $F(b) - F(a) = b^3/3 - a^3/3$. I am sure you would not mind if I formulate this result in the above form to calculate the definite integral of $f(x) = x^2$:

$$F_a(b) = \int_a^b x^2 dx = F(b) - F(a) = \frac{b^3}{3} - \frac{a^3}{3}.$$

A little practice certainly cannot hurt now:

Exercise 6.17 (a) Determine the content of the area enclosed by the graph of $f(x) = x^2$ as $[1, 3]$ passes with the *x-axis*. (b) Determine the location $t > 1$ such that the graph of f from (a) encloses an area with content $7/3$ as $[1, t]$ passes with the *x-axis*.

Surely you have noticed some regularity in the formation of the previous definite integrals. You see what I am getting at when I reveal that $\int_a^b x^3 dx = b^4/4 - a^4/4$ and $\int_a^b x^4 dx = b^5/5 - a^5/5$ apply. So the *rough rule* for finding the definite integral of a general **power function** $p_i(x) = x^i$ with respect to $[a, b]$ is, *in a sense*, the *reverse of the* rough rule of differentiating (see the first section in this chapter) such functions. It is: Increase the exponent by one and divide by this new exponent $x^{i+1}/(i + 1)$ – then put b and a into the resulting term and form the difference.

Integration of the Power Function $p_i(x) = x^i$, Where $i \in \mathbb{N} \cup \{0\}$

The definite integral of the **power function** $p_i(x) = x^i$ with respect to $[a, b]$ is given by

$$\int_a^b x^i dx = \frac{b^{i+1}}{i+1} - \frac{a^{i+1}}{i+1} = \frac{b^{i+1} - a^{i+1}}{i+1}.$$

Of course, the calculation of definite integrals is not limited to power functions. I think it is already time to tell you what is generally understood by the **integrability of** functions and what a **definite integral of** a real-valued function is:

Integrability and Definite Integral

A function f is said to be **integrable** on $[a, b]$ if for each subdivision of $[a, b]$ into n subintervals $I_{j,n} = [x_{j,n}, x_{j+1,n}]$ whose lengths $x_{j+1,n} - x_{j,n}$ go towards 0 for $n \to \infty$ and for each choice of points $\widetilde{x}_{j,n} \in I_{j,n}$, where $j \in \{0, \ldots, n-1\}$, the limit of sequence

$$\mathcal{F}_n = f(\widetilde{x}_{0,n})(x_{1,n} - x_{0,n}) + \cdots + f(\widetilde{x}_{j,n})(x_{j+1,n} - x_{j,n}) + \cdots + f(\widetilde{x}_{n-1,n}) \times (x_{n,n} - x_{n-1,n})$$

exists for $n \to \infty$ and is equal in each case. In this case the limit is called

$$\int_a^b f(x)dx = \lim_{n\to\infty} \mathcal{F}_n = \lim_{n\to\infty} \sum_{j=0}^{n-1} f(\widetilde{x}_{j,n})(x_{j+1,n} - x_{j,n})$$

(read: "integral from a to b over f") the **definite integral of f with respect to** $[a, b]$.

I suspect that this definition is not necessarily immediately as clear as daylight and intuitively perfectly comprehensible to the average person skimming over it for the first time. Even though you almost certainly share my view that this is a clear exposition of the notion of integrability of functions, I now feel compelled to provide some more detailed explanations.

The integral sign only appears here when a special limit value is named. Thus I must – as always, if it is about limit-values – pay attention, that this limit-value *exists* (is a *real* number) and is *unambiguously fixable* (no choice of several candidates exists). On closer inspection, you can then see what kind of limit it is. Similar to my (somewhat simplified) example with $f(x) = x^2$, the $\mathcal{F}_n$ are sums of area contents of rectangles of side lengths $x_{j+1,n} - x_{j,n}$ (the length of the interval $I_{j,n}$) and $f(\widetilde{x}_{j,n})$ (a function value of f at a point $\widetilde{x}_{j,n}$ from $I_{j,n}$). These area values are thus $f(\widetilde{x}_{j,n})(x_{j+1,n} - x_{j,n})$. The sums $\mathcal{F}_n$ are approximations to the contents of the area enclosed by the graph of f with the *x-axis* as it passes through $[a, b]$. But here I consider *all subdivisions* of $[a, b]$ into n subintervals $I_{j,n}$ of length $x_{j+1,n} - x_{j,n}$, which are closer and closer to 0 as n increases. Even more: At the same time, for each such subdivision, I look at *all* possibilities to choose a function value of f at a point $\widetilde{x}_{j,n}$ from $I_{j,n}$. Now, if for all such choices of subdivisions and for all choices of locations the resulting approximate contents $\mathcal{F}_n$ in the limit for $n \to \infty$ lead to a *definite real* limit $\lim_{n\to\infty} \mathcal{F}_n$, then f *is* called integrable on $[a, b]$, and the limit itself is called the definite integral of f *with* respect to $[a, b]$. The hope, then, is that this will be approximated arbitrarily precisely for all approximate contents $\mathcal{F}_n$ thus formed.

Leibniz was faced with the (mathematical) world-changing question of how he could then best designate this limit value. The limit value arises from sums. Therefore it is obvious to use a similar sign for the limit value as the sum sign $\sum$: $\int$. Now you know the *meaning of the* **integral sign** $\int$. The individual terms $f(\widetilde{x}_{j,n})(x_{j+1,n} - x_{j,n})$ of the sums $\mathcal{F}_n$ are areas of rectangles. As n gets larger and larger, the *difference* $x_{j+1,n} - x_{j,n}$ should get smaller and smaller, and you will have chosen *almost all* x as a $\widetilde{x}_{j,n}$ in this process. Therefore, it seems reasonable to use symbols in the limit label for this that intuitively reflect this process: $f(x)\,dx$. Finally, one puts *a at the* bottom and b at the top of the $\int$-sign, so that one does not forget over which interval one wants to integrate f. Thus, one arrives at the overall natural notation $\int_a^b f(x)dx$ for the limit of $\mathcal{F}_n$ for $n \to \infty$. By the way, f is then called the **integrand** (also: **integrand function**), x the **integration variable** and a, b are the **integration constants**.

You may still have the slightly unsettling feeling that integrability of functions is a difficult thing. After all, one would have to consider *all* subdivisions of subintervals $I_{j,n}$ with smaller and smaller lengths and at the same time *every* selection of values of $f(\widetilde{x}_{j,n})$ at locations $\widetilde{x}_{j,n} \in I_{j,n}$ and then check whether the limit $\int_a^b f(x)dx$ of $\mathcal{F}_n$ always exists and is then also unique. Indeed: proceeding in this way is often far too laborious. Now you can see why Master Newton and Master Leibniz probably – unfortunately I cannot ask them personally anymore – had their justified thoughts about the integration of functions. The approach just described was more or less the state of science *before* their fundamental discovery mentioned at the beginning. This led to a general machinery, which – as I will show you below – simplifies the integration of functions considerably.

In addition, I have another very positive news for you: Almost all functions that you normally encounter are integrable. For example, **monotone functions** are *always integrable*. **Differentiable functions** – indeed, a more general class of functions called **continuous functions** – are also *always integrable*. Very roughly speaking, these are functions whose graph can be drawn without putting down the writer. Somewhat more precisely, this means that (definitional) gaps, jumps in the function values, and places where the function values oscillate excessively in some way may not occur. Very many functions are continuous, and the *class of all integrable functions* is fortunately *even larger*. You have to do some tinkering to find a function that is not integrable.

Example 6.31

I will just show you the following example of a function f that is *not* integrable on [0,1]:

$$f(x) = \begin{cases} 0, & \text{falls } x \in [0,1] \cap \mathbb{Q}, \\ 1, & \text{falls } x \in [0,1] \cap (\mathbb{R} \backslash \mathbb{Q}). \end{cases}$$

$\mathbb{Q}$ denotes here, as in Sect. 1.2, the rational numbers.

Of course, this function in particular has no chance of being monotonic or even continuous, because to draw its graph I would have to move the writer back and forth between 0 and 1 an infinite number of times.

I cannot do this, and therefore I now move myself in a positive direction by *assuming* from now on – even if I do not mention this explicitly every time – *always* that the function(s) of consideration are integrable on the corresponding intervals. I now first name natural properties of the definite integral of such functions: ◀

Properties of the Definite Integral

(a) $\int_a^b f(x)dx = \int_a^c f(x)dx + \int_c^b f(x)dx$ *(Additivity of Intervals)*

(b) $\int_a^b f(x)dx = -\int_b^a f(x)dx$ *(Change of bounds.)*

(c) $\int_a^a f(x)dx = 0$ *(Empty Integral)*

(d) If $f(x) \geq 0$ for all $x \in [a, b]$, then:

$$\int_a^b f(x)dx \geq 0 \quad (Positivity)$$

To see these properties, I should recall that the definite integral of f *with* respect to $[a, b]$ is the limit of sums $\mathcal{F}_n$ whose terms are of the form $f(\widetilde{x}_j)(x_{j+1,n} - x_{j,n})$. Property (a) says that the definite integral with respect to $[a, b]$ is the sum of the definite integrals with respect to the two subintervals $[a, c]$ and $[c, b]$. This becomes insightful by looking at sums of the construction $\mathcal{F}_n$ for $[a, c]$ and $[c, b]$ and making clear that their sum is of the form $\mathcal{F}_n$ for $[a, b]$. To see b), first notice that when we pass $[a, b]$ in reverse order, the terms $f(\widetilde{x}_j) \times (x_{j,n} - x_{j+1,n}) = -f(\widetilde{x}_j)(x_{j+1,n} - x_{j,n})$ appear in the formation of the sum $\mathcal{F}_n$. The minus sign is preserved here in the formation of the limit and therefore it occurs when the integration limits are swapped in (b). However, (c) follows from (a) and (b):

$$\int_a^a f(x)dx \overset{a)}{=} \int_a^b f(x)dx + \int_b^a f(x)dx \overset{b)}{=} \int_a^b f(x)dx - \int_a^b f(x)dx \overset{c)}{=} 0.$$

Further, I can see d) as follows: From $f(x) \geq 0$ it follows that all terms $f(\widetilde{x}_j)(x_{j+1,n} - x_{j,n})$ are also ≥ 0, because it is also $x_{j+1,n} - x_{j,n} > 0$. So $\mathcal{F}_n \geq 0$ – and therefore the limit for $n \to \infty$ is also ≥ 0.

Now it occurs to me that (d) implies the following: If $f(x) \leq 0$ for all $x \in [a, b]$, then $\int_a^b f(x)dx \leq 0$. This implies that the definite integral can sometimes *become negative*. For example, all function values of $f(x) = x$ for $x \in [-1, -3]$ are non-positive and I calculate according to what I showed you at the beginning,

$$\int_{-3}^{-1} xdx = \left((-1)^2 - (-3)^2\right)/2 = -4.$$

If a function in the considered interval changes sign, even 0 can be the value of the determined integral:

$$\int_{-1}^{1} xdx = \left(1^2 - (-1)^2\right)/2 = 0.$$

You can see that the indefinite integral generally *does not* give the **absolute area of** the area enclosed by the graph of f with the *x-axis,* but only its **oriented area**. The sign of the integrand f also determines the sign of the integral in the various subintervals. The determined integral thus represents the summation of the corresponding *signed* area contents. Because of

$$\int_{-1}^{0} xdx = -1/2 \quad \text{and} \quad \int_{0}^{1} xdx = 1/2$$

I am therefore no longer surprised that I have calculated $\int_{-1}^{1} xdx = 0$ above, because this must be so because of property (a). If I want to know the absolute area of the area enclosed by the graph of f with the *x-axis*, I should form the definite integral of the (concatenated) function $|f| = |\cdot| \circ f$:

$$\int_{a}^{b} |f(x)|\, dx.$$

Practically, because of the specification of the magnitude$|\cdot|$ (see the first section in this chapter), this means that instead of integrating f at all points in $[a, b]$ where f is not positive, I integrate the function $-f$. In Fig. 6.18, the graph of $|f|$ for a given f is shown as a broader curve. I will look at the concrete example associated with this figure in a bit more detail below. Here I first note that the absolute area in the above example is calculated as follows:

$$\int_{-1}^{1} |x|\, dx = \int_{-1}^{0} -xdx + \int_{0}^{1} xdx = -\int_{-1}^{0} xdx + \int_{0}^{1} xdx = -(-1/2) + 1/2 = 1.$$

Here, at the second "=" sign, I just loosely dragged the **factor** -1 in front of the integral.

However, I should now explain this "loose" arithmetic operation in more detail: You know from the first section of this chapter that the derivative rules for the **factor product** $\lambda \cdot f$, defined by $(\lambda \cdot f)(x) = \lambda f(x)$, and the **sum of functions** $f + g$, defined by $(f + g)(x) = f(x) + g(x)$, have often simplified the computation of derivatives for us. I will now formulate similar *rules to simplify the calculation of integrals.*

The definite integral of $\lambda \cdot f$ with respect to $[a, b]$ is the limit of sums $\mathcal{F}_n$, where the individual terms are of the construction $\lambda f(\widetilde{x}_j)\left(x_{j+1,n} - x_{j,n}\right)$. So I can factor λ out of $\mathcal{F}_n$ and factor out in the limit for $n \to \infty$. The definite integral of $f + g$ *with* respect to $[a, b]$ is the limit of sums $\mathcal{F}_n$, where the individual terms are of the form

$$\left(f(\widetilde{x}_j) + g(\widetilde{x}_j)\right)\left(x_{j+1,n} - x_{j,n}\right) = f(\widetilde{x}_j)\left(x_{j+1,n} - x_{j,n}\right) + g(\widetilde{x}_j)\left(x_{j+1,n} - x_{j,n}\right)$$

Are. So I can consider two sums of the form $\mathcal{F}_n$ and consider them separately in forming the limit. In this way, the following two simple integration rules arise.

Integration Rules for the Factor Product $\lambda \cdot f$ and the Sum $f + g$

The definite integral of the factor product $\lambda \cdot f$ with *respect to* $[a, b]$ is the product of λ with the definite integral of f with respect to $[a, b]$:

$$\int_a^b (\lambda \cdot f)(x)dx = \lambda \int_a^b f(x)dx.$$

The definite integral of the sum $f + g$ of functions f and g *with respect to* $[a, b]$ is equal to the sum of the definite integrals of f and g with respect to $[a, b]$:

$$\int_a^b (f + g)(x)dx = \int_a^b f(x)dx + \int_a^b g(x)dx.$$

Finally, I can now calculate the definite integral even for somewhat more difficult functions than the power functions:

Example 6.32

For example, if I am interested in the definite integral of the function $h(x) = 2x - x^2$ with respect to [0,2], I can apply the sum and factor rule and first obtain

$$\int_0^2 2x - x^2 dx = \int_0^2 2x dx + \int_0^2 -x^2 dx = 2\int_0^2 x dx - \int_0^2 x^2 dx.$$

Thus I have expressed the definite integral of h by definite integrals of functions – $f(x) = x$ and $g(x) = x^2$ – for which I already know them:

$$\int_0^2 x dx = 2^2/2 - 0^2/2 = 2 \quad \text{and} \quad \int_0^2 x^2 dx = 2^3/3 - 0^3/3 = 8/3.$$

Thus, the result is $\int_0^2 h(x)dx = \int_0^2 2x - x^2 dx = 2 \cdot 2 - 8/3 = 4/3$. One interpretation of this result is that the area enclosed by the straight line defined by $2x$ and the **normal parabola** for $x \in [0,2]$ has content 4⁄3.▪ ◄

Example 6.33

Next, I calculate the definite integral of the cubic polynomial $f(x) = 1/4\, x^3 + 3/4\, x^2 - 3/2\, x - 2$ (see Fig. 6.10) in terms of $[-4, -1]$:

$$\int_{-4}^{-1} f(x) dx =$$

$$\frac{1}{4}\int_{-4}^{-1} x^3 dx + \frac{3}{4}\int_{-4}^{-1} x^2 dx - \frac{3}{2}\int_{-4}^{-1} x dx - 2\int_{-4}^{-1} 1 dx \quad = \frac{1}{4}\left(\frac{(-1)^4 - (-4)^4}{4}\right) + \frac{3}{4}$$
$$\times \left(\frac{(-1)^3 - (-4)^3}{3}\right) - \frac{3}{2}\left(\frac{(-1)^2 - (-4)^2}{2}\right) - 2((-1) - (-4)) \quad = \frac{81}{16}.$$

Here I again first used the sum and factor rule and then the above knowledge about the determined integrals of the power functions 1, x, x^2 and x^3. It is quite gratifying, and perhaps even a little surprising, that I am thus able to determine the content of the relatively complicated bounded area highlighted in gray in Fig. 6.10. The number 81⁄16 is the absolute content of this surface, because $f(x) \geq 0$ holds for all $x \in [-4, -1]$. The situation is different with the definite integral of this f *with respect* to $[-4, 2]$ – for this I calculate by analogous procedure as just now: $\int_{-4}^{2} f(x)dx = 0$. Thus, the oriented area computed with respect to $[-4, 2]$ has the value 0. The absolute area over $[-4, 2]$, on the other hand, is 81⁄8, because I determine the gray area in Fig. 6.18 as follows:

$$\int_{-4}^{2} |f(x)| dx = \int_{-4}^{-1} f(x)dx - \int_{-1}^{2} f(x)dx = \frac{81}{16} - \left(-\frac{81}{16}\right) = \frac{81}{8}. \quad ■$$

◄

Exercise 6.18 (a) Calculate the oriented and absolute area of the area enclosed by the graph of $f(x) = x^3 - x$ over $[-1, 1]$ with the *x-axis*. (b) Find the definite integral $\int_{-1}^{1} f(x)dx$ for the two functions $f = p$ and $f = f_t$ from Exercise 6.10.

Exercise 6.19 For $t > 0$, find an explicit representation of the integral $\int_0^t -x^2 + 2x dx$. Take the explicit result to be a *function* F_0 *in* t and, for fun, compute the derivative of F_0 with respect to t. Do you notice anything?

Using the factor and sum rule allows me to calculate certain integrals now for a larger class of functions than the power functions:

Integration of Polynomials

The definite integral of a **polynomial function** p **of degree** n, $p(x) = a_n x^n + \cdots + a_i x^i + \cdots + a_1 x + a_0$ with respect to $[a, b]$ is given by

$$\int_a^b p(x)dx = a_n\left(\frac{b^{n+1} - a^{n+1}}{n+1}\right) + \cdots + a_i\left(\frac{b^{i+1} - a^{i+1}}{i+1}\right) + \cdots + a_1\left(\frac{b^2 - a^2}{2}\right) + a_0(b - a).$$

Great – but of course I am *not* satisfied with just integrating polynomials. I already told you above that the class of all integrable functions is quite extensive. Therefore it would be especially desirable to have a *general machinery for integration*, which is applicable for very many of these functions simultaneously, so to speak. A mathematical dream would come true, if this machinery were such, that certain integrals could be calculated in a similarly simple way as for polynomials. I may reveal it here already in advance: Such a dream became reality a long time ago.

In order to approach the procedure together with you, I will look again at an example from the beginning of this chapter. Above I explained to you that

$$F_a(b) = \int_a^b x^2 dx = \frac{b^3}{3} - \frac{a^3}{3}$$

is the definite integral of $f(x) = x^2$ with respect to $[a, b]$. I *deliberately* use here *again* the designation $F_a(b)$ from the beginning of this section, because I am about to take the upper limit of integration as a *variable of the function* F_a. So now I relocation the fixed location b by a variable – say t – so that no confusion with x can *happen*. What the heck, you might say – in principle it still says the same thing:

$$F_a(t) = \int_a^t x^2 dx = \frac{t^3}{3} - \frac{a^3}{3}.$$

True. *However*, my interpretation is now different: I now consider F_a to be a function in the variable t: $F_a(t) = t^3/3 - a^3/3$. With this function F_a I can do everything I usually do with functions: Determine zeros, draw graphs, . . ., and in particular I can *compute the derivative of* the F_a with respect to t here for fun. Since $a^3/3$ is a constant, I calculate the derivative F_a' of F_a as follows:

$$F'(t) = \frac{1}{3}(t^3)' - \left(\frac{a^3}{3}\right)' = \frac{1}{3}3t^2 = t^2.$$

The crucial *observation* here is that the derivativeF_a' of F_a coincides with the function f for which I am forming the integral (the integrand function): $F_a'(t) = f(t)$. Since you did Exercise 6.19, you already know this phenomenon for another function. "Coincidence" you may say – please read on in a relaxed manner.

This phenomenon apparently occurs for any power function $f(x) = x^i$. By setting $F_a(t) = \int_a^t x^i dx$ and then differentiating by the variable, I see this:

$$F_a'(t) = \left(\int_a^t x^i dx\right)' = \left(\frac{t^{i+1}}{i+1}\right)' - \left(\frac{a^{i+1}}{i+1}\right)' = (i+1)\frac{t^i}{i+1} = t^i = f(t).$$

Now for any polynomial function $p(x) = a_n x^n + \cdots + a_i x^i + \cdots + a_1 x + a_0$, I can observe this phenomenon – differentiating $\int_a^t p(x)dx$ I get p:

$$a_n \overbrace{\left(\frac{t^{n+1} - a^{n+1}}{n+1}\right)'}^{t^n} + \cdots + a_i \overbrace{\left(\frac{t^{i+1} - a^{i+1}}{i+1}\right)'}^{t^i} + \cdots + a_1 \overbrace{\left(\frac{t^2 - a^2}{2}\right)'}^{t} + a_0 \overbrace{(t-a)'}^{1} = p(t).$$

So for the f considered so far, I *observe* that the **integral function of** f

$$F_a(t) = \int_a^t f(x)dx$$

is differentiable and its first derivative F_a' coincides with the integrand function f: $F_a'(t) = f(-t)$. This *observation* was the *starting point of* the development of a *general machinery of integration*. The function F_a is sometimes called an **indefinite integral of** f, because the upper limit of integration t is variable here and therefore *not fixes*. Sometimes – especially when the lower limit is *no longer of* too much interest – this is referred to as $\int^t f(x)dx$ or $\int^t f$, depending on the advanced degree of personal laziness in writing.

Since there are many integrable functions, one can consider for which functions f *of* a more general kind the integral function $F_a(t) = \int_a^t f(x)dx$ could be differentiable, so that additionally $F_a' = f$ is valid. Since f *should* ideally be of a quite general kind, I must now **remember** the procedure from the first section of this chapter for proving **ddifferentiability** by forming the **difference quotients** and the **differential quotient**. So

I proceed as in the investigations there: Thus I consider first for a given (integrable) function f the difference quotients of the integral function

$$\frac{F_a(t+h)-F_a(t)}{h}=\frac{\int_a^{t+h} f(x)dx-\int_a^t f(x)dx}{h}.$$

By using the properties (b) and (a) of the definite integral for the numerator appearing here, I obtain for it

$$\int_a^{t+h} f(x)dx+\int_t^a f(x)dx=\int_t^a f(x)dx+\int_a^{t+h} f(x)dx=\int_t^{t+h} f(x)dx,$$

and thus the difference quotients of the integral function are given as $\int_t^{t+h} f(x)dx/h$. Unfortunately, I cannot further simplify the integral $\int_t^{t+h} f(x)dx$ occurring here for functions f *of a* very general kind. However, for any function f that is not too crazy – namely, for f whose graph I can draw without putting down the writer – I fortunately succeed. I called these f ***continuous*** functions above. For continuous f, the representation as a rectangular area is

$$\int_t^{t+h} f(x)dx=f(\xi)h$$

possible. Here ξ (pronounced "xi") is a location in $[t, t+h]$. Unfortunately, this ξ is not specified in general – all I know is that there is such a location. In Fig. 6.19 I have tried to represent this **mean value theorem of the integral calculus.** The difference quotients of the integral function are thus:

$$\frac{F_a(t+h)-F_a(t)}{h}=\frac{\int_t^{t+h} f(x)dx}{h}=\frac{f(\xi)h}{h}=f(\xi).$$

The location ξ depends in general on f and the choice of t and h. Now, if h goes towards 0, $h \to 0$, then $t + h$ approaches more and more t, and $\xi \in [t, t+h]$ has no other chance than to go towards t for $h \to 0$: $\lim\limits_{h\to 0}\xi=t$. Thus *it seems plausible* that the value of f in ξ, $f(\xi)$, goes against the value of f in t, $f(t)$, for $h \to 0$: $\lim\limits_{h\to 0} f(\xi)=f(t)$. This is also true here. Only for even too crazy functions could the latter limit possibly not exist – no difficulty arises here, because I already decided above to consider *continuous* functions f. Strictly speaking, the validity of $\lim\limits_{h\to 0} f(\xi)=f(t)$ under the condition $\lim\limits_{h\to 0}\xi=t$ is even just a form of mathematical description of *the continuity of functions* – but that leads too far today. Now, I can determine the differential quotient of the integral function of *f*:

$$\lim_{h \to 0} \frac{F_a(t+h) - F_a(t)}{h} = \lim_{h \to 0} f(\xi) = f(t),$$

and I realize that the integral function F_a is differentiable. *Crucially*, the phenomenon I *observed* above only for polynomial functions is also valid for the large class of continuous functions *f*: The first derivative F_a' of the integral function F_a then coincides with *f*: $F_a' = f$.

Differentiability of the Integral Function of Continuous Functions
For any continuous function *f* on $[a, b]$, $F_a(t) = \int_a^t f(x)dx$, $t \in [a, b]$ is a differentiable function whose first derivative coincides with *f*: $F_a'(t) = f(t)$, $t \in [a, b]$.

If I compute the derivative of the integral function F_a of a continuous function *f* (with respect to *t*), I get the function *f* itself:

$$F_a'(t) = \left(\int_a^t f(x)dx \right)' = f(t).$$

Thus, differentiating cancels integrating. Even more: If I consider an integral function F_a, then F_a' is the corresponding integrand function (if it is continuous):

$$\int_a^t F_a'(x)dx = F_a(t).$$

Integrating therefore cancels out the differantiation. Now, you see why integrating and deriving can be seen as the *inverse of each other*.

This is – I am sure you agree – highly interesting, but I should not lose sight of the fact that I wanted to introduce you to a general machinery for integrating continuous functions. Besides, I promised you, that the calculation of such indefinite integrals should often go similarly easy as for polynomials. For this purpose, I first briefly make clear to myself that the definite integral of continuous functions *f* with respect to $[a, b]$ can be calculated with the integral function F_a by *merely evaluating it at the point b*:

$$\int_a^b f(x)dx = F_a(b).$$

Another problem here is that the integral function F_a is unfortunately not described very explicitly. $F_a(t) = \int_a^t f(x)dx$ is rather formally represented as an integral with variable *t* in the upper integration limit. One might turn away disappointed at this point – were it not for our *observation* that $F'_a = f$ holds. *This now helps us decisively.* To get away from the formal representation of F_a as an integral, to calculate the definite integral of f I might – as a test – try to find *some function F* whose derivative gives f: $F' = f$. I would thus find such an F by *reversing the derivative process, so* to speak. My quiet hope is that such an F would not be too different from F_a, and that this F *would* be more likely to be represented explicitly, so that I could conveniently use it to calculate the definite integral of f. A differentiable function F with this property – $F' = f$ – is called a **primitive** of f. I do not think I should continue writing now without listing a few examples of primitives:

Example 6.34

The function $F(x) = x^3/3$ is certainly a primitive of $f(x) = x^2$, because $F'(x) = 1/3 (x^3)' = (1/3)3x^2 = x^2 = f(x)$. The function $G(x) = x^3/3 + 15$ is also a primitive of $f(x) = x^2$, because **constants** *vanish when they are compute the derivative ofd* and thus: $G'(x) = (1/3x^3)' + (15)' = x^2 = f(x)$. The function $P_i(x) = x^{i+1}/(i+1)$ is a primitive of $p_i(x) = x^i$, because $P_i'(x) = 1/(i+1)(x^{i+1})' = 1/(i+1)(i+1)x^i = x^i = p_i(x)$. The function

$$F(x) = 1/16x^4 + 1/4x^3 - 3/4x^2 - 2x + \pi$$

is a primitive of the cubic polynomial already considered above

$$f(x) = 1/4x^3 + 3/4x^2 - 3/2x - 2,$$

because

$$F'(x) = 1/16(x^4)' + 1/4(x^3)' - 3/4(x^2)' - 2(x)' + (\pi)' = 4/16x^3 + 3/4x^2 - 3/2x - 2. \blacksquare$$

These are all functions for which I have already told you above how to calculate their definite integrals. To document that I am really serious, I guess I need to discuss a few more examples of primitives of non-polynomial functions: ◀

Example 6.35

The function $F(x) = \exp.(x) = e^x$ is a primitive of $f(x) = e^x$, because $F'(x) = (e^x)' = e^x = f(x)$. The natural logarithm $\ln(x)$ is -as I showed you in the last section- a primitive function of $f(x) = 1/x$, because $\ln'(x) = 1/x$ holds for all $x > 0$. If you turn to Example 6.15, you will see that $F(x) = (x^2 + x - 1)/e^x$ *is* a primitive function of $f(x) = (-x^2 + x + 2)/e^x$ – there I called F "k" and showed $F' = f$. The function $F(x) = -cos(x) + 4$ is a primitive of $f(x) = \sin(x)$, because in the last section I showed you $F'(x) = sin(x) = f(x)$ *as* part of the discussion of the connection between convexity and the sign of the second derivative. An example of the delicate kind is the function $g(x) = (1 - x^2)/(x^2 + 1)^2$, for which I know the primitive function $G(x) = x/(x^2 + 1)$. The graph of G is shown in Fig. 6.12. In Example 6.26 I denoted this G by f, and already there I showed you $G' = g$.

You can see that I have made things quite simple for myself in these examples. After all, I have mainly considered functions f here for which I have seen before that they occur as derivatives of some other function F: $F' = f$. ◀

Exercise 6.20 (a) Show that $H(x) = -1/3\, x^3 + x^2 - 12$ is a primitive function of $h(x) = -x^2 + 2\,x$. (b) Using the information to solve Exercise 6.8b, determine a primitive function F of $f(x) = (2x - 1)\, e^{x^2 - x}$. Do you find any other primitive functions of f that are different from F?

You must have noticed it just now: if F *is* a primitive of a given function f, i.e. $F' = f$, then I can add any constant c to F and get further primitives of f: $(F + c)' = F' + (c)' = F' + +0 = F' = f$. Perhaps somewhat surprisingly, this way even yields *all the* primitive functions of f. To see this, I look at two primitives – let us call them F and G – of the same function f. The first derivative of their difference $F - G$ vanishes for all x:

$$(F - G)'(x) = F'(x) - G'(x) = f(x) - f(x) = 0$$

and therefore $F - G$ is constant.

Primitive Functions Differ only by Additive Constants
Two primitive functions F and G *of* the same function differ only by an additive constant c: $F(x) = G(x) + c$ for all x.

Primitives push the gate wide open for a general machinery for the integration of continuous functions. I *choose* – with the aim of calculating the definite integral $\int_a^b f(x)dx$ – *some* primitive F of f. The somewhat unwieldy representation of the integral function $F_a(t) = \int_a^t f(x)dx$ should not make me forget that F_a is also a primitive of the continuous function f, for it holds, after all: $F_a' = f$. Consequently, F_a and F differ only by a constant c: $F_a(t) - F(t) = c$ for all t. I decided to call the variable t again – and not x, for example – because that is what I have been doing for F_a all along. In this scenario, I can even easily calculate the constant c by *setting* $t = a$: $c = F_a(a) - F(a) = \int_a^a f(x)dx - F(a) = 0 - F(a) = -F(a)$. Now, for the sake of clarity, I will write F on the other side: $F_a(t) = F(t) - F(a)$ for all t. At the special location $t = b$ we get $F_a(b) = F(b) - F(a)$. The determined integral $\int_a^b f(x)dx = F_a(b)$ can be easily calculated with *the* help of *my chosen primitive function* F by *simply* determining the function value of F in b and a and calculating the difference:

$$\int_a^b f(x)dx = F(b) - F(a).$$

The fundamental discovery was thus that for any continuous function f I can calculate the definite integral with the help of one of its primitives F essentially in exactly the same way as in my three simple input examples (Figs. 6.16 and 6.17 (right))! So that you will not forget this statement, I formulate the result in the following **main theorem of differential and integral calculus**, which describes the *central* connection between deriving ($F' = f$) and integrating ($\int f = F$):

Main Theorem of Differential and Integral Calculus
If f is a continuous function on $[a, b]$ and F is an arbitrary primitive function of f, then the definite integral of f *with respect to* $[a, b]$ can be calculated as follows: $\int_a^b f(x)dx = F(b) - F(a)$.

In summary, the *general machinery* now developed *for integrating continuous functions* is as follows:

- Find *any* primitive function F of f.
- Calculate $F(b)$ and $F(a)$ by "substituting" the digits b and a into F.
- The difference $F(b) - F(a)$ is the definite integral of $f$his $[a, b]$.

For the latter difference, two abbreviated *spellings* are often used:

$$[F(x)]_a^b = F(x)\Big|_a^b = F(b) - F(a).$$

Now I could already calculate certain integrals for many continuous functions f explicitly.

Example 6.36
For example, I had shown you above that $F(x) = x^3/3$ and $G(x) = x^3/3 + 15$ are primitive functions of $f(x) = x^2$. Thus I now calculate

$$\int_1^{10} x^2 dx = F(10) - F(1) = \frac{10^3}{3} - \frac{1^3}{3} = \frac{999}{3} = 333$$

and

$$\int_{-1}^{1} x^2 dx = [F(x)]_{-1}^{1} = \frac{1^3}{3} - \frac{(-1)^3}{3} = \frac{1}{3} + \frac{1}{3} = \frac{2}{3},$$

but also

$$\int_{1}^{10} x^2 dx = \frac{x^3}{3} + 15\bigg|_1^{10} = \left(\frac{10^3}{3} + 15\right) - \left(\frac{1^3}{3} + 15\right) = \frac{999}{3} = 333$$

and

$$\int_{-1}^{1} x^2 dx = G(1) - G(-1) = \left(\frac{1^3}{3} + 15\right) - \left(\frac{(-1)^3}{3} + 15\right) = \frac{1}{3} + \frac{1}{3} = \frac{2}{3}. \quad ■$$

As you can see, it is *completely irrelevant which primitive function* I use for the calculation, because the constant – in the example this is the number 15 – occurs with a positive and negative sign and therefore cancels out of the calculation. This is always the case. ◀

Example 6.37

$F(x) = 1/16\, x^4 + 1/4\, x^3 - 3/4\, x^2 - 2\,x + \pi$ is a primitive function of $f(x) = 1/4\, x^3 + 3/4\, x^2 - 3/2\, x - 2$ and I compute the above definite integral $\int_{-4}^{-1} f(x)dx = [F(x)]_{-4}^{-1} = F(-1) - F(-4)$ now as

$$\frac{1}{16}(-1)^4 + \frac{1}{4}(-1)^3 - \frac{3}{4}(-1)^2 - 2(-1)$$
$$- \left(\frac{1}{16}(-4)^4 + \frac{1}{4}(-4)^3 - \frac{3}{4}(-4)^2 - 2(-4)\right) = \frac{81}{16}.$$

Since $(e^x)' = e^x$ applies, I calculate

$$\int_1^3 e^x dx = [e^x]_1^3 = e^3 - e^1 = e\left(e^2 - 1\right)$$

and

$$\int_{-1}^{1} e^x dx = [e^x]_{-1}^{1} = e^1 - e^{-1} = e - \frac{1}{e}.$$

Because of $\ln'(x) = 1/x$ for all $x > 0$, I find.

$$\int_1^2 \frac{1}{x}dx = \ln(x)\Big|_1^2 = \ln(2).$$

Since $(-\cos(x))' = \sin(x)$, I compute

$$\int_0^\pi \sin(x)dx = [-\cos(x)]_0^\pi = (-\cos(\pi)) - (-\cos(0)) = 2$$

and

$$\int_{-\pi}^{\pi} \sin(x)dx = [-\cos(x)]_{-\pi}^{\pi} = -\cos(\pi) - (-\cos(-\pi)) = 1 - 1 = 0. \blacksquare$$

◀

Exercise 6.21

(a) Calculate $\int_0^2 -x^2 + 2xdx$.

(b) Calculate $\int_1^2 (2x-1)e^{x^2-x}dx$.

(c) Calculate $\int_1^2 (1-x^2)/(x^2+1)^2 dx$.

(d) Determine $\int_0^\pi (\cos(x) - \sin(x))e^x dx$, using what you learned in Exercise 6.7b.

(e) For $x > 0$, determine a primitive function F of $f(x) = 1/x^2$ and then find $\int_1^2 1/x^2 dx$. Note: Read again below Exercise 6.7.

So you see, once I know a primitive F of f *explicitly*, calculating the definite integral of f *with* respect to $[a, b]$ is relatively easy – all I have to do is work out $[F(x)]_a^b = F(b) - F(a)$. However, I am not always in this happy situation. Unfortunately for us at the moment, for example, with the integrals.

$$\int_a^b xe^x dx, \quad \int_1^7 \ln(x)dx, \quad \int^t \sin^2(x)dx, \quad \int_a^b 6x(3x^2+1)^{71}dx, \quad \int_1^2 \frac{\ln(x)}{x}dx$$

is the case. Of course, I can formally write down a truncated function here, because the integral functions – for example $\int_1^t \ln(x)dx$ – are such functions. Only this helps heartily little, because I know then the function F, in which I would like to insert the integration borders, unfortunately *not* yet *explicitly*. For your information, I tell you that there are indeed continuous integrands f for which it is not possible to give a **closed expression** by elementary functions for one of their primitives. For example, I cannot explicitly write down any of the primitives $\int_a^t e^{-x^2} dx + c$ of $f(x) = e^{-x^2}$ as a closed expression. The same happens when I want to **calculate** the length of an **elliptic arc** (see end of Chap. 4) – this leads to so-called **elliptic integrals**, which are not expressible by elementary functions, which is very annoying. But that is the way it is.

Fortunately, however, in addition to the rules already formulated above (integrating the factor product and the sum of functions), there are the following two integration rules that I can sometimes use to find closed-form expressions for primitive functions. Roughly speaking, I get them by integrating the differantiation rules called **product and chain rule in** the first section of this chapter.

If I consider two differentiable functions f and g, the product rule says that $(f \cdot g)' = f' g + g' \cdot f$ holds. The idea now is to integrate the functions of both sides. For $(f \cdot g)'$ I know an explicit primitive function $f \cdot g$, while I might want to find one for $f' g$. Since $f' g = (f \cdot g)' - g' \cdot f$ holds, it suffices to find an explicit primitive for $g' \cdot f$ for this purpose,

$$\int^t f' \cdot g = \int^t (f \cdot g)' - \int^t g' \cdot f = (f \cdot g)(t) - \int^t g' \cdot f,$$

and the hope is that, at least sometimes, it will be relatively easy to do so.

Product Integration: Partial Integration
If f and g are functions differentiable on $[a, b]$, the following is true.

$$\int_a^b f'(x)g(x)dx = f'(x)g(x)|_a^b - \int_a^b f(x)g'(x)dx$$

Example 6.38

A typical example is finding a primitive of $h(x) = x\,e^x$. I set $f(x) = e^x$ and $g(x) = x$ and recognize $f'(x) = e^x$ and $g'(x) = 1$. Thus $h(x) = f'(x)\,g(x)$ holds here. If I now apply partial integration, I get

$$\int^t \overbrace{e^x x}^{=f'(x)g(x)} dx = \overbrace{e^t t}^{=f(t)g(t)} - \int^t \overbrace{e^x 1}^{=f(x)g'(x)} dx.$$

The indefinite integral on the right side can be written explicitly $\int^t e^x dx = e^t$, and thus it follows that $F(t) = t\,e^t - e^t = (t - 1)\,e^t$ is a primitive function of f. I can now check this by calculating the derivative of F. The product rule indeed gives:

$$F'(t) = (t-1)'e^t + (t-1)(e^t)' = 1e^t + (t-1)e^t = te^t = h(t).$$

If I now want to calculate a certain integral of h, this is done as usual:

$$\int_a^b xe^x dx = (x-1)e^x \Big|_a^b = (b-1)e^b - (a-1)e^a.$$

For example, I calculate

$$\int_1^2 xe^x dx = e^2. \quad \blacksquare$$

◀

Example 6.39

Another example is the calculation of the definite integral $\int_1^7 \ln(x)dx$. Here it seems difficult at first glance to do anything promising – I cannot seem to simplify the **natural logarithm** ln any further or decompose it trickily. At this point you begin to understand why some people call *differentiation a technique, but integration an art.* After some trial and error, you might get the idea of *artificially* taking $g(x) = \ln(x)$ to be the product of two functions $g(x) = 1 \cdot \ln(x)$. If I now apply partial integration, I get for $t > 0$:

$$\int^t \overbrace{1 \cdot \ln(x)}^{=(x)'g(x)} dx = \overbrace{t\ln(t)}^{=tg(t)} - \int^t \overbrace{x\frac{1}{x}}^{=x\ln'(x)} dx.$$

The indefinite integral on the right-hand side can be written explicitly $\int^t x \; 1/x dx = \int^t 1 dx = t$, and thus it follows that $F(t) = t\ln(t) - t = t(\ln(t) - 1)$ is a primitive function of $\ln(t)$. Applying the product rule confirms this:

$$F'(t) = (t\ln(t))' - (t)' = (t)'\ln(t) + t(\ln(t))' - 1 = 1\ln(t) + t1/t - 1 = \ln(t).$$

Using the primitive function F, I can now calculate the definite integral mentioned above:

$$\int_1^7 \ln(x)dx = [x\ln(x) - x]_1^7 = 7 \cdot \ln(7) - 7 - (1 \cdot \ln(1) - 1) = 7 \cdot \ln(7) - 6.$$

I have shown the corresponding area in grey in Fig. 6.9 (right). ◀

Example 6.40

Another nice example is finding a primitive of $f(x) = \sin^2(x) = \sin(x)\,\sin(x)$. With partial integration I first get

$$\int^t \overbrace{\sin(x)\sin(x)}^{=(-\cos(x))'\sin(x)}\, dx = -\cos(t)\sin(t) - \int^t \overbrace{-\cos^2(x)}^{=-\cos(x)(\sin(x))'}\, dx.$$

At first glance, I do not like this at all, because on the right, the integrand $-\cos^2(x)$ appears, which is just as complicated as the integrand I started from: $\sin^2(x)$. In fact, sometimes at this point you end up in a dead end and have to think of something else. Here, however, this is fortunately not the case. I remember $\cos^2(x) = 1 - \sin^2(x)$ and calculate like this

$$\int^t \sin^2(x)dx = -\cos(t)\sin(t) + \int^t 1dx - \int^t \sin^2(x)dx.$$

Thus $F(x) = 1/2\,(-\cos(x)\,\sin(x) + x)$ is a primitive function of $f(x) = \sin^2(x)$.▪ ◀

Exercise 6.22 Use partial integration to determine all the primitive functions of $f(x) = \sin(x)\,x$ and $g(x) = x^2e^x$.

The second integration rule, which can also sometimes be used to determine primitives, is called the **substitution rule**. Here – at least in the standard application for which I will first show you examples – a complicated integrand is *relocationd* by a simpler integrand whenever possible. I assume here that for a simple function f I know a primitive F and the complicated integrand is of the form $g' \cdot (f \circ g)$ – so f *is* here **concatenated** with g and this is multiplied by g'. This looks very special – but in fact occurs more often than is commonly thought. By the **chain** rule from the first section, $(F \circ g)'(x) = g'(x)\,F'(g(x))$ – so it follows because $F' = f$: $(F \circ g)'(x) = g'(x)\,f(g(x))$. Thus $(F \circ g)$ is a primitive function of $g' \cdot (f \circ g)$ and I calculate the integral with respect to the more complicated integrand $g' \cdot (f \circ g)$ as follows:

$$\int^t g' \cdot (f \circ g) = (F \circ g)(t) = F(g(t)) = \int^{g(t)} f.$$

I see that this integral is obtained by integrating the simpler integrand f, just adjusting the integration limit(s) there.

Substitution Rule

If f is continuous and g is differentiable on $[a, b]$, then

$$\int_a^b g'(x)f(g(x))dx = \int_{g(a)}^{g(b)} f(\overline{x})d\overline{x}.$$

An example may be worth a thousand words:

Example 6.41

I consider the polynomial function $p(x) = 6x\,(3x^2 + 1)^{71}$ (of degree 143). I see that here for $g(x) = 3\,x^2 + 1$ just $g'(x) = 6$ x appears *as* multiplicative term in p. Further, I see that using the power function $f(\overline{x}) = \overline{x}^{71}$ I can write down the polynomial p as $g'\ (f \circ g)$. Indeed, if I relocation the variable $\overline{x}$ with $g\ (x)$, it follows $\overline{x} = 3x^2 + 1$. Further, I recall that I have no trouble finding a primitive for the power function f: $F(\overline{x}) = 1/72\overline{x}^{72}$. So I calculate

$$\int^t 6x\left(3x^2+1\right)^{71}dx = \int^{g(t)} \overline{x}^{71}d\overline{x} = 1/72\overline{x}^{72}\Big|^{3t^2+1} = 1/72\left(3t^2+1\right)^{72}.$$

In fact, $P(x) = 1/72(3x^2 + 1)^{72}$ is a primitive function of p, which you can verify by differentiating P with the chain rule.▪ ◀

Example 6.42

Since I want to avoid leaving you alone with one of the integrals mentioned further above, I now turn to the particular integral $\int_1^2 \ln(x)/xdx$. Using our general machinery, I first try to find a primitive of ln(x)/x. To do this, I first realize that $g'(x) = 1/x$ is the derivative of $g(x) = \ln(x)$. This should be exploited. To do this, I define $f(\overline{x}) = \overline{x}$, because the above integrand can be written as $g' \cdot (f \circ g)$ – indeed, if I substitute $\overline{x} = g(x) = \ln(x)$ in the definition of f and multiply by $g'(x) = 1/x$ at the same time, I get the integrand $1/x \cdot \ln(x)$. For f I have no problems finding a primitive, since my second example in this section already showed in principle that $F(\overline{x}) = 1/2\overline{x}^2$ has the property $F' = f$. So now I calculate a primitive of the integrand ln(x)/x for $t > 0$:

$$\int^{t} \frac{\ln(x)}{x} dx = \int^{\ln(t)} \bar{x} d\bar{x} = \left[\frac{\bar{x}^2}{2}\right]^{\ln(t)} = \frac{\ln^2(t)}{2}.$$

Finally I calculate

$$\int_1^2 \ln(x)/x dx = \left[\ln^2(x)/2\right]_1^2 = \ln^2(2)/2 - \ln^2(1)/2 = \ln^2(2)/2. \quad \blacksquare$$

◀

Example 6.43

At the end of Chap. 4 it was difficult for me to calculate the **area of an ellipse** around the center (0|0) with major axis radius r and minor axis radius b using the means there. I would like to do this now with the help of the integral calculus. Such an ellipse has, as seen there, the representation $x^2/r^2 + y^2/b^2 = 1$. If I resolve this to y, I get $y = b\sqrt{1 - x^2/r^2}$ for the points on the ellipse with non-negative y-values. To calculate the area F of the ellipse I have to solve the integral

$$I = \int_{-r}^{r} b\sqrt{1 - \frac{x^2}{r^2}} dx = 2b \int_0^r \sqrt{1 - \frac{x^2}{r^2}} dx$$

which represents one half of F. Considering the parameter representation of the ellipse from Chap. 4, that is in particular $x = r \sin(t)$, I can relocation x *in* the integral by multiplying the resulting integrand by the derivative $x' = r \cos(t)$ and adjusting the integration limits ($0 = r\sin(0)$ and $r = r\sin(\pi/2)$):

$$I = 2b \int_0^{\frac{\pi}{2}} r\cos(t)\sqrt{1 - \frac{(r\sin(t))^2}{r^2}} dt = 2rb \int_0^{\frac{\pi}{2}} \cos(t)\sqrt{1 - \sin^2(t)} dt = 2rb \int_0^{\frac{\pi}{2}} \times \left(1 - \sin^2(t)\right) dt = 2rb\left(\frac{\pi}{2} - \int_0^{\frac{\pi}{2}} \sin^2(t) dt\right).$$

Here I used that $\cos(t) = \sqrt{1 - \sin^2(t)}$ and $\sqrt{z} \cdot \sqrt{z} = z$ apply. Now I can apply my knowledge from Example 6.40 and calculate

$$\int_0^{\frac{\pi}{2}} \sin^2(t)dt = \frac{1}{2}\left(-\cos\left(\frac{\pi}{2}\right)\sin\left(\frac{\pi}{2}\right) + \frac{\pi}{2}\right) - \frac{1}{2}(-\cos(0)\sin(0) + 0) = \frac{\pi}{4},$$

so in total $I = 2rb\pi/4 = rb\pi/2$. As area of the ellipse F *with* major axis radius r and minor axis radius b I thus get

$$F = 2 \cdot I = \pi rb. \quad \blacksquare$$

This was also quite a nice example to show that I can also read the rule in the last box from right to left and then apply it as well. I must then – as I showed you in Example 6.43 – multiply the derivative of the term to be relocationd to the integrand and adjust the integration limits. It takes some experience to recognize which term to relocation in this way, so that the resulting integrand is easily solvable. This approach – in contrast to the method in the other examples above – is sometimes called the **second substitution rule**.

I think it is now really time for me to relax a little after these laborious descriptions and give you the reins one last time in this chapter: ◀

Exercise 6.23 Using the above substitution rule, determine all the primitive functions of $f(x) = 3x^2(x^3 + 4)^{88}$ and $g(x) = (2x - 1)e^{x^2 - x}$.

Exercise 6.24 Determine the contents of the area shown in gray in Fig. 6.12.

Fundamentals of Probability Theory

Probability theory has always been the most mysterious, almost mystical part of mathematics for me. This is not meant to scare you off, on the contrary, I want to arouse your curiosity. The term "mysterious" in no way refers to the hard mathematical facts one has to deal with in probability theory, but rather to the astonishing fact that the results calculated in this way really have something to do with the real world. Or are you not amazed when on election night, only a few minutes after the polling stations have closed and only a very few constituencies have been counted, you can already find out the overall result to within 1%? (Only heretics suspect that the remaining votes are usually not counted at all and the election official determines the final result after a few hours on the basis of the first projection).

Another example of the mysterious nature of probability theory: When a coin is tossed that has "heads" on one side and "coats of arms" on the other, it is rightly assumed that the appearance of both sides is equally probable; since in mathematical probability theory one does not calculate with percentages as in colloquial language, but with numbers between 0 and 1, the two tosses "heads" and "coats of arms" each have the probability 1⁄2. One tacitly assumes here that the coin never stops on the edge; I refrain from saying that this would also be quite "improbable".

Now suppose you had already tossed the coin 999 times, and *each time* "heads" had appeared; how likely is it that on the thousandth toss "heads" will appear again? Intuitively, you certainly want to answer, "Pretty unlikely!", but probability theory says that, again, with probability 1⁄2, "heads" will appear, because each toss is an *independent event* from the other tosses; the coin has "no memory", so to speak. If you (understandably) do not want to believe this, imagine that at the very moment you are about to make your thousandth toss, your friend (if you still have friends after studying mathematics) enters the room and you ask him about the probability of this toss. Since he, like the coin itself,

G. Walz et al., *Foundations of Mathematics: A Preparatory Course*,
https://doi.org/10.1007/978-3-662-67809-1_7

knows nothing about the first 999 tosses, he will surely answer "1⁄2", and he would be right!

Now, just to confuse you completely at the end of these introductory remarks, let me tell you this: If you are asked before the first roll about the probability of throwing "heads" 1000 times in a row, the correct answer is: "2^{-1000}"; this is a number that has about 300 zeros after the decimal point before a valid digit appears.

If you are about to fire this book into the corner or at least close it, then wait a moment! I believe I can explain to you in the following pages why probabilities are the way they are and enable you to deal with simple to moderately difficult probabilistic problems; and you would not be asked to do much more than that in college.

7.1 Combinatorics

One of the clearest situations in probability theory is when there are only a finite number of, let us say n, possible outcomes for an experiment (in which case a coin toss is also an experiment) and all of them are equally probable. Since the total probability, i.e. the probability that *any* outcome takes place *at all* (which is clear), is equal to 1 by definition, i.e. the sum of all *n individual probabilities* must be 1, each of these individual probabilities must be equal to 1⁄n.

Experiment with Finitely Many Equally Probable Outcomes
If a trial has n distinct outcomes, all of which are equally likely, then each outcome occurs with probability 1⁄n.

I already mentioned a first example above: in the coin toss there are only two possible outcomes, consequently each of these two has probability 1⁄2. Another standard example of elementary probability theory is the roll of a standard die; since there are six different outcomes here (namely the six numbers of eyes), each one of these six possible numbers of eyes occurs with probability 1⁄6.

Now comes an important consideration: Even when picking six balls at random from a drum of 49 balls, i.e., the Saturday number lottery or lotto, every single six-ball selection is equally likely, so you can easily calculate the probability of a particular outcome as the inverse of the total number of all possibilities if, yes, you know how many total possibilities there are to pick 6 out of 49 balls.

This is exactly what combinatorics, the study of the number of combinations, is all about: One wants to find out how many possibilities there are to choose from a certain number of things a smaller number according to certain rules; the reciprocal of this number of possibilities then just gives the probability of encountering a certain possibility.

Oh my goodness! I just read the last paragraph again and hardly understood it myself, how might it be for you? It cannot go on like this, I have to be specific now; so: The model of thinking just described with the lottery drum is indeed a very classical one, but in earlier times the balls were not placed in glass drums but in urns, which is why we also speak of urn models. In addition, one must still distinguish according to which rules one may take out the balls. All this is described in the following, therefore somewhat more detailed text box:

Urn Models

By an **urn model** one understands the following thought experiment: In a vessel (the urn) n distinguishable balls are contained, of which k pieces are chosen randomly (i.e. without looking) one after the other. We now ask for the number of possibilities this selection can produce, distinguishing the following selection rules: If each ball is put back into the urn after its removal, this is called **selection with putting back**; if it is not put back, this is called **selection without putting back**.

In each of the two cases, a distinction is also made as to whether the order of removal is important (**selection with consideration of the order**) or not (**selection without consideration of the order**).

In total, therefore, the following four urn models can be distinguished:

- Selection with backspacing with attention to the order
- Selection with backspacing without regard to the order
- Selection without backspacing with consideration of the order
- Selection without backspacing without consideration of the order

In each of the four cases, you can say exactly how many possibilities there are in total, and thus calculate the probability of a particular choice as the reciprocal of that number. I shall in what follows derive for you, as far as is possible and necessary within the scope of this book, the expression of the formula in each case, but in any case I shall make it plausible. In what way the balls are made distinguishable is, of course, quite irrelevant; for example, they can be coloured differently, given letters or numbers; since I, as a mathematician, am innately best at dealing with numbers (although my wife sometimes doubts this too), I will assume for the following examples that the n balls in the urn are numbered consecutively.

I start with what I think is the simplest case, namely the *selection with putting back with attention to the order*: For the selection of the first ball I obviously have exactly n *possibilities*. I note the number of this ball and then put it back into the urn. Therefore, when I draw the second ball, I again have all n possibilities. And here is the thing: for *every one of* the n possibilities for choosing the first ball, there are n possibilities for the second, so in total there are n times n, that is, n^2 different possibilities for choosing the first two balls. And so it goes on, of course, if I draw a third ball, I again have n possibilities, so in total there can then be n^3 different situations, and since there is no reason to stop after three balls, we already have a general formula for the first case:

Selection with Backspacing with Attention to the Sequence
To select k balls from a set of *nballs* with reclining and with respect to the order, there are

$$n^k$$

possibilities. The probability of making a particular choice is therefore

$$\frac{1}{n^k}.$$

So here the good old exponentiation with natural numbers from the first chapter is called for again. For example, one calculates that for the selection of four balls from an urn with 15 balls, there are a total of $15^4 = 50{,}625$ different possibilities; not bad, eh?

Of course, these results are not limited to pure spherical calculation, which is only a thought model. With a little ability of abstraction, one can solve also quite other situations with it. Here is an example:

Example 7.1 There are five number wheels on a vault, each of which can be set to one of the numbers from 1 to 99. How many possibilities are there to form a number combination from this?

The correct urn model here is (who is surprised?) the selection with putting back (since one can set for example the second cog to the same number as the first one, so this is available again) and with attention to the order. Thus there are

$$99^5 = 9{,}509{,}900{,}499,$$

so almost 10 billion possibilities. Good luck cracking the vault!

Exercise 7.1 From a deck of 32 different cards, a card is drawn three times in a row and then returned to the deck. How likely is it to draw Queen of Hearts three times?

Next I turn to the *selection with laying back without regard to the order*. Here, too, once a ball has been drawn, it is put back into the urn, but now the order in which the individual balls were drawn is no longer important when considering the final result; for example, the moves 3-6-2 and 6-3-2 would be counted as the same result, since they both produced the balls 2, 3 and 6 once each.

If you find it hard to imagine this because you prefer playing dice games to drawing balls, then perhaps the following will help. The problem of selecting two pieces from a set of six numbers (balls) with repetition without regard to order is exactly the same as the problem of achieving a particular roll result with two indistinguishable dice: Throwing a die is exactly equivalent to selecting a number between 1 and 6, and since the dice are supposed to be indistinguishable, the results of rolling a two with the "first" die and a five with the "second" die, and of rolling a five with the "first" die and a two with the "second" die, for example, are also considered to be the same event.

Before I can get serious about calculating how many possibilities there are in this situation, I am afraid I have to introduce some new notation, because simple exponentiation is no longer enough. Do you remember the definition of factorial given in the first chapter? If not, it is not so bad, I will repeat it here: if m is a natural number, then $m!$ ("m factorial") is defined as

$$m! = 1 \cdot 2 \cdot 3 \cdots (m-1) \cdot m.$$

As if that was not bad enough, we now need a new term, for the definition of which I have to resort to the factorial, namely the binomial coefficient. This combines two natural numbers, say m and l, and is defined as follows:

Binomial Coefficient

If m and l are two natural numbers with $l \leq m$, the **binomial coefficient** is

$$\binom{m}{l},$$

pronounced "m over l", defined as

$$\binom{m}{l} = \frac{m!}{l! \cdot (m-l)!}. \tag{7.1}$$

You should not ask yourself or me, why this was defined in this way; it just turned out, that exactly this combination of two numbers is needed in many formulas, and therefore this abbreviated notation was introduced.

For example, I will use the binomial coefficient to calculate the number of choices in the context under consideration. Before that, however, I want to point out a way to save work and also sources of error when calculating the binomial coefficient: Since $l \leq m$ is assumed, both l and $m - l$ are numbers between 0 and m, thus both $l!$ and $(m - l)!$ are included as factors in $m!$, and therefore the following identities can be derived from the definition (7.1):

$$\binom{m}{l} = \frac{(l+1)\cdot(l+2)\cdots(m-1)\cdot m}{(m-l)!} = \frac{(m-l+1)\cdot(m-l+2)\cdots(m-1)\cdot m}{l!}.$$

To compute $\binom{m}{l}$ one can use any of the representations, one should always take the one that requires less computation; this again depends on the size of m and l. For example

$$\binom{7}{3} = \frac{(7-3+1)\cdot(7-3+2)\cdot 7}{3!} = \frac{5\cdot 6\cdot 7}{2\cdot 3} = 35$$

and

$$\binom{4}{2} = \frac{3\cdot 4}{2} = 6.$$

Perhaps you have been wondering all along that I have not designated the natural numbers involved with n and k for the definition of the binomial coefficient, as I usually do. This is simply because in the context of urn models, in which we are right now – I hope you had not forgotten this – n and k are otherwise occupied, namely as the number of balls. The connection is now made:

Selection with Backspacing without Regard to the Sequence

To select k balls from a set of n *balls* with backspacing but without regard to order, there is

$$\binom{n+k-1}{k}$$

possibilities. The probability of making a particular choice is therefore

$$\frac{1}{\binom{n+k-1}{k}}.$$

Example 7.2 I promised in the introduction to calculate how many different rolls there are with two indistinguishable dice. We had already agreed that the correct urn model for this is selection with laying back and without regard to order. Consequently there are

$$\binom{7}{2} = \frac{6 \cdot 7}{2} = 21$$

different litters.

Exercise 7.2 There are seven balls in a drum, which are labelled with the letters A, B, E, H, M, N, T. The balls are drawn one after the other. You now draw five balls one after the other, write down the letter and put the drawn ball back into the drum. What is the probability of being able to form the word "MATHE" from the letters you have written down?

By the way, if you are desperate for more practice problems, do not worry, they will be coming. But I will do them after I have presented the two remaining urn models, and I will mix them up to make it harder!

Now I first treat the selection of k out of n balls *without putting them back*; this also means that I must now always assume $k \leq n$, because since I do not put back a ball that has already been drawn, I cannot draw more than *n balls in* total.

Let us look at the *selection without putting back with respect to the order*: For the selection of the first ball I obviously have exactly n possibilities. But since this is not put back, there are still $n - 1$ possibilities for the second ball, $n - 2$ for the third, and so on. If you now think this through further, you will find that there are still $n - k + 1$ possibilities for the *kth* ball, and since you now have to multiply these individual values, you get the number of possibilities to *select* k balls from n:

$$n \cdot (n-1) \cdot (n-2) \cdots (n-k+2) \cdot (n-k+1).$$

This can be written more elegantly using the factorial notation, and I will do this in the next notebox.

Selection Without Backspacing with Consideration of the Sequence
To select k balls from a set of *nballs* without reclining, but with respect to the order, there is

(continued)

$$\frac{n!}{(n-k)!} = n \cdot (n-1) \cdot (n-2) \cdots (n-k+1) \tag{7.2}$$

possibilities. The probability of making a particular choice is the inverse of this, i.e.

$$\frac{(n-k)!}{n!}.$$

Example 7.3 In a horse race with 12 horses participating, someone randomly bets on the top 5 finishers and wants to know the probability of being right. In order to determine this, one can apply the urn model just discussed, because it is a matter of selection without laying back, since a horse that has already run through the finish cannot run in again, and the order is of course to be observed in horse betting as hardly anywhere else. The sought-after probability is thus calculated according to the above formula with $n = 12$ and $k = 5$ as follows

$$\frac{(12-5)!}{12!} = \frac{1}{8 \cdot 9 \cdot 10 \cdot 11 \cdot 12} = \frac{1}{95{,}040} \approx 0.0000105. \quad ■$$

Exercise 7.3 There are seven balls in a drum, which are labelled with the letters A, B, E, H, M, N, T. You now draw five balls one after the other and place them in this order. What is the probability that the word "MATHE" is formed?

Before I go on to the last of the four urn models, I would like to point out a special case of selection without laying back with respect to the order, which is often presented as a separate case by mistake, or even to emphasize it. Surely you remember, that by definition was set: $0! =$ The reason for this was and is that many formulas remain correct even for 0, and this is also true for (7.2) or the probability derived from it as a reciprocal. If one sets $k = n$, which means that one takes *all* balls out of the urn, then (7.2) becomes

$$\frac{n!}{(n-n)!} = \frac{n!}{0!} = n!.$$

So there are $n!$ possibilities to take out n balls while respecting the order; this is also called the **complete draw** and is also interpreted simply as the number of possibilities to arrange n balls or other things.

Arrays
There are $n!$ possibilities to arrange n distinguishable elements of a set.

Here too is a small example: The chess club *Checkmate 07* would like to take a group picture of its current members; the five ladies are to line up in the front row and the nine men in the back row. How many possibilities are there for arranging the people in the picture?

The little trick here is to first treat the arrangement of the ladies and the arrangement of the men as separate problems. But this is easy, because since the five ladies are presumably distinguishable, there are for their arrangement $5! = 120$ possibilities and for the same reason one can find $9! = 362{,}880$ arrangements for the men. But since for *each* arrangement of the ladies *all* arrangements of the gentlemen can now be played through, I have to multiply the two values just determined to find the total possible cases. The required number of possibilities is therefore

$$5! \cdot 9! = 120 \cdot 362{,}880 = 43{,}545{,}600.$$

So the reigns will have to go through quite a while until they have tested all possible setups.

But now I finally come to the last of the four urn models, which one could jokingly call "selection without everything", namely it is the *selection without putting back and without paying attention to the order*. So you choose k balls out of a set of n, and once you have chosen a ball you do not put it back, and you do not care in the end in which order you drew the balls. Also for this I need the binomial coefficient introduced above, actually it was introduced just for this situation many years ago. The following applies:

Selection Without Backspacing Without Consideration of the Sequence
To select k balls from a set of n *balls* without reclining and without considering the order, there is

$$\binom{n}{k} = \frac{n!}{k!(n-k)!}$$

possibilities. The probability of making a particular choice is therefore

$$\frac{1}{\binom{n}{k}} = \frac{k!(n-k)!}{n!}.$$

You have probably been waiting all this time for the lottery numbers to finally come in; well, here they are:

Example 7.4 The number lottery, for example "6 out of 49", is just the standard example for the selection without putting back without consideration of the order: A ball once drawn is not put back again in the lottery, and the order in which the balls are drawn is not important (otherwise it would be even more difficult to achieve a main prize!). Thus for the selection of 6 out of 49 balls there are exactly

$$\binom{49}{6} = \frac{49!}{6!43!} = 13{,}983{,}816$$

possibilities, so the probability of hitting a particular one of them is

$$\frac{1}{13{,}983{,}816} \approx 0.0000000715.$$

At the latest now you know why very few mathematicians play the lottery. ▪

If you do not want to play the lottery, you can at least play cards; again, the formula for the number of picks can be helpful:

Example 7.5 I would like to clarify the problem of how many different ways there are to distribute the 32 cards of a Skat game among the three players (and the Skat).

For this purpose I assume (contrary to the international rules of Skat, but in accordance with the model) that 10 cards are initially dealt to the first player. If the deck is well shuffled, i.e. if we exclude the presence of Aunt Erna, who always picks up only once and does not shuffle, then there are for this

$$\binom{32}{10}$$

options. The second player also receives 10 cards, but now there are only 22 to choose from; consequently, for the second player's hand there are

$$\binom{22}{10}$$

possibilities. And since I do not want to bore you for the life of me, I will say right now, without further ado, that for the third player's hand there are only

$$\binom{12}{10}$$

possibilities. But if all three players have their cards, the Skat (i.e. the remaining two cards) is clearly determined, so there is no longer a choice for this. The number of possibilities for the distribution of the cards, which I asked for at the beginning, is now obtained by multiplying the individual possibilities just determined; this results in

$$\binom{32}{10} \cdot \binom{22}{10} \cdot \binom{12}{10} = \frac{32! \cdot 22! \cdot 12!}{10! \cdot 22! \cdot 10! \cdot 12! \cdot 10! \cdot 2!} = \frac{32!}{10! \cdot 10! \cdot 10! \cdot 2!} \approx 2.753 \cdot 10^{15}. \quad \blacksquare$$

That was quite exhausting at my age; so I am going to take a moment and let you take the field:

Exercise 7.4

(a) A football club has 15 active players. How many possibilities does the coach have to form a team of 11 players out of this? (Even if it hurts my son, who is an active footballer, I assume here for simplification that every player can take every position, including that of the goalkeeper).

(b) The coach from part a) has resigned because he cannot work with a team of 15 all-rounders. The new coach analyses the potential players more closely and finds that there are seven defenders, six strikers and two goalkeepers (he has given up the midfield). How many options does the coach have to form a team from this consisting of five defenders, five strikers and a goalkeeper?

I had already threatened to formulate a few more exercises in which you will have to decide for yourself which of the urn models formulated above can be applied; now the time has come:

Exercise 7.5

(a) In the classic football pool you have to bet on a draw (0), home win (1) or away win (2) in a total of 11 matches. How many different picks are possible here?

(b) You have the choice of playing either the "5 out of 25" lottery or the "4 out of 20" lottery. Which has a higher chance of winning the jackpot?

(c) At a bus stop, four passengers board the bus and find seven empty seats. How many possibilities do they have to distribute themselves on four of these seats?

(d) In a sports tournament, the 12 participating teams must be divided into three groups of four teams each. How many options does the organiser have for this?

7.2 Relative Frequency and Classical Definition of Probability

In this chapter I have already written about probabilities for about 10 pages and calculated them in certain cases. However, I have not yet *defined* the term properly, and that is not surprising, because such a definition is not so easy, at least if the probability defined in this way is to coincide with the calculated numbers. This starts at the very beginning, for example, when I assumed that a certain side of a coin lies on top after the toss with the "probability 1/2". This may correspond to some life experience, but it cannot be proven by any mathematical theorem in the world.

To get closer to the concept of probability, it is a good idea to look at the so-called frequency of events:

Random Experiment and Random Event

A **random event**, usually **referred to as** an **event for** short, is the result of a random experiment. A random **experiment,** in turn, is an experiment that can (at least theoretically) be repeated any number of times and whose outcome is unpredictable.

Such a random event is, for example, the toss of a coin or the drawing of balls from an urn, in which case "coat of arms up" or "2 and 5 drawn" are associated random events.

With the help of these terms I can now define the frequency(ies) of random events:

Absolute and Relative Frequency

A random experiment is repeated n *times* and the random event A occurs exactly a *times.*

Then the number is called

$$H_n(A) = a$$

the **absolute frequency of** the event A and the number

$$h_n(A) = \frac{a}{n}$$

the **relative frequency** of event A.

Of greater significance here is certainly the relative frequency $h_n(A)$, because it indicates in which fraction of all n cases the event A has occurred.

Since n *is* a positive number and a is an integer between 0 and n, $h_n(A)$ is a real number between 0 and 1, more precisely $h_n(A)$ is equal to one of the numbers

$$0, \ \frac{1}{n}, \ \frac{2}{n}, \ \ldots, \ \frac{n-1}{n}, \ 1.$$

For example, if event A occurs 864 times in a trial performed 2000 times, then

$$H_{2000}(A) = 864$$

and

$$h_{2000}(A) = \frac{864}{2000} = 0.432.$$

Believe it or not, but in order to be able to give you a realistic numerical example here, I sat down while writing these lines and flipped a coin 100 times; the event $W =$ coat of arms above occurred exactly 47 times. So for this event

$$H_{100}(W) = 47$$

and

$$h_{100}(W) = \frac{47}{100} = 0.47.$$

Maybe coin tossing is too boring for you? In that case, I invite you to throw the dice:

Exercise 7.6 When rolling a standard die, the six numbers of points occurred with the following absolute frequencies:

Roll of the dice	1	2	3	4	5	6
Frequency	137	140	127	120	140	133

Calculate the relative frequencies of the following events in this random experiment:

A_1 : throw of a 5

A_2 : throw an even number

A_3 : throw an odd number not divisible by 3 ■

I had bragged a bit above about having gone to the trouble of flipping a coin 100 times. That is quite a respectable feat for an aging math professor, but of course it is still not enough to make a firm statement about the frequency to expect for W to occur.

Fortunately, there were and are far more industrious people than me who have taken on many thousands of flips of a coin, in earlier times actually at the gaming table, today also gladly as a computer simulation.

The result, and *this* is indeed one of the mysteries of probability theory, is the same everywhere in the world and at all times of mankind: after a certain number of throws, the relative frequency of the throw of the crest settles at 0.5. Put another way: In pretty much half of all cases, the coin at the top will show the coat of arms. Mind you: No one can predict how the next toss, the one after that, the third toss, ... will turn out, but one can predict with great certainty that after 100,000 tosses one will have seen the coat of arms on top about 50,000 times.

Before you ask me, I have *no* idea why that is! I can, with some effort, calculate the most convoluted probabilities of compound, conditional, and otherwise misshapen events, but I really cannot tell you why this darn coin behaves the way it does.

It is no different with more complex systems; for radioactive materials, for example, one can predict fairly accurately how much of the material initially present will still be there after a certain unit of time and how much will have decayed, these are relatively simple equations whose results nevertheless agree amazingly well with reality. But of *a given* single particle, no man or formula in the world can tell whether it will decay in the next second or not.

Probability is therefore something that can only be calculated "in the big picture"; even if – let it be said for the mathematical purists among you – it puts the cart before the horse, one can intuitively define the probability of an event as that value to which the relative frequencies of that event settle for very large numbers of trials. However, you cannot calculate the probability of an event in this way, you only get estimated values; moreover, this is only limited to relatively simple experiments such as coin tosses or dice games, because, for example, determining the probability of a total meltdown in a nuclear power plant by relative frequencies of this event, i.e. with the help of a sufficiently long series of experiments, is not something I consider advisable.

There is a slightly better definition of probability that I do not want to withhold from you. It is also called the **classical definition of probability**, and it goes back to the great mathematician and probability theorist Pierre S. de Laplace, after all. But even great people are not perfect, and this definition is not either, because it is only applicable to random experiments with *finitely many* outcomes, and only those where every single outcome has the same single probability. So, in principle, this is exactly the situation I assumed in the first section on combinatorics. In this respect, this definition is not so wrong, at least as a first approach, which is why I am now also dealing with it for a while.

Classical Definition of Probability
A random experiment has N different equally probable outcomes, in exactly k *of* which event A occurs. Then one calls the number

$$p(A) = \frac{k}{N}$$

is the **(classically defined) probability of** the event A. Sometimes we also say that k is the number of **cases favorable** to A.

The letter p stands for probability and should be easier to understand than this definition. To make this again more understandable, here is a first example:

Example 7.6 There are 20 balls in an urn, numbered from 1 to 20. What is the probability for the event U of drawing a ball whose number is odd but not divisible by 3?

In this simple case, you simply count the favorable cases: They are numbers 1, 5, 7, 11, 13, 17, 19, i.e., seven pieces, consequently

$$p(U) = \frac{7}{20} = 0.35. \quad \blacksquare$$

Exercise 7.7 In a raffle, a total of 500 tickets are sold; among them is one grand prize, 25 high-value prizes, 140 consolation prizes, and the rest are studs. What is the probability of event G, which is to win anything at all?

One of the great advantages of the above definition of probability is that it can be used to do "real math"; I will show you two examples of this.

If, as in the definition, the experiment has N outcomes, of which the event A occurs in exactly k cases, this means that A does not occur in exactly $N - k$ cases. The event "*A does* not occur" is usually called the **counter-event** of A and is denoted by $\overline{A}$; since it occurs in exactly $N - k$ *of* the N cases, it has probability

$$p(\overline{A}) = \frac{N-k}{N} = 1 - \frac{k}{N}.$$

But since k/N was just the probability of $p(A)$, we derived the following rule of calculation:

Probability of Counter Event

The probabilities of an event A and the counter event $\overline{A}$ are related as follows:

$$p(\overline{A}) = 1 - p(A).$$

This I want to understand as a first example for the mentioned "calculating with probabilities". To illustrate it, I take up again Example 7.6. The counter-event $\overline{U}$ of the event U defined there consists just in drawing a number, which is either even or divisible by 3 (or both). The favorable cases for this are numbers 2, 3, 4, 6, 8, 9, 10, 12, 14, 15, 16, 18, 20, that is, 13 of them. The probability of $\overline{U}$ is thus

$$p(\overline{U}) = \frac{13}{20} = 0.65,$$

so just $1 - p(U) = 1 - 0.35$.

For the second example of "calculating with probabilities" we need two events, say A *and* B, and a new conceptualization:

Incompatible Events

Two events A and B are called **incompatible** if they never occur simultaneously, i.e. if the occurrence of B is possible if A occurs, and vice versa.

For example, when rolling with a standard die, the events are

$$A = \text{throw of a 1}$$

and

$$B = \text{throw of an even number}$$

are incompatible.

From two (or more) events one can construct a new one by celebrating the occurrence of at least one of them already as a new event; formally, this new event is called the sum of the two initial events and symbolized by the sign "∪"; if this makes you feel unpleasant associations with set theory, this is quite justified (the associations, not necessarily the fact that they are unpleasant): In fact, the whole of modern probability theory is built axiomatically with the help of set theory, by conceiving of events as sets and defining their probability as the "measure" of that set, in a sense its size. I will take a step towards this axiomatic view in the next section we are heading towards, but in order to spare you (and me) the introduction of full set theory, I will limit myself to what is probabilistically necessary. Unfortunately, this has the small disadvantage that a few terms (like here the sum of events) are introduced, whose full background or their name cause remain hidden, but I think you can live with that at this point.

Sum of Events

If A and B are two events, then one can define a new event $\boldsymbol{A \cup B}$, which is called the **sum of** A **and** B. This event occurs when *at least one of* the two events A or B occurs.

The sum of the two events defined above would thus be

$$A \cup B = \text{throw of a 1 or an even number},$$

hence

$$A \cup B = \text{throw of a } 1, 2, 4 \text{ or } 6.$$

The probability of event $A \cup B$ can be seen directly here: Since it occurs in four out of six possible cases.

$$p(A \cup B) = \frac{4}{6} = \frac{2}{3}.$$

But this result can also be obtained in another way: Event A, i.e. the throw of the number 1, obviously has the probability 1⁄6, while event B, i.e. the throw of an even number, has the probability 1⁄2. Since both events are incompatible, i.e. can never occur at the same time, I can try to add these probabilities, and using the rules for fractions I get:

$$p(A) + p(B) = \frac{1}{6} + \frac{1}{2} = \frac{1}{6} + \frac{3}{6} = \frac{2}{3},$$

thus the identical result. This is true not only in this example, but in general, and this should be noted – as a second rule of calculation for probabilities:

Sum of Incompatible Events

If A and B are incompatible events, then the probability of their sum can be calculated as follows:

(continued)

$$p(A \cup B) = p(A) + p(B). \tag{7.3}$$

So, in catchphrase form, we can say, "The probability of the sum is equal to the sum of the probabilities." But, let us remember once again, this only applies to incompatible events!

Exercise 7.8 Let the six sides of a cube be printed with the numbers 1, 1, 3, 3, 4, 5. We consider the three events

$$A = \text{throw of an even number,}$$
$$B = \text{throw of a number divisible by 3,}$$
$$C = \text{throw of the number 5.}$$

Show that these three events are pairwise incompatible, that is, that every possible pair of two that can be formed from these events is incompatible, and calculate in two different ways the probability of the union of each two of these events.

7.3 Axiomatic Definition of Probability

I think by now you have progressed so far into mathematical thinking that I can confront you with an axiomatic definition – do not worry, I am with you. According to the Mathematics Dictionary, an axiom is "a statement which, because of its content, is fundamental and considered to be evident, and therefore requires no proof". In other words, an axiom is simply slapped down as the basis of a theory without further justification or even motivation, and then conclusions are drawn from it and the whole theory is built up.

In the case of probability theory, this was done at the beginning of the twentieth century by the Russian mathematician A.N. Kolmogoroff; he was guided by the rules of calculation given above, which must obviously apply to "reasonable" probabilities, and then wrote down what properties a mapping from the set of events to that of the real numbers must possess in order to be called a "probability". The result was the following:

Axiomatic Definition of Probability

On the set of all events that can occur as a result of a random experiment (this can now also be an infinite number), define a function p that assigns a real number to each

(continued)

event. This function or its values is called **(axiomatically defined) probability** if it has the following properties:

1. For any event A, $0 \leq p(A) \leq 1$, so the function p only takes values between 0 and 1.
2. The so-called **sure event**, which combines all possible events and is traditionally denoted by Ω ("omega"), has the value 1: $p(\Omega) = 1$.
3. For each pair of incompatible events A and B, the following applies

$$p(A \cup B) = p(A) + p(B).$$

4. Property (3) also holds for the sum of any number of events, including infinitely many, if they are pairwise incompatible.

I think I have earned myself a piece of chocolate for now, and you a few explanatory remarks to make this definition more understandable to you:

(a) There are many abstrue functions in mathematics which, according to this definition, represent a "probability", but which have little to do with what is commonly understood by it; if this only confirms your prejudice about mathematics, I cannot change it now, unfortunately. In this book, however, I will deal only with "normal" probabilities, and only with those that allow only finitely many events. By the way, if you doubt that there are any such ones with infinitely many events at all, why not play the game (just invented by me) "Think of any real number!" with a partner? Since there are infinitely many numbers, this random experiment also has just as many events.

(b) Back to the analysis of our axioms: Property (1) says that a probability is always a number between 0 and 1, so (2) says that the sure event Ω has the maximum probability. One more word about this "sure event": This consists, as already written, of the summary of *all possible* events and thus occurs with certainty; examples of this are the throwing of one of the numbers 1, 2, 3, 4, 5, 6 in dice or the drawing of one of the numbers 1,2, …, 49 in the standard lotto.

(c) The opposite of the sure is the **impossible event**, symbolized by the $\emptyset$ sign. Examples include rolling a seven in dice, drawing a negative number in the lottery, or writing a math book free of typos.

(d) Perhaps you are surprised that one has to demand (4) in addition to (3), although it looks as if one can derive this from (3). Well, I can only say here, without being able or wanting to prove this in this framework, that this derivation is *not* possible, if we are dealing with infinitely many events, which I had admitted.

Above I advertised that one has computational rules for axiomatically defined probabilities, and it is now time to make good on that promise; to make it clear that these are in addition *to* axioms as rules, I will just keep counting.

Rules for Calculating Probabilities
In addition to the above axioms, the following rules apply to probabilities:

5. The impossible event has probability zero: $p(\emptyset) = 0$.
6. If $\overline{A}$ is the counter-event to A, then $p(\overline{A}) = 1 - p(A)$.
7. If an event A *is* part of another event B, i.e. if the occurrence of A implies that of B *in* all cases, then $p(A) \leq p(B)$ holds.

Incidentally, since the impossible event is just the counter-event to the event Ω, (5) just follows from (6) and the axiom $p(\Omega) = 1$, but it is customary to write down this property separately anyway.

In order to illustrate the way of speaking used in (7), I again make use of the dice game and define the events

$$A = \text{throw of a 2}$$

and

$$B = \text{throw of an even number.}$$

Surely A is part of B, for 2 is an even number; the second way of speaking is also made clear by this, for the event A, having rolled a 2, implies that an even number has been rolled, that is, B. And the asserted inequality is also true here, for $p(A) = 1/6$ is certainly smaller than $p(B) = 1/2$.

But now it is high time for a more detailed example:

Example 7.7 This time we roll the dice with a dice from the secret workshop of Al Capone. For rolling the numbers 2–6, the dice hasthe following probabilities:

$$p(6) = p(5) = \frac{1}{5} \quad \text{and} \quad p(4) = p(3) = p(2) = \frac{1}{6}.$$

(a) What is the probability of rolling a one with this dice?

To answer this, all I have to do is note that the sum of all probabilities, which is, after all, the certain event, must be one. Since the sum of the given five probabilities is just

$$\frac{2}{5}+\frac{3}{6}=\frac{9}{10}$$

is, remains only

$$p(1)=\frac{1}{10}$$

left.

(b) What is the probability of *not* rolling *a* six with this dice?

This is quite simple: rolling "no 6" is just the counter-event of rolling a six, which in turn has probability 1/5 according to the specification. Thus

$$p(no\ 6)=1-\frac{1}{5}=\frac{4}{5}.$$

(c) What is the probability of rolling an odd number with this dice?

Since the rolls of the three odd numbers are incompatible (you can only roll either a one or a three or a five), to answer this question according to axiom (3) you only have to add up the three individual probabilities; the following follows

$$p(\text{odd number})=p(1)+p(3)+p(5)=\frac{1}{10}+\frac{1}{6}+\frac{1}{5}=\frac{7}{15}. \quad \blacksquare$$

I have saved the real "banger" among the calculation rules for probabilities for the end, namely the formula for the probability of a sum of events which are not necessarily incompatible. In order to be able to formulate this now, I need one last new term in this section, namely that of the product of events:

Product of Events

If A and B are two events, then one can define a new event $A \cap B$, which is called the **product of** A **and** B. This event occurs when *both A and B* occur.

The product of the two events

$$A=\text{throw of a 1 or 2}$$

and

$$B = \text{throw of an even number}$$

would be the event

$$A \cap B = \text{throw of a 2,}$$

because only in this case both A and B have occurred.

Now I am in a position to give you the following formula, which is certainly the most frequently used rule of calculation when dealing with probabilities:

Sum of any Events

For any (i.e. not necessarily incompatible) events A and B, *the* following applies

$$p(A \cup B) = p(A) + p(B) - p(A \cap B). \tag{7.4}$$

The abundance of rules you have learned in the meantime really cries out for detailed examples, and since we are "through" with the material of this section, I will also only cover all the rest with them.

Example 7.8 In a random experiment, a total of four events A, B, C, D are possible, all of which are supposed to be pairwise incompatible. The following data are known:

$$p(A \cup B) = \frac{13}{21}, \quad p(A \cup C) = \frac{8}{15}, \quad p(B \cup C) = \frac{17}{35}.$$

Now calculate the individual probabilities of the four events.

Since the events are inconsistent, I can use the special formula (7.3); I first write down what I know: it is

$$\begin{aligned} p(A) + p(B) &= p(A \cup B) = \frac{13}{21}, \\ p(A) + p(C) &= p(A \cup C) = \frac{8}{15}, \\ p(B) + p(C) &= p(B \cup C) = \frac{17}{35}, \end{aligned}$$

and this is a beautiful linear system of equations with three lines to calculate the three unknown individual probabilities. Using the methods you learned in the chapter on linear algebra, you can easily find the solutions

$$p(A) = \frac{1}{3}, \quad p(B) = \frac{2}{7}, \quad p(C) = \frac{1}{5}$$

out. As there is to be only *one* more event altogether, namely D, this is just the counter-event of these three. It follows

$$p(D) = 1 - (p(A) + p(B) + p(C)) = 1 - \left(\frac{1}{3} + \frac{2}{7} + \frac{1}{5}\right) = 1 - \frac{86}{105} = \frac{29}{105}. \quad \blacksquare$$

And because it is so nice, here is another example:

Example 7.9 In a plastics processing plant, it is found that 5% of the parts produced have deformations and 8% have color inaccuracies; 3% of all parts even have both defects. What is the probability that a randomly selected part will have

(a) at least one of the two errors,
(b) at most one of the two errors,
(c) no mistake

With such "text tasks" one should first translate the percentages into probabilities. I denote the event that a randomly selected part exhibits a deformation with V, correspondingly with F the colour fidelity. Then holds, since $p\%$ are just *p/100*,

$$p(V) = \frac{5}{100} = 0.05, \quad p(F) = \frac{8}{100} = 0.08 \quad \text{sowie} \quad p(V \cap F) = \frac{3}{100} = 0.03.$$

The latter, since "both errors" means just the simultaneous occurrence of V and F. Now one only needs to apply the mentioned calculation rules appropriately:

(a) This is the event $V \cup F$ and according to (7.4) is

$$p(V \cup F) = p(V) + p(F) - p(V \cap F) = 0.05 + 0.08 - 0.03 = 0.1.$$

(b) This is the counter-event to "having both faults at the same time", that is, to $V \cap F$. Thus

$$p(\text{has at most one error}) = 1 - p(V \cap F) = 0.97.$$

(c) Again, the quickest way to get there is to use the counter-event argument; this is now $V \cup F$, hence

$$p(\text{no error}) = 1 - p(V \cup F) = 0.9. \quad \blacksquare$$

Exercise 7.9 In a grammar school the foreign languages English and Latin are offered, each pupil has to learn at least one language. Out of a total of 50 students, 35 learn English and 25 learn Latin. What is the probability that a randomly selected pupil will

(a) learned both languages,
(b) only learned English,
(c) only learned one language?

Exercise 7.10 Show that formula (7.3) for the sum of incompatible events is included as a special case in formula (7.4).

7.4 Conditional Probabilities

This last section will be quite short, I promise you, but it is necessary, because you cannot close the chapter on probability theory without having talked about conditional probabilities. I will now introduce you to this concept, and I will start, as usual, with an example: The probability of rolling a 6 on the standard dice is, as you already know, equal to 1⁄6. But suppose someone were to guarantee you that an even number would fall in any case (that would be the "condition"), we should assume that this would increase the probability of a 6 to 1⁄3, since there would then only be three numbers to choose from.

This is indeed the case, and to be able to calculate this in general, one uses the formula for the conditional probability:

Conditional Probability

I assume that *A* and *B* are two events of the same random experiment and that event *B* does not have probability zero: $p(B) > 0$. Then $p(A|B)$, defined by

$$p(A \mid B) = \frac{p(A \cap B)}{p(B)} \tag{7.5}$$

the **probability of the event** *A* **under the condition** *B* or in short **the conditional probability of** *A* with **respect to** *B*.

So the probability of the product of *A* and *B* is divided by that of *B*, and since this is non-zero by premise, nothing can happen in the process.

I take up the initial example again: Let *A* be the event "throw of a 6" with $p(A) = 1/6$ and let *B* be the event "throw of an even number" with $p(B) = 1/2$. In order to apply formula

(7.5), I still need the probability of the product $A \cap B$ of A and B. *This consists of just* throwing a 6 and thus p(A ∩ *B*) = *1⁄6*. This just consists of the roll of a six, and so $p(A \cap B) = 1/6$. Overall, we get as the probability of the roll of a six under the condition that an even number is rolled:

$$p(A \mid B) = \frac{\frac{1}{6}}{\frac{1}{2}} = \frac{1}{3}$$

in accordance with "intuition".

In general, rolling dice is very good for illustrating the conditional probability formula, since you can interpret the results quite well; almost every textbook I know does it this way, and why should I object?

Example 7.10 I roll a standard die and take as condition B the "rolling of an odd number". So it is $p(B) = 1/2$. Now I consider three different dice events A_1, A_2, A_3 and calculate their probability as well as the probability of these events under the condition B.

(a) Let A_1 be the event "throw of 1, 2 or 3". Then $A_1 \cap B =$ "throw of a 1 or 3" and thus

$$p(A_1) = \frac{1}{2} \quad \text{and} \quad p(A_1 \mid B) = \frac{\frac{1}{3}}{\frac{1}{2}} = \frac{2}{3}.$$

So if you know that an odd number will be thrown (event B), the probability of event A_1 increases.

(b) Let A_2 be the event "throw of 2, 3 or 4". Then $A_2 \cap B =$ "throw of a 3" and thus

$$p(A_2) = \frac{1}{2} \quad \text{and} \quad p(A_2 \mid B) = \frac{\frac{1}{6}}{\frac{1}{2}} = \frac{1}{3}.$$

So if you know that an odd number will be thrown, the probability of event A_2 decreases.

(c) Let A_3 be the event "throw of 5 or 6". Then $A_3 \cap B =$ "throw of a 5" and thus

$$p(A_3) = \frac{1}{3} \quad \text{and} \quad p(A_3 \mid B) = \frac{\frac{1}{6}}{\frac{1}{2}} = \frac{1}{3}.$$

So, if you know that an odd number is thrown, this does not affect the probability of event A_3.⍰

To test the correctness or plausibility of a formula, one should always look at special cases; therefore, in the case of formula (7.5), I will conclude by testing two such special cases:

(a) If A and B are incompatible events, then $A \cap B$ is impossible, and so $p(A \cap B) = 0$. But this also means that $p(A \mid B) = 0$, and this is just as well, since this is the probability of A occurring under the condition that B *has* already occurred; but if, as assumed, A and B are incompatible, then A *can* no longer occur if B has already occurred.
(b) If $A = B$, then $A \cap B = B$ too, and so $p(A \mid B) = 1$. This is fine too, because if I set the condition that *leg kick* and $A = B$, then of course A must occur too.

In order to practice the handling of the conditional probability a little bit, I would like to ask you to work on the last exercise.

Exercise 7.11 A company receives its vendor parts from two different companies, 75% from company A and 25% from company B. A random sample has shown that the probability of a randomly sampled part being intact and coming from company B is $p = 0.23$. What percentage of the parts supplied by firm B are intact?

Descriptive Statistics

Winston Churchill is often credited with saying that he did not trust any statistics that he had not falsified himself. Although this is well formulated, it has one or two weaknesses, because firstly, Churchill was a soldier, politician and historian – but not a statistician who understood anything about the matter. And second, much more seriously, despite a long search, no evidence has been found that he ever uttered a sentence of this kind; indeed, it is now assumed that Goebbels disseminated it during the course of the Second World War and blamed it on Churchill in order to discredit him as a habitual liar.

This may be comforting for Churchill, but it will do little to change the rather dubious reputation of statistics. After all, it was Benjamin Disraeli, a predecessor of Churchill's in the office of British Prime Minister, who opined that there are three kinds of lies: Lies, damned lies, and statistics. That does not sound very confidence-inspiring. But why do statistical statements have such a bad reputation? There may be at least two reasons. On the one hand, of course, it has always happened and continues to happen that statistical material is used to lie and cheat, that data is either falsified or prepared in such a way that it gives a false impression. There is not much I can do about that; the world is what it is. On the other hand, many people are not at all familiar with statistics, and the tables and graphs they are confronted with are more frightening than insightful. And that is what I want to do something about in this chapter.

So here I will show you how to display collected data, how to transform long tables into clearer tables and even clearer graphs, and how to calculate certain parameters such as the mean and the standard deviation from a set of data. But first I want to say a few words about what statistics, and descriptive statistics in particular, is all about.

G. Walz et al., *Foundations of Mathematics: A Preparatory Course*,
https://doi.org/10.1007/978-3-662-67809-1_8

8.1 Introduction

What is a "statistic"? Usually, it means some more or less clearly arranged columns of numbers in the form of tables or graphs, but why does one actually take the trouble to compile such statistics? As a rule, the aim is to obtain information about a certain section of reality, data that tell the user something about the current situation and perhaps also allow one or two conclusions to be drawn about the future. There are statistics on unemployment figures and sales, on eating habits and voting behaviour, in fact on almost all variables in which anyone is interested.

But that leads to a problem. For example, you cannot ask *all* drivers in Germany about the number of times they swear while waiting at a red light, just as you cannot determine what *all* German citizens eligible to vote think about the government: there are simply too many of them, and the eventual value of such statistics would be far exceeded by the cost of producing them. For this reason, one usually restricts oneself to *random samples*, which – in contrast to a total survey that covers everything – are limited to a section of the entire possible data. So you do not ask all motorists how upset they are at red lights, but only 100 or 1000, and you do not ask all eligible voters, but only a certain manageable number. This means that you have solved the problem of the excessive costs of a total survey simply by not carrying out a total survey, but you have obviously got yourself into another problem, because a sample does not include all motorists or eligible voters: How then can you infer the situation of everyone else from the results of the sample? The set of all objects of interest to me is also called the *population*, and the question is how to draw conclusions from a sample to the population.

This, however, is the task of *inductive* or *inferential* statistics, which will not concern me here. What is needed to accomplish this task, however, is reasonable information about the properties of the sample at hand, because one cannot expect to learn anything about the population if one already knows nothing about the sample. So the first thing is to present and describe the properties of the sample, and that is why we talk about "descriptive" statistics. I will deal with this in the following pages.

Descriptive Statistics
The aim of descriptive statistics is to clearly present and arrange the empirically collected data of a sample by means of tables, graphs and key figures.

You will see in the course of this chapter what exactly is meant by *key figures.* But before I show you how to deal with the empirically collected data I mentioned, I would like to encourage you to be a little cautious and suspicious about the survey. In the U.S., there was once a survey of public attitudes toward free speech that was conducted in two different ways. One half of the sample was asked, "Do you think the U.S. should prohibit public attacks on democracy?" while the other half was asked, "Do you think the U.-S. should not allow public attacks on democracy?" You can see the results in Table 8.1.

Table 8.1 Results of the surveys

	Prohibit (%)	Do not allow (%)
Yes	54	75
No	46	25

Although both questions actually want to know the same thing, the respondents were not quite so petty in dealing with freedom of speech in the supposedly friendlier formulation. As one can see, the outcome of a survey may depend heavily on the wording of the question. I should mention that this example and many others can be found in Walter Krämer's book "So lügt man mit Statistik".

So much for the psychological side of statistical surveys. In the next sect. I will tell you something about the handling of *characteristics*.

8.2 Representation Methods

Obviously, as a rule it is not possible to represent complete aggregates within the framework of a statistic, and it would not be particularly interesting to gather any drivers or voters in a table. What is really interesting are certain properties of the objects to be studied, which can be determined in various ways: You can ask the opinions of motorists, the size of apartment buildings can be measured, the number of potholes in a state highway can be counted. Such characteristic properties of an object are usually called a *feature*, while the objects whose features are collected are called *feature carriers*. You should be reasonably careful with the term characteristic; the characteristic is the property itself, such as "age of a person", "color of a car", or "number of ex-wives". The concrete value of a characteristic, on the other hand, is called the *characteristic value*.

Example 8.1

You yourself are the characteristic carrier of many different characteristics, including the characteristic "Height in centimeters". If, for example, you are 180 cm tall, this characteristic has the characteristic value 180 for you.

Now there are obviously very different types of characteristics, because it does make a difference whether I am interested in the colour of a car or rather in its average petrol consumption. Characteristics whose expressions cannot be meaningfully expressed in numerical values are called *qualitative characteristics*, all others *quantitative characteristics*, which is of course because their expressions correspond to numerical values, i.e. they are quantifiable. The usual distinction is, however, somewhat more precise and is usually expressed with the aid of the following three *scales*, whereby here the term scale is simply to be understood as a classification.

The least demanding is the *nominal scale*. Its characteristic expressions are basically only qualitative in nature and often consist of simple adjectives. For example, the characteristic "color" has the expressions "yellow" or "red" and the characteristic "state of health" has the expressions "sick" and "healthy". However, they do not always have to be adjectives, as the example of the characteristic "Favourite sport" shows, which can have the characteristics "football", "swimming" or "none", among others.

While it is not possible to make any meaningful comparisons (except for equality) between expressions of a nominal scale, the situation is different for the *ordinal scale* or *rank scale*. Here, the characteristic values can be ordered and sorted, even though it would not really make sense to calculate the distances between the different values. Examples are the characteristics "Exam grades" and "Ranking place of a university". Surely grade 1 is better than grade 4, but the distance of three between the two grades has no real meaning, the grades only give a ranking of the different performances.

The most accurate way to do this is to use a *metric scale*. In this case, you are dealing with sensible real numbers for the characteristic values that actually say something, and the differences between different values also have a concrete meaning, since they express the differences between the characteristic values in numbers. If, for example, the characteristic "height in centimetres" is concerned, person A may have the characteristic 180, while person B has the characteristic 175, which of course means, among other things, that A is 5 cm taller than B: the numerical difference between the characteristics describes the actual difference between the characteristics.

The three scales are worthy of being immortalized in their own box.

Statistical Scales

A distinction is made between the following statistical scales:

(a) Nominal scale: Characteristic expressions are distinguishable, but only qualitative properties are named to distinguish them, usually verbally.
(b) Ordinal scale or ranking scale: The proficiencies allow a comparison, so that a sorting is possible.
(c) Metric scale: The expressions are real numbers and the differences between the expressions describe the difference of the respective characteristic expressions.

It is quite nice to be able to assign different characteristics to different statistical scales, but it is about time to have a look at one or the other concrete statistic.

Example 8.2

A university wants to know more about its new students. Since it is possible to study with a wide variety of school-leaving qualifications nowadays, the highest previous school-leaving qualification from a general school is registered by each newcomer, whereby for the sake of simplicity I will restrict myself to the classic german schools of secondary school, Middle schooland High school. In addition, however, great importance is attached to ergonomically acceptable seating, which is why the university administration is also

Table 8.2 Tabular presentation

Feature carrier	Height	Graduation
1	185	Middle school
2	177	Secondary school
3	163	Secondary school
4	175	High school
5	162	Middle school
6	177	High school
7	174	High school
8	168	High school
9	190	Middle school
10	175	High school
11	163	Secondary school
12	172	Middle school
13	184	Secondary school
14	184	High school
15	180	Middle school
16	175	Secondary school
17	172	High school
18	168	High school
19	190	Middle school
20	166	High school
21	180	High school
22	172	Secondary school
23	184	Middle school
24	167	High school
25	172	Middle school
26	185	High school
27	174	High school
28	166	High school
29	175	Middle school
30	168	High school

interested in the height of its students. Thirty new students were interviewed; the result can be found in Table 8.2, with the body height given in centimetres without decimal places.

You can see that the students in their role as feature carriers are not listed with their name, but with the sequence number, since the name does not contribute anything to the matter. Furthermore, even with the basically manageable number of 30 characteristic carriers – i.e. also 30 data records – it is noticeable that the table does not necessarily look very clear.

The idea now is to separate the characteristics summarized in this one table from each other and at the same time only read from the table what is really of interest.

Table 8.3 Compressed representation

Sequence number	Graduation	Absolute frequency	Relative frequency
1	Secondary school	6	0.2
2	Middle school	9	0.3
3	High school	15	0.5

So now I am only looking at the characteristic "school-leaving qualification" for a while. Do you really need to know that the student with the number five went to Middle school, while the one with the number six went to High school? No one really cares that much; what you really want to know is the number of graduates from each type of school. I therefore count how many students come from which school and obtain Table 8.3 from this.

This already looks much more manageable. There are three different types of school-leaving qualifications, and Table 8.3 lists how often which school-leaving qualification occurs among the 30 students surveyed: this is called the *absolute frequency*. In addition, the *relative frequency* is also given, which describes what proportion of all participants have this or that school-leaving qualification. In this case, the relative frequencies are calculated from $6/30 = 0.2$; $9/30 = 0.3$ and $15/30 = 0.5$.

So that is how easy it is to summarize the data for the characteristic "school-leaving qualification" in a summary table, and of course you can do that with any nominally scaled characteristic, since no one will stop you from counting. I will now note the two new terms again.

Absolute and Relative Frequencies

If n characteristic carriers – i.e. n data records – and the characteristic values $A_1, \ldots, A_j$ are given, the absolute frequency n_i describes the number of characteristic carriers that have the value A_i. The following always holds

$$n_1 + \cdots + n_j = n.$$

The relative frequency of characteristic expression A_i is calculated by

$$h_i = \frac{n_i}{n}.$$

It always applies

$$h_1 + \cdots + h_j = \frac{n_1}{n} + \cdots + \frac{n_j}{n} = \frac{n_1 + \cdots + n_j}{n} = 1.$$

Now this is a rather formal description of a simple fact. In Example 8.2, I have three characteristic values for the school-leaving qualification, so here $j = 3$. The number of characteristic carriers is 30, so $n = 30$. The absolute frequencies, which can be determined by simply counting, are $n_1 = 6$, $n_2 = 9$, $n_3 = 15$, and as soon as these absolute frequencies are divided by the total number 30 of all available data sets, the relative frequencies given are obtained. The fact that $h_1 + h_2 + h_3 = 0.2 + 0.3 + 0.5 = 1$ then also applies need not surprise anyone.

However, in the case of a nominally scaled characteristic such as school-leaving certificate, one need not leave it at the tabular representation, for after all many people prefer a picture to any table. The appropriate tool for this purpose is the *pie chart.*

Example 8.3

If one imagines the set of all characteristic carriers as a circle, one can enter the individual shares of the characteristic values in this circle. In the case of school-leaving qualifications, half of the area of the circle is occupied by pupils, a share of 0.3, i.e. 30%, is claimed by Middle school pupils, and Secondary school pupils hold a share of 0.2, i.e. 20% (see Fig. 8.1).

So much for nominal scaled characteristics. I should also mention that you can represent ordinally scaled characteristics using exactly the same methods, which is why I can save myself a separate treatment and move straight on to metrically scaled characteristics.

Example 8.4

I take another look at the student table from Example 8.2. So far I was only interested in the characteristic "school-leaving qualification", now I turn to the metrically scaled characteristic "body height". To get a better overview, I count how often which height occurs and summarize the results in a new table. So that the table does not become too wide, I use the following abbreviations: af denotes the already discussed *absolute frequency*, rf the likewise known *relative frequency*, with cf. mean the *cumulated frequency*, whose meaning I will explain shortly, and with ref. correspondingly the *relative cumulated frequency.*

As you can see, I have listed the occurring heights by height, which will prove beneficial in a moment. In contrast to the original table, I can now immediately read that, for example, four freshmen are 175 cm tall without having to search long, and I also know without further effort that this corresponds to a proportion of 0.1333, or 13.33%. In addition, however, I have also cumulated the frequencies that occur, and that means that I have added up, for example, how often a value has occurred that is less than or equal to 175: namely exactly 19 times. So I get the cumulative frequency of a characteristic value in such an ordered table by simply adding up the number of all absolute frequencies that have occurred up to this table entry. And the same applies to the relative cumulative frequency, which corresponds to the sum of the usual relative frequencies that have occurred up to this row. Therefore, we can immediately see from Table 8.4 that the proportion of freshmen with a height less than or equal to 175 cm is 0.6333, and this means that 63.33% of all freshmen are no taller than 175 cm.

Fig. 8.1 Circular diagram

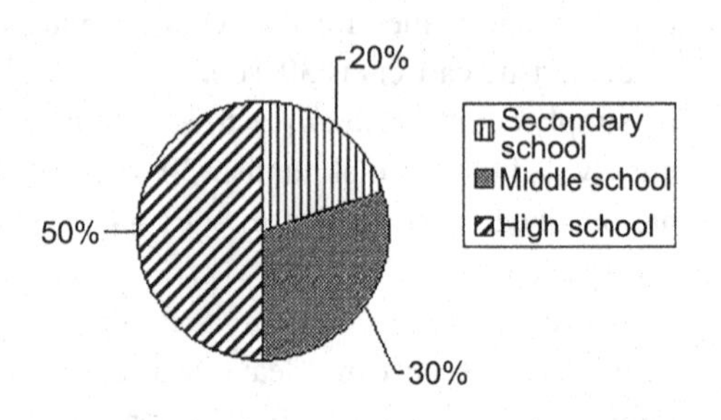

Table 8.4 Table of body sizes

Sequence Number	Height	af	rf	cf	ref
1	162	1	0.0333	1	0.0333
2	163	2	0.0667	3	0.1
3	166	2	0.0667	5	0.1667
4	167	1	0.0333	6	0.2
5	168	3	0.1	9	0.3
6	172	4	0.1333	13	0.4333
7	174	2	0.0667	15	0.5
8	175	4	0.1333	19	0.6333
9	177	2	0.0667	21	0.7
10	180	2	0.0667	23	0.7667
11	184	3	0.1	26	0.8667
12	185	2	0.0667	28	0.9333
13	190	2	0.0667	30	1

Of course, the cumulative frequency must end up with the total number of all characteristic carriers, and the last entry in the relative cumulative frequencies must be a 1, because after all, the table should contain the data of 100% of all characteristic carriers.

I will now briefly summarize the new terms.

Cumulative and Relative Cumulative Frequencies

If n characteristic carriers – i.e. n data sets – of a metrically scaled characteristic with the characteristic values $A_1, \ldots, A_j$ are given and if $A_1 < A_2 < \cdots < A_j$, the cumulative frequency K_i describes the number of characteristic carriers that have a value less than or equal to A_i. The following applies

(continued)

$$K_i = n_1 + \cdots + n_i,$$

where $n_1, \ldots, n_j$ are the absolute frequencies of the characteristic values.

The relative cumulative frequency k_i of the characteristic expression A_i is calculated by

$$k_i = \frac{n_1}{n} + \cdots + \frac{n_i}{n} = \frac{K_i}{n},$$

that is, as the sum of the relative frequencies $h_1 = n_1$ /n to $h_i = n_i$ /n.

It always applies

$$K_j = n \quad \text{und} \quad k_j = 1,$$

that is: the last cumulative frequency is equal to the number of characteristic carriers, the last relative cumulative frequency is equal to 1.

In Table 8.4, for example, $K_5 = 9$ and $k_5 = 0.3$, because 9 freshmen are no taller than 168 cm and this number corresponds to a proportion of 30% of all freshmen.

In contrast to nominally scaled characteristics, metrically scaled characteristics allow a number of other ways of representation, which I would not like to deprive you of. In order to summarize the absolute frequencies in a graph, one often uses *bar charts* and *column charts* or *histograms*, which do not differ much from each other.

Example 8.5

I refer back to Table 8.4, which I would now like to represent in a bar chart. The method is very simple. On the horizontal axis you enter the possible characteristic values, in my case these are the natural numbers from 162 to 190, whereby you can also take the liberty of starting, for example, at 160 and ending at 195 or 200.

The vertical axis is labelled with the possible absolute frequencies, in our case these are the natural numbers from 1 to 4. However, the axis labelling starts at 0, because there are also body sizes such as 169 or 183, which do not occur at all. As soon as you have drawn a thin bar above each body size, the height of which corresponds to the associated absolute frequency, you have already created a bar diagram, as you can admire in Fig. 8.2.

I do not need to say much more about the corresponding column chart; it is created in exactly the same way as a bar chart, except that you use thick columns that connect to each other instead of thin bars. You can see how this looks in Fig. 8.3.

Of course, no one will prevent you from using the vertical axis to represent numbers between 0 and 1 or percentages between 0% and 100%; in this case, your bar or column chart illustrates relative frequencies rather than absolute ones.

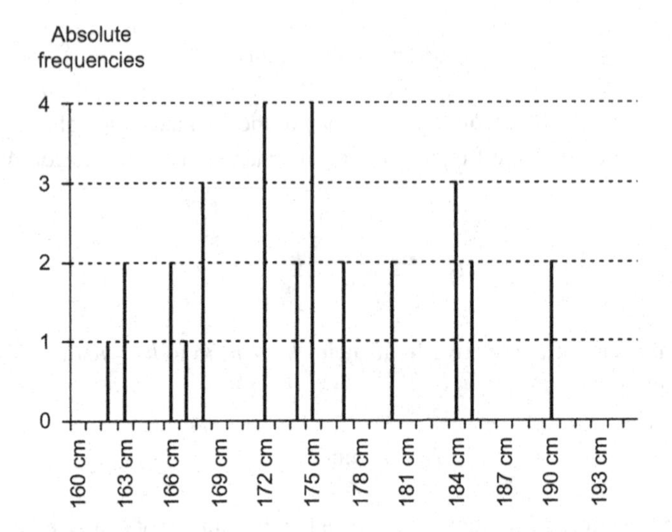

Fig. 8.2 Bar diagram for Table 8.4

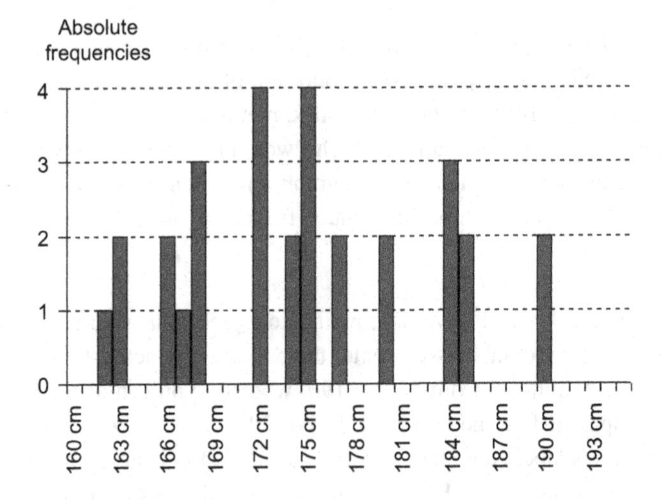

Fig. 8.3 Bar chart for Table 8.4

Bar and Column Diagrams

Both absolute and relative frequencies can be represented graphically using bar and column charts.

Bar charts are also popular to abstract a bit from the individual results of a survey. Consider our example of body sizes: Is it really necessarily desirable to know how many

Table 8.5 Classes of body sizes

Sequence Number	Height	af	rf	cf	ref
1	160 to under 170	9	0.3	9	0.3
2	170 to under 180	12	0.4	21	0.7
3	180 to 190	9	0.3	30	1

newcomers are exactly 175 cm tall? To get an overview of the size ratios, a somewhat coarser grid is obviously sufficient.

Example 8.6
The height data from Table 8.4 can be condensed by *grouping* certain ranges of body heights into *classes*. Here it is advisable to proceed in steps of 10 cm. I divide the possible values between 160 and 190 into three classes: 160 to under 170, 170 to under 180 and 180 to 190. Using the same abbreviations as in Table 8.4, this results in Table 8.5.

So instead of giving all the individual values, such a class table only lists how often characteristics occur within a particular class. Thus, according to Table 8.5, there are 12 beginners whose height is at least 170 cm but less than 180 cm, and the relative frequency of this size range within the 30 beginners is 0.4. Of course, this fact can also be illustrated again with the aid of a bar chart, about which I probably need say no more.▯

A few words about the classification. Of course, the lowest value of the first class will be the smallest occurring characteristic or a number just below it; there, opinions actually differ a little. One should proceed accordingly with the highest value of the last class. In order not to convey a false picture, the class widths – for example, the 10 cm from Example 8.6 – should be the same, and of course the classes must have neither gaps nor overlaps.

I have talked enough about representation options now; it is time to turn to position measures. But first, two exercises.

Exercise 8.1
From the current production of a screw company 25 screws are taken, whose diameter is measured in millimeters. This results in the following list:

6.0; 6.1; 6.0; 6.0; 5.9;
5.9; 5.8; 5.9; 6.0; 6.1;
6.0; 6.0; 6.1; 5.9; 5.8;
5.9; 6.0; 6.0; 5.7; 6.1;
6, 2; 6, 2; 6, 0; 5, 8; 5, 9.

(a) Which different values occur?
(b) Create a table of absolute, relative, cumulative absolute, and cumulative relative frequencies.

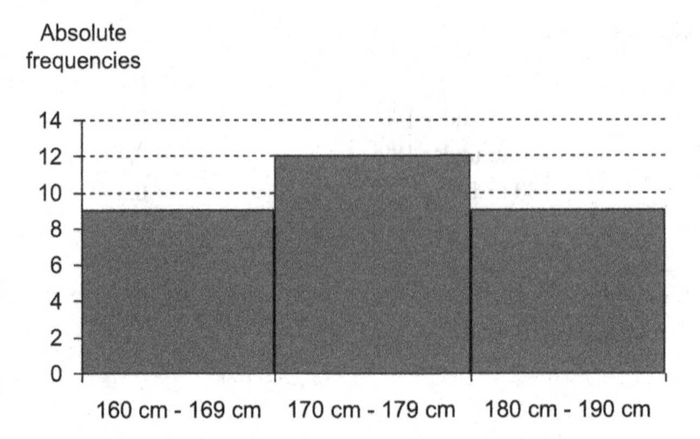

Fig. 8.4 Bar chart for Table 8.5

(c) Plot the absolute frequencies on a bar graph.

Exercise 8.2
Create a new class table with class width 5 from the data in Table 8.4. Then graph the absolute frequencies from this class table using a bar chart (Fig. 8.4).

8.3 Position and Spreading Dimensions

As you may have noticed, the number of actual calculations has been kept within narrow limits so far. I have made tables out of data and graphs out of tables, but I have really only calculated relative frequencies, and that was a simple exercise in dividing. It does not get any worse than that now, but there are a few so-called *statistical measures* that can be obtained from statistical surveys, which often provide a short and succinct summary of the statistical material. I will start with the *position measures modal value*, *arithmetic mean,* and *median*.

The modal value is a very simple matter, requiring nothing more than a certain patience in counting. If you take another look at Table 8.4, you will see that the values 172 and 175 each occur 4 times, while the other values occur at most three times. That is why in this case 172 and 175 are called the *modal values of* the height characteristic: they are simply the values that occur most frequently in my table.

Modal Value
If the characteristic values of a characteristic are given, the characteristic value that occurs most frequently is a modal value of the characteristic. A characteristic can have several modal values.

There is nothing more to be said about this except one thing: Of course, modal values can also be determined for nominally scaled characteristics, as you can see, for example, in Table 8.3. The most frequent school-leaving qualification there was that of the High school, which is why the modal value is "High school".

I still have not calculated anything yet, but that is about to change.

Example 8.7

For an evening party among statisticians, five bags of potato chips are bought, the contents of which supposedly weigh 250 g each. Of course, the statisticians present first weigh the respective contents and obtain the results 251, 248, 250, 243, 254. In order to now obtain an average weight, they calculate the average of the measured values, thus adding the values and dividing by the number of values. This gives

$$\frac{251 + 248 + 250 + 243 + 254}{5} = \frac{1246}{5} = 249.2.$$

So, on average, a bag is missing 0.8 g.

This is how it always works. If you have a number of expressions of a metrically scaled characteristic, you calculate the average value of these expressions, the *arithmetic mean*, by adding the values of the expressions and then dividing by the number of values.

Arithmetic Mean

If n values $x_1, \ldots, x_n$ of a metrically scaled characteristic are given, their arithmetic mean$\overline{x}$ is calculated according to the formula

$$\overline{x} = \frac{x_1 + x_2 + \cdots + x_n}{n}.$$

This is not very difficult, and we will take a quick look at what the mean value is for the heights from Table 8.2.

Example 8.8

Given are the 30 body sizes from Table 8.2, which I will of course not list again now. Their mean value is then calculated from:

$$\frac{185 + 177 + 163 + 175 + \cdots + 174 + 166 + 175 + 168}{30} = \frac{5243}{30} = 174.77.$$

Thus, the mean height among freshmen is 174.77 cm.

But that is not all. You will recall that I introduced the principle of class formation at the end of the previous section. Can you then also determine an arithmetic mean if you have nothing more than a table already divided into classes? Indeed you can. In Table 8.5 there were 9 values between 160 and 170, 12 between 170 and 180 and finally 9 more between 180 and 190. From this table alone I obviously cannot reconstruct the individual values, and I do not want to, because otherwise I could have saved myself the class formation. I therefore assume that the values within a class scatter to some extent around the *class midpoint*, and simply pretend that within a class this class midpoint comes out every time. Of course, this is only an approximation, but you cannot have everything. So for calculating the arithmetic mean, I may approximate 9 times 165 cm, 12 times 175 cm, and then 9 times 185 cm again. This then gives the mean value:

$$\bar{x} = \frac{9 \cdot 165 + 12 \cdot 175 + 9 \cdot 185}{30} = \frac{5250}{30} = 175.$$

Not quite the same value as before, but not bad at all for an approximation.

It cannot hurt to put this approach into a formula for once.

Arithmetic Mean for Class Division

If the individual expressions of a characteristic are not known, but only a class division is given, the arithmetic mean is calculated according to the formula:

$$\bar{x} = \frac{n_1 \cdot x_1 + n_2 \cdot x_2 + \cdots + n_k \cdot x_k}{n}.$$

Here k denotes the number of classes, $x_1, \ldots, x_k$ the respective class midpoints of the classes, $n_1, \ldots, n_k$ the absolute frequencies of the feature expressions in the respective class, and $n = n_1 + \cdots + n_k$ the total number of these expressions.

As important as the arithmetic mean is – in some situations it is not to be used because it does not adequately reflect the actual conditions. The following example shows what I mean.

Example 8.9

In earlier times, there was often a lord of the manor who ruled a manageable region and essentially lived off the labor of his peasants. I now assume that a lord of the manor could tyrannize 100 peasants. Even under optimistic assumptions, we can assume that each individual peasant owned only a single thaler, while the laird owned considerably more, say 1000 talers. The mean value from the 101 existing possessions then amounts to

$$\overline{x} = \frac{100 \cdot 1 + 1 \cdot 1000}{101} = 10.89.$$

But this mean has nothing to do with reality, because each of the peasants would have licked his fingers for the nearly 11 talers he would never see in his life, while the landowner would have complained bitterly about his poverty in view of 10.89 talers.

In such cases, where there are outliers such as the value 1000, which stand out from the large amount of other values and distort the mean value, one likes to use the *median*: the value that is in the middle of the characteristic values ordered by size. So, to determine the median, you need to sort the characteristics by size, then identify the value that 50% of the values do not *exceed* and 50% of the values do not *fall below*. In Example 8.8, we have 101 values that can be easily arranged in the form

$$1 \leq 1 \leq \cdots \leq 1 \leq 1000$$

can be arranged. The value that lies exactly in the middle is, of course, the fifty-first, i.e. 1. Consequently, the historical wealth statistics have the median 1, which describes the actual and predominant wealth much better than the arithmetic mean.

Determining the median is easy, provided your characteristic values are in sorted form. If you have an odd number of values, simply take the middle one, i.e. the fifth of nine, the tenth of nineteen or the fifty-first of one hundred and one. Generally, if n is odd, you take the median of n values to be the value numbered $\frac{n+1}{2}$, because that value is right in the middle. Unfortunately, there is no real median value for even n: Which value should be exactly the median for four values? For four values, for example, one helps oneself by looking at the second and the third, because these two frame the middle, so to speak; for ten values, one looks at the fifth and the sixth for the same reason; and for n values, one takes the values with the numbers $n/2$ and $n/2 + 1$. And since the median is supposed to be a number and not, say, two, one calculates the arithmetic mean from these two values that are almost in the middle.

Median

If n characteristic values of a metrically scaled or ordinally scaled characteristic are available in ordered form, i.e. $x_1 \leq x_2 \leq \cdots \leq x_n$, the median $\tilde{x}$ is determined according to the following formula:

$$\tilde{x} = x_{\frac{n+1}{2}}, \quad \text{if } n \text{ is odd}$$

respectively

(continued)

$$\tilde{x} = \frac{x_{\frac{n}{2}} + x_{\frac{n}{2}+1}}{2}, \quad \text{if } n \text{ is odd.}$$

Example 8.10

I take another look at the body sizes from Tables 8.2 and 8.4. There, $n = 30$, which is why I have to resort to the more awkward formula. So the fifteenth as well as the sixteenth value are to be determined from the values ordered by size.

Table 8.4 shows that $x_{15} = 174$ and $x_{16} = 175$. From this, the median $\tilde{x} = \frac{174+175}{2} = 174.5$ is calculated.

The ratios discussed so far are called *position measures* because they indicate approximately where the essential value of the characteristic values lies – be it a greatest value or a mean value. However, once you have found a mean value, you may also be interested in how far away from this mean value the individual characteristic values are: Should one expect large distances, or can one assume that the individual values are more or less close to the mean? In short, one would like to know how far the expressions are scattered, which is why one also likes to speak of *scatter measures*.

The simplest of all scattering measures is the *range*, which is not worth more than a few words.

Range

The range of expressions of a characteristic is the difference between the largest and smallest expression of the characteristic.

For the often mentioned body sizes, the range is 190–162 = 28, since the largest body size is 190 and the smallest 162. There is nothing more to be said about this, especially since the range has little to do with the mean value.

The situation is different with the *mean absolute deviation*. Its principle is simple. As usual, I assume that there are some characteristic values of a metrically scaled characteristic, which I label $x_1, x_2, \ldots, x_n$. From these characteristics, I can then calculate a mean value as discussed – whether the arithmetic mean $\overline{x}$ or the median $\tilde{x}$, I do not care at the moment; in any case, I denote the mean value of whatever kind by x. But now each value $x_1, \ldots, x_n$ has a certain distance from this x: For x_1 the distance is $|x_1 - x|$, for x_2 of course $|x_2 - x|$ and so on. Since I am not interested in individual distances, but in an average distance of the existing expressions from their mean, I calculate the arithmetic mean from the individual distances, which leads me to the expression

$$d = \frac{|x_1 - x| + |x_2 - x| + \cdots + |x_n - x|}{n}$$

It is called the *mean absolute deviation,* and depending on which mean I had previously taken as a basis, one speaks of a deviation with respect to the arithmetic mean or of a deviation with respect to the median.

Mean Absolute Deviation
If n expressions $x_1, \ldots, x_n$ of a metrically scaled characteristic are given, the size

$$\frac{|x_1 - \overline{x}| + |x_2 - \overline{x}| + \cdots + |x_n - \overline{x}|}{n}$$

Is denoted as the mean absolute deviation with respect to the arithmetic mean $\overline{x}$, and the quantity

$$\frac{|x_1 - \widetilde{x}| + |x_2 - \widetilde{x}| + \cdots + |x_n - \widetilde{x}|}{n}$$

as the mean absolute deviation with respect to the median $\widetilde{x}$.

A little example cannot hurt.

Example 8.11
Let me remind you of the five chip bags whose mean I had already calculated in Example 8.7. The values for the content weights were 251, 248, 250, 243, 254, and the arithmetic mean was 249.2, just below the promised value of 250. Sorting the values now by size gives 243, 248, 250, 251, 254, which is why the median is exactly 250. Now I can calculate the two mean absolute deviations. With regard to the arithmetic mean, the result is a deviation of

$$\frac{1}{5} \cdot (|251 - 249.2| + |248 - 249.2| + |250 - 249.2| + |243 - 249.2| + |254 - 249.2|)$$
$$= \frac{14.8}{5} = 2.96.$$

On the other hand, with respect to the median, we have a mean absolute deviation amounting to

$$\frac{|243-250|+|248-250|+|250-250|+|251-250|+|254-250|}{5}=\frac{14}{5}$$
$$=2. \quad \blacksquare$$

It must be admitted that both forms of mean absolute deviation are not very popular in statistics. The reason is that they are difficult to apply in so-called inferential or inductive statistics, and what is difficult to apply is better left alone, provided there is a useful substitute for it. And there is indeed one: the *dispersion* or *variance* and the *standard deviation* derived from it. It avoids the annoying absolute values by switching to the positive squares, i.e. instead of $|x_1 - \overline{x}|$ it uses $(x_1 - \overline{x})^2$. And there is another difference that you will see immediately.

Dispersion and Standard Deviation

If n expressions $x_1, \ldots, x_n$ of a metrically scaled characteristic are given, the size

$$s^2 = \frac{(x_1-\overline{x})^2 + (x_2-\overline{x})^2 + \cdots + (x_n-\overline{x})^2}{n-1}$$

Is denoted as variance or dispersion. The size

$$s = \sqrt{\frac{(x_1-\overline{x})^2 + (x_2-\overline{x})^2 + \cdots + (x_n-\overline{x})^2}{n-1}}$$

is called the standard deviation.

You can see the differences from the mean absolute deviation. First, you take squares instead of amounts, which you compensate for a bit by then taking the root. And second, you no longer divide by n, but by $n - 1$. For an explanation of this somewhat odd change, I will have to put you off until later times, when you may be looking into inductive statistics. At this time, I can only anticipate that the slightly different denominator has something to do with what is known as the fidelity of expectations. In practical arithmetic, it usually does not make a huge difference anyway whether you divide by n or by $n - 1$, because for large values of n, it hardly changes the result at all.

Let us take another look at calculating the dispersion and standard deviation with an example.

Example 8.12

The example of the body sizes from Table 8.2 is not yet completely exhausted. In Example 8.8 I had already calculated the arithmetic mean $\overline{x} = 174.77$ of the body sizes. Now I want to determine the standard deviation s. The following applies to the scatter

$$
\begin{aligned}
s^2 &= \frac{(185-174.77)^2+(177-174.77)^2+\cdots+(168-174.77)^2}{29} \\
&= \frac{1817.367}{29} = 62.67.
\end{aligned}
$$

You have to remember that my table has 30 values listed, so I have to divide by 29. This then gives the standard deviation

$$s = \sqrt{62.67} = 7.92. \quad \blacksquare$$

With the introduction of the dispersion and the standard deviation, we have now already approached the border to conclusive statistics, which, however, is not the subject of this bridge course. I may therefore leave you with the following exercises for your further studies.

Exercise 8.3
Determine the arithmetic mean, median, dispersion, and standard deviation for the screw diameters from Exercise 8.1.

Exercise 8.4
For Table 8.4 of body sizes, determine the mean absolute deviation with respect to the arithmetic mean and the mean absolute deviation with respect to the median.

Exercise 8.5
Determine the arithmetic mean for the class table from Exercise 8.2.

Complex Numbers

An important topic of the first chapters was root extraction, many examples and exercises were given, and somehow you were probably given the feeling that you could now already calculate roots from all real numbers.

That was a lie.

Well, at least if it came across that way, it was not the full truth. That comes now: Up to this point, you can calculate roots from all *positive* real numbers and, of course, from zero, because $\sqrt{0} = 0$. But you cannot yet calculate roots from *negative numbers*, and this is not due to you, but to the fact that in the real world there are simply no roots from negative numbers: For example, there is no real number x with the property

$$x \cdot x = -1.$$

To realize such a thing we must enter the realm of complex numbers, and I ask you to follow me there now; never fear, I am with you – if that is any consolation to you.

If not, perhaps the following will comfort you: A few weeks ago I was with my children in an amusement park in southern Germany and on this occasion I also had to ride the "highest roller coaster in Europe" (statement of the operator, whom I believe to the letter) ("Dad, you're not afraid, are you?").

After this experience, there is nothing scary about complex numbers, believe me!

G. Walz et al., *Foundations of Mathematics: A Preparatory Course*,
https://doi.org/10.1007/978-3-662-67809-1_9

9.1 The Imaginary Unit *i* and the Set of Complex Numbers

The introduction of complex numbers stands and falls with the introduction of a number, which multiplied with itself results in -1. In the real, such a thing does not exist, I had pointed that out again in the introduction. On the other hand, it would be very nice to have such a thing, because already the solution of the simple looking equation

$$x^2 + 1 = 0$$

requires the existence of such a number.

At this point, mathematicians do not do it any differently than – for example – humanities scholars: If something is needed that does not yet exist, it is defined:

Imaginary Unit

An **imaginary unit** is that number i (which does not occur in the real, therefore "imaginary") which has the property

$$i^2 = -1$$

has. In other words, it is

$$i = \sqrt{-1}.$$

If you are not quite comfortable with this definition, I can understand that; however, I can reassure you that calculating with this newly defined number i will lead to correct results and – for example when solving differential equations – results that are consistent with the real world.

But we are not there yet. Maybe you are thinking now: Fine, now we can calculate the root of -1, but what about root of -2, of -3, etc.? Should we introduce a new letter for each negative number and rename the solution?

This is certainly not necessary, as the first application of the new number i now *show* you that you can calculate the root of *any negative* number, if you only master the one from -1: For example, if you are faced with the task of calculating $\sqrt{-9}$, you can set up the following chain of equations:

$$\sqrt{-9} = \sqrt{(-1) \cdot 9} = \sqrt{-1} \cdot \sqrt{9} = i \cdot \sqrt{9} = 3i.$$

So with that, you have determined the root of −9; and since this procedure does not seem to depend on the specific choice −9, you now know how to – calculate the root of any negative number: It is

$$\sqrt{-a} = \sqrt{(-1) \cdot a} = \sqrt{-1} \cdot \sqrt{a} = i \cdot \sqrt{a}.$$

In general, the number i is the key to the introduction of a new range of numbers beyond the known real numbers, the set of complex numbers:

Complex Numbers
If i *is* the complex unit defined above and a and b are *arbitrary* real numbers, then a number of the form

$$a + ib$$

a **complex number.**
The set of all complex numbers is denoted by $\mathbb{C}$, i.e.

$$\mathbb{C} = \left\{a + bi;\ a, b \in \mathbb{R}, i = \sqrt{-1}\right\}.$$

The number a is called the **real part**, the number b the **imaginary part of** the complex number $a + i\,b$.

Examples of complex numbers are $1 + 2i$, $-\sqrt{3} - i$ and $\pi + \pi\,i$. But i itself is also a complex number, for it can be written in the form $0 + 1i$; and finally every real number x is *also* a complex one, for it can be written in the form $x + 0$. The complex numbers are thus an extension of the real numbers, just as the real numbers were themselves an extension of the rational numbers, and the rational numbers were an extension of the integers.

In the next section, I will show you how to calculate with complex numbers. Before that, however … well, you know:

Exercise 9.1 Which of the following expressions are complex numbers?

(a) $-2 - 3i$,
(b) i^2,
(c) the solutions x of the equation $x^2 + 2 = 0$.

9.2 Basic Arithmetic Operations for Complex Numbers

In this section I will show you how to add, subtract, multiply and divide complex numbers, starting of course with the two simplest arithmetic operations, addition and subtraction.

If z_1 and z_2 are complex numbers, then by definition they have the representation $z_1 = a_1 + ib_1$ and $z_2 = a_2 + ib_2$. The sum $z_1 + z_2$ is therefore simply equal to

$$z_1 + z_2 = (a_1 + ib_1) + (a_2 + ib_2),$$

where I have put mathematically unnecessary brackets to clarify the composition of this expression. In this form, the result is not quite recognizable as a complex number yet, but of course you can rearrange it to do so: It is

$$z_1 + z_2 = (a_1 + ib_1) + (a_2 + ib_2) = (a_1 + a_2) + i \cdot (b_1 + b_2),$$

a complex number of the purest water.

So we have already solved the addition of complex numbers (which was not difficult) and the subtraction is not worse: If you replace the plus sign between z_1 and z_2 by a minus sign, you get

$$z_1 - z_2 = (a_1 + ib_1) - (a_2 + ib_2) = (a_1 - a_2) + i \cdot (b_1 - b_2),$$

no problem either.

Thus, for example

$$(3 - 2i) + (1 + 4i) = 4 + 2i$$

and

$$(-2 + 4i) - (-3 + 3i) = 1 + i.$$

I would not even bother you with exercises on this topic, we are out of that stage. Instead, let us move on to the next higher basic arithmetic, multiplication: What is the product of two complex numbers? Here, too, you do not have to be shy and you do not have to use any tricks, you just multiply out and drag along the ominous number i, "forgetting", so to speak, what kind of unimaginable object is hidden behind it, and calculate blithely away: Simple multiplication out results in

$$z_1 \cdot z_2 = (a_1 + ib_1) \cdot (a_2 + ib_2) = a_1a_2 + ib_1a_2 + ia_1b_2 + i^2b_1b_2. \tag{9.1}$$

That does not look much like a complex number yet, but we get that: Now "remember" that yes i is not some parameter, but that $i^2 = -1$. If we substitute this into (9.1), we get

$$z_1 \cdot z_2 = a_1 a_2 - b_1 b_2 + i(b_1 a_2 + a_1 b_2), \tag{9.2}$$

where I immediately excluded the factor i from the two mixed terms.

So this is how you multiply complex numbers. It is up to your personal taste whether you want to remember the formula (9.2) or whether you want to take the way just shown (i.e. multiply out and then set $i^2 = -1$) – the result is the same.

As a small example I calculate the product

$$(3 - 2i) \cdot (1 + 4i) = (3 + 8) + i(-2 + 12) = 11 + 10i.$$

I saved the best for last: Division of complex numbers, which is the fourth basic arithmetic. Here we approach the problem very slowly and see first of all how to calculate the reciprocal value of a complex number, i.e. what

$$\frac{1}{a + ib}$$

is. To calculate this complex number, there is a simple trick, which I will now show you without too much talk: You first expand the fraction with the number $a - ib$ and thus get

$$\frac{1}{a + ib} = \frac{a - ib}{(a + ib)(a - ib)}.$$

If you now think that this has only made things worse, I can understand you, but I can assure you that everything is about to change for the better: If you multiply out the denominator, you get

$$\frac{1}{a + ib} = \frac{a - ib}{(a + ib)(a - ib)} = \frac{a - ib}{a^2 + b^2} \tag{9.3}$$

and this denominator is a *positive real number*. But you can always divide by such a number, and so we have solved the problem. If we now divide the last fraction in (9.3) into two parts, we get a complex number of purest water, which is the result of the division. As a rule of thumb:

Reciprocal of a Complex Number

For any complex number $z = a + ib \neq 0$ is

(continued)

$$\frac{1}{z} = \frac{1}{a+ib} = \frac{a}{a^2+b^2} - i \cdot \frac{b}{a^2+b^2}.$$

For example

$$\frac{1}{2+3i} = \frac{2}{13} - i \cdot \frac{3}{13}$$

and

$$\frac{1}{1-i} = \frac{1}{2} + i \cdot \frac{1}{2}.$$

If you do not believe me – which, to be honest, I would always advise you to do – you can check the results by multiplying the value in the denominator on the left by the result on the right; the result must be 1.

In the first example this is also true, because

$$(2+3i)\left(\frac{2}{13} - i \cdot \frac{3}{13}\right) = \frac{4}{13} + \frac{6i}{13} - \frac{6i}{13} - i^2 \cdot \frac{9}{13} = \frac{13}{13} = 1,$$

since, as should be well known by now, $-i^2 = 1$. Please test the second example yourself using the same method.

So far we can only calculate reciprocals of complex numbers, but what about general division; so how do we calculate values of the form

$$\frac{c+id}{a+ib}?$$

Well, you do it exactly as just shown with the reciprocal (which is why I did this), so you expand with the number $a - ib$. This leads first to

$$\frac{c+id}{a+ib} = \frac{(c+id)(a-ib)}{(a+ib)(a-ib)} = \frac{(c+id)(a-ib)}{a^2+b^2},$$

and if you still do the multiplication in the numerator, you get the desired result, which I will formulate in a rule box in a moment:

Division of Complex Numbers
If $z_1 = c + id$ and $z_2 = a + ib$ are arbitrary complex numbers with $z_2 \neq 0$, then

$$\frac{z_1}{z_2} = \frac{ac + bd + i(ad - bc)}{a^2 + b^2}. \tag{9.4}$$

That takes care of the four basic arithmetic operations; what comes next is clear:

Exercise 9.2 Given the complex numbers $z_1 = 2 + 5i$, $z_2 = -1 - 2i$ *and* $z_3 = 1 - 3i$. Calculate the following complex numbers:

$$z_1 \cdot z_2, \quad \frac{z_1}{z_2 \cdot z_3}, \quad \frac{z_1 + z_2}{z_2 - z_3}. \quad \blacksquare$$

9.3 The Gaussian Number Plane and the Trigonometric Form of Complex Numbers

Perhaps you have been wondering all along where this new set of numbers, the complex numbers, should be placed on the number line. As a reminder: The number line is the set of all real numbers and, conversely, every real number has its place on the number line. In other words, the number line is full to the brim with real numbers, and there is no room left for other things, such as the complex numbers.

This is also completely correct and yet not a problem: It is an idea that is as simple as it is ingenious (although I personally believe that *truly* ingenious ideas are always simple, though not necessarily vice versa), which goes back to the great C.F. Gauss, to open up a second dimension here, so to speak, and to represent the set of complex numbers in a plane, the so-called **Gaussian number plane.**

For this purpose, one interprets the real and imaginary parts of the respective number as its two coordinates in the plane, i.e. one draws the number $a + ib$ *at* the position (a, b) in the coordinate system. In Fig. 9.1 you can see three complex numbers and their position in the Gaussian plane.

Now, where has the real number line actually got to? Well, the set of real numbers is just the set of all complex numbers with $b = 0$; accordingly, the good old real number line can be found as a horizontal coordinate axis in the Gaussian number plane.

Of course, the Gaussian number plane is not only used to simply draw the complex numbers, but can also be used for "graphical arithmetic", especially for extracting roots from complex numbers, which would otherwise be a rather difficult undertaking. To do

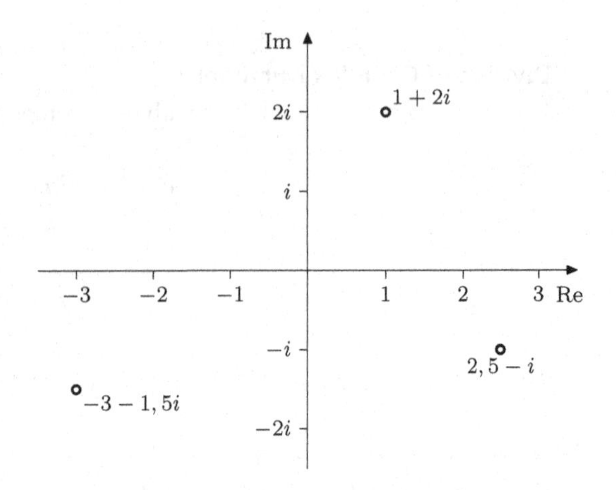

Fig. 9.1 Some complex numbers in the Gaussian number plane

this, you need a different representation of complex numbers, called trigonometric form. This is based on the fact that a point in the plane (and every complex number is now regarded as such) can be described unambiguously not only by its Cartesian coordinates (i.e. the values a and b), but also by specifying how far it is from the zero point and what angle it makes – for example – with the positive real axis.

That was quite a sentence monstrosity, one of the "grammatically correct but in no way understandable" variety. So I would rather state that again in mathematically precise form right away:

The distance of a complex number from the zero point of the Gaussian number plane is also called its absolute value, because this distance can also be interpreted as the absolute value (i.e. the length) of the vector pointing from the zero point to this number.

Be that as it may, this absolute value is calculated in any case according to the theorem of our colleague Pythagoras as follows:

Absolute Value

The **absolute value of** the complex number $z = a + ib$ is the real number

$$|z| = \sqrt{a^2 + b^2}.$$

For example

$$|2 + 3i| = \sqrt{2^2 + 3^2} = \sqrt{13}$$

and

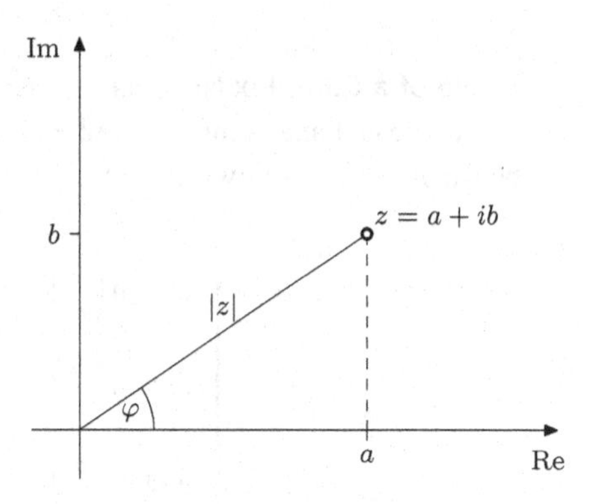

Fig. 9.2 Complex number and associated angle

$$|-3-4i| = \sqrt{(-3)^2+(-4)^2} = \sqrt{9+16} = 5.$$

Now, how do we calculate the angle that a complex number, or more precisely the vector just mentioned, makes with the positive axis? To do this, it is best to take a look at the complex number $z = a + ib$ in Fig. 9.2: You can see that the angle φ can be interpreted as an acute angle in a right triangle, whose opposite cathetus *is* b and whose adjacent cathetus *is* a. Consequently, the tangent of this angle is equal to

$$\tan\varphi = \frac{\text{opposite cathetus}}{\text{adjacent cathetus}} = \frac{b}{a}.$$

If you are not familiar with the tangent anymore or not so much yet, then please just trust me and accept the statement I just made. It is correct. However, I am not interested in the tangent of φ, but in φ itself. But now I get this angle quite simply by applying to both sides the inverse function of the tangent, which is the arc tangent. Thus

$$\varphi = \arctan\left(\frac{b}{a}\right).$$

If you do not find a key with the inscription “ARCTAN” on your calculator, look for “INV TAN” or “TAN^{-1}”, all this means the same, namely the inverse function of the tangent, the arc tangent.

So we have clarified the calculation of the angle, but only if the complex number in question is in the first quadrant. In all other cases, you still have to add certain angle numbers, the derivation of which would simply lead too far here; I therefore list the calculation of these angles without comment in a small table:

Angle of a Complex Number
The angle φ that a complex number $z = a + ib$ makes with the positive real axis can be calculated as follows:

$$\varphi = \begin{cases} \arctan\left(\frac{b}{a}\right), & \text{if } a > 0 \text{ and } b \geq 0, \\ \arctan\left(\frac{b}{a}\right) + 360^\circ, & \text{if } a > 0 \text{ and } b < 0, \\ \arctan\left(\frac{b}{a}\right) + 180^\circ, & \text{if } a < 0, \\ 90^\circ, & \text{if } a = 0 \text{ and } b > 0, \\ 270^\circ, & \text{if } a = 0 \text{ and } b < 0, \\ 0^\circ, & \text{if } a = 0 \text{ and } b = 0. \end{cases}$$

For example, the angle of $2 + 3i$ is equal to

$$\varphi = \arctan\frac{3}{2} = 56.31^\circ$$

and that of $-3 - 2$ is equal to

$$\varphi = \arctan\frac{-2}{-3} + 180^\circ = \arctan\frac{2}{3} + 180^\circ = 213.69^\circ.$$

Thus we have the two ingredients (admittedly, I have again secretly read in the foreign dictionary) of the desired trigonometric form of a complex number together and can now write it down formally:

Trigonometric Form of a Complex Number
The **trigonometric form of** the complex number z is the representation

$$z = |z| \cdot (\cos\varphi + i\sin\varphi).$$

Where $|z|$ is the absolute value of z and φ is the angle calculated according to the above definition.

By the way, the representation $a + ib$ *of* a complex number is also called its **normal form** to distinguish it from the trigonometric form.

Again, of course, a few examples: I use for this the two complex numbers, whose angles I had already calculated above, of which I have to determine only the absolute value (one gets older!).

The absolute value of $z = 2 + 3i$ is

$$|z| = \sqrt{2^2 + 3^2} = \sqrt{13} \approx 3.6056,$$

I had calculated the angle above to be 56.31°. Thus the trigonometric form of this number is equal to

$$z = \sqrt{13} \cdot \left(\cos 56.31^\circ + i \sin 56.31^\circ\right).$$

To check, you can fire up your calculator and work out that

$$3.6056 \cdot \left(\cos 56.31^\circ + i \sin 56.31^\circ\right) = 3.6056 \cdot (0.5547 + i \cdot 0.8321) = 2 + i \cdot 3$$

is.

The absolute value of the second example number, $z = -3 - 2i$, is also $\sqrt{13}$; this therefore has the trigonometric form

$$z = 3.6056 \cdot \left(\cos 213.69^\circ + i \sin 213.69^\circ\right).$$

I am entrusting the control bill to you this time.

Frankly speaking, I had toyed with the idea of astonishing you at this point with the fact that I do *not* set *any* exercises on this topic. But I very quickly decided against it, because especially the trigonometric form is very important and should therefore be practiced.

Exercise 9.3 Determine the trigonometric form of the following complex numbers:

(a) $z = 1 - i$,
(b) $z = -5 - 3i$.

9.4 Powers and Roots of Complex Numbers

In the last section, I already talked about the fact that the trigonometric form of complex numbers makes some arithmetic operations easier to perform or even possible in the first place; this is what I want to show you now.

Addition and subtraction are so easy to perform in the normal form of complex numbers that they do not require any further simplification. The situation is somewhat different for multiplication and division, both of which can be somewhat complex in the normal form and are therefore always available for simplifications. The trigonometric form offers such a simplification:

Multiplication and Division of Complex Numbers in Trigonometric Form

There are

$$z_1 = |z_1| \cdot (\cos \varphi_1 + i \sin \varphi_1)$$

and

$$z_2 = |z_2| \cdot (\cos \varphi_2 + i \sin \varphi_2)$$

two complex numbers in trigonometric form.

Then their product is

$$z_1 \cdot z_2 = |z_1| \cdot |z_2| \cdot (\cos(\varphi_1 + \varphi_2) + i \sin(\varphi_1 + \varphi_2)) \tag{9.5}$$

and – if $z_2 \neq 0$ – their quotient

$$\frac{z_1}{z_2} = \frac{|z_1|}{|z_2|} \cdot (\cos(\varphi_1 - \varphi_2) + i \sin(\varphi_1 - \varphi_2)). \tag{9.6}$$

By the way, the two formulas are not that difficult to derive, you just do the multiplication or division in normal form and use the addition theorems for sine and cosine, but I will spare you and me that here anyway.

So you multiply complex numbers by multiplying their absolute value (which is easy, since they are positive real numbers), and adding their angles; the same is true for division.

Again, I want to support this with a small example. Since I cannot think of any complex pizza calculation or similar, I will leave it at a simple numerical example. I use the two numbers already used above as an example for the trigonometric form

$$z_1 = 2 + 3i = 3.6056 \cdot \left(\cos 56.31^\circ + i \sin 56.31^\circ\right)$$

and

$$z_2 = -3 - 2i = 3.6056 \cdot \left(\cos 213.69^\circ + i \sin 213.69^\circ\right).$$

Multiplication of the two absolute values and addition of the angles results here in

$$z_1 \cdot z_2 = 13 \cdot \left(\cos 270^\circ + i \sin 270^\circ\right),$$

hence

$$z_1 \cdot z_2 = 13 \cdot (0 - i) = -13i,$$

which can be confirmed by direct multiplication of the two numbers according to formula (9.2).

To calculate a quotient of the two numbers, I must first divide the absolute values; this gives 1, and in total I get

$$\frac{z_1}{z_2} = 1 \cdot \left(\cos\left(-157.38^\circ\right) + i \sin\left(-157.38^\circ\right)\right) = -0.9230 + i \cdot 0.3846.$$

Well, admittedly, that was not too thrilling yet, a simple multiplication or division of complex numbers can still be done quite well in the normal form.

The trigonometric form becomes a little more attractive when it comes to repeated multiplication, i.e. exponentiation of a complex number. For example, if you want to calculate $(3 + 7i)^5$ (I do not know why you should do this, but mathematicians have rarely bothered to apply their results, that is what others are for), you can of course do this using the formula (9.2) for multiplying complex numbers, but you will have some work to do. You can do it much more elegantly by repeatedly applying the formula (9.5) for multiplying in trigonometric form. This gives:

Exponentiation of Complex Numbers

If

$$z = |z| \cdot (\cos \varphi + i \sin \varphi)$$

is a complex number and n is a natural number, then

$$z^n = |z|^n \cdot (\cos n\varphi + i \sin n\varphi).$$

So you exponentiate a complex number by exponentiating its absolute value (did I mention that this is a simple operation since the absolute value is a positive real number?) and multiplying its angle by *n*.

By the way, as a special case of this rule for complex numbers whose absolute value is equal to 1, we obtain the formula

$$(\cos\varphi + i\sin\varphi)^n = (\cos n\varphi + i\sin n\varphi),$$

which is also called **de Moivre's formula**, named after a French mathematician named – you would not believe this – de Moivre, who lived from 1667 to 1754.

As a first example of this new exponentiation rule, I calculate the number already mentioned above $(3 + 7i)^5$. The trigonometric form of $3 + 7i$ *is* (you better recalculate it, the next edition of this book can still use improvements!)

$$3 + 7i = \sqrt{58} \cdot \left(\cos 66.8014^\circ + i\sin 66.8014^\circ\right)$$

and thus

$$(3 + 7i)^5 = \sqrt{58}^5 \cdot \left(\cos 334.0070^\circ + i\sin 334.0070^\circ\right) = 23027.9907 - 11228.0199i.$$

But since we know that the real and imaginary parts of the number $(3 + 7i)^5$ must be integers, since they are generated by repeated multiplication of integers when calculating with the normal form, I prefer to round this up to $23.028–11.228i$, and this result in turn can be confirmed by explicit multiplication.

As a second example, I calculate the fourth power of i, i^4. The trigonometric form of i is – strictly according to the rule –

$$i = 1 \cdot \left(\cos 90^\circ + i\sin 90^\circ\right),$$

and thus

$$i^4 = 1^4 \cdot \left(\cos 360^\circ + i\sin 360^\circ\right) = 1 + i \cdot 0,$$

so $i^4 = 1$. Admittedly, this can also be seen more quickly directly, because

$$i^4 = \left(i^2\right)^2 = (-1)^2 = 1,$$

but this was only meant to illustrate the correctness of the potentization formula.

More examples will now follow in the do-it-yourself process:

Exercise 9.4 Calculate the following complex numbers and state them in normal form:

(a) $(-1 + 2\,i)^5$,

(b) $\left(\frac{1}{\sqrt{2}} + \frac{i}{\sqrt{2}}\right)^8$.

So far we have already seen some of the advantages of the trigonometric form, but I am only now showing you the real strength of this approach (great, is not it, first stall a few pages, and only then come out with the big advantage! Well, that is called "marketing" elsewhere). What I want to say is this: So far, I have been able to show you some techniques – that is, performing the basic arithmetic operations – that are easier to do in the trigonometric form than in the standard form. Now, however, comes something that is only possible in the trigonometric form, i.e. that is practically impossible to perform in the standard form: taking the square root of complex numbers.

Before I explain this to you in more detail, however, I would like a drum roll . . .? Well, if need be, I will do it myself. Because what follows now is worth such an effort: While in the real, so to speak, one never knows exactly how many different *nth* roots there are from a given number x – that depends on whether n is even or odd and, of course, above all on whether x is positive or negative – here in the complex there is a very simple rule without any exception: If n is any natural number and z any complex number different from zero, then there are exactly n different *nth* roots from z. Period, done, off, no exceptions necessary and allowed!

So from each complex number you can draw exactly three different third roots, exactly seven different seventh roots, exactly nineteen different nineteenth roots, and, if you absolutely want to, even two hundred and forty-third different two hundred and forty-third roots.

And it gets even better: There is even a simple formula to calculate these roots. I will give it to you right now:

Roots of Complex Numbers

Let z be any nonzero complex number in the trigonometric form

$$z = |z| \cdot (\cos\varphi + i\sin\varphi)$$

and n is any natural number.

There are exactly n *nth* roots from z, which I will denote by $\zeta_0, \zeta_1, \ldots, \zeta_{n-1}$. These roots are calculated as follows: It is

(continued)

$$\zeta_k = \sqrt[n]{|z|} \cdot \left(\cos\left(\frac{\varphi + k \cdot 360^\circ}{n} \right) + i \cdot \sin\left(\frac{\varphi + k \cdot 360^\circ}{n} \right) \right) \tag{9.7}$$

for $k = 0, 1, \ldots, n - 1$.

By the way, ζ is the Greek letter "zeta", i.e. the Greek "z"; after all, the painstakingly endured 5 years of Greek lessons must pay off at some point.

The formula (9.7) just given is a true marvel of mathematics: It "produces" all possible roots of a complex number without further inquiry and/or case distinction. The small disadvantage of such a miracle formula is that one still has to get used to dealing with it. This is best done with a few examples:

First, I will take the third roots from the number

$$z = 8 \cdot \left(\cos 21^\circ + i \cdot \sin 21^\circ \right)$$

Since I am interested in the third roots, here $n = 3$, and thus the index k runs through the values 0, 1, 2. Fortunately, z is already given in trigonometric form, so that I can read off the absolute value 8 directly, and for the same reason the angle $\varphi = 21°$ is easily recognizable.

I get the absolute value of all three third roots by taking the third root from $|z|$, i.e. from eight, which gives two without further effort. To determine the first of the three third roots, ζ_0, I now have to set $\varphi = 21°$, $n = 3$ and $k = 0$ in the above formula. This gives

$$\zeta_0 = 2 \cdot \left(\cos 7^\circ + i \cdot \sin 7^\circ \right),$$

and if one still calculates this, one receives the normal form

$$\zeta_0 = 2 \cdot (0.9925 + i \cdot \sin 0.1219) = 1.985 + i \cdot 0.2438.$$

Likewise one calculates the two further third roots to

$$\zeta_1 = 2 \cdot \left(\cos 127^\circ + i \cdot \sin 127^\circ \right) = -1.2036 + 1.5972i$$

and

$$\zeta_2 = 2 \cdot \left(\cos 247^\circ + i \cdot \sin 247^\circ \right) = -0.7815 - 1.8410i.$$

This first example was – precisely *because* it was the first one – a little atypical, since the number from which roots were to be drawn was already given in trigonometric form. Of course, this is not usually the case, and so I will now work through a "serious" example, namely the third roots of

$$-1+i$$

The absolute value of this number is equal to $\sqrt{(-1)^2+1^2}=\sqrt{2}$, the angle is

$$\arctan\left(\frac{1}{-1}\right)+180^\circ=-45^\circ+180^\circ=135^\circ.$$

This gives the following three third roots:

$$\begin{aligned}\zeta_0 &= \sqrt[6]{2}\cdot\left(\cos 45^\circ+i\sin 45^\circ\right)=0.7937+0.7937i,\\ \zeta_1 &= \sqrt[6]{2}\cdot\left(\cos 165^\circ+i\sin 165^\circ\right)=-1.0842+0.2905i,\\ \zeta_2 &= \sqrt[6]{2}\cdot\left(\cos 285^\circ+i\sin 285^\circ\right)=0.2905-1.0842i.\end{aligned}$$

Perhaps you wonder about the sixth root, which occurs in the absolute value? Well, according to the rule, I had to take the third root from the absolute value of $-1+i$, this in turn is the (second) root of 2, so the total is the sixth root.

At the risk of boring you (which would be a good sign, by the way, because then you would already get it), I will now give a third example: I calculate the second roots (i.e. square roots) from -1.

Any more questions, Watson? Well, you may be puzzled by the fact that there is no "*i*" *to* be found here, so that the number cannot necessarily be identified as a complex number. Nevertheless, it is one, because at the very beginning of the chapter we agreed that the real numbers are contained as a subset in the complex ones; they are distinguished precisely by the fact that their imaginary part is zero.

But that does not matter at all, I claimed above that the root calculation formula works without any case distinction and exception, and I will show you that now with the example $z=-1$.

The absolute value of this number is 1. You can either see this with the naked eye or calculate it strictly according to the formula $\sqrt{(-1)^2+0}$. It is similar with the angle: Since the number -1 lies on the negative real axis, it naturally includes the angle 180° with the positive axis; if you do not see this directly, you can of course also use the table above. In any case you get the trigonometric form

$$-1=1\cdot\left(\cos 180^\circ+i\sin 180^\circ\right).$$

Now let us start: Obviously I do not have to care about the absolute value, it is and remains 1, it is just about the angles: For $k=0$ I get

$$\zeta_0 = \cos\frac{180^\circ}{2} + i\sin\frac{180^\circ}{2} = \cos 90^\circ + i\sin 90^\circ = 0 + i = i.$$

So the first square root of -1 is i. Great, that should not surprise anyone, since that is how the imaginary unit i had just been defined. At least, we have confirmed that the square root formula works in this case, too.

But what is the second square root of -1? Of course, you can guess this too, but you can also calculate it, and that is exactly what I want to do now: For $k = 1$ (and this is already the last value, because for $n = 2$ we have $n - 1 = 1$) the formula is

$$\zeta_1 = \cos\frac{540^\circ}{2} + i\sin\frac{540^\circ}{2} = \cos 270^\circ + i\sin 270^\circ = 0 + i(-1) = -i,$$

whereas you can take these values of sine and cosine from any collection of formulas or table.

So the second square root of -1 is $-$i, and this can be verified directly, because

$$(-i)^2 = (-1)^2 \cdot i^2 = -1.$$

Exercise 9.5 Calculate

(a) all second roots from $-2 + 3i$,
(b) all third roots from 8,

and give the result in normal form.

9.5 Complete Solution of Quadratic and Biquadratic Equations

Probably already in your school days, but at the latest in the third chapter of this book, you have seen how to calculate the solutions of a quadratic equation. As a reminder: The two solutions of the quadratic equation

$$x^2 + px + q = 0$$

are the numbers

$$x_1 = -\frac{p}{2} + \sqrt{\frac{p^2}{4} - q}$$

and

$$x_2 = -\frac{p}{2} - \sqrt{\frac{p^2}{4} - q},$$

where up to now, since we were only dealing with real numbers, there was always the addition *if the radicand $p^2/4 - q$ is not negative, otherwise the equation has no solution.*

Exactly this addition can be omitted now, because by now it is clear how to calculate roots of negative numbers; at the very beginning of the chapter I pointed out that the root of a negative number $-a$ can be calculated as follows:

$$\sqrt{-a} = \sqrt{(-1) \cdot a} = \sqrt{-1} \cdot \sqrt{a} = i \cdot \sqrt{a}.$$

Applying this to the above case of the *$p - q$ formula*, we obtain the following solution formulas of quadratic equations:

Complete Solution of Quadratic Equations
The quadratic equation

$$x^2 + px + q = 0$$

has the solutions

$$x_1 = -\frac{p}{2} + \sqrt{\frac{p^2}{4} - q}$$

and

$$x_2 = -\frac{p}{2} - \sqrt{\frac{p^2}{4} - q},$$

if the radicand $p^2/4 - q$ is not negative. Otherwise the solutions are the complex numbers

$$x_1 = -\frac{p}{2} + i \cdot \sqrt{-\left(\frac{p^2}{4} - q\right)}$$

and

(continued)

$$x_2 = -\frac{p}{2} - i \cdot \sqrt{-\left(\frac{p^2}{4} - q\right)}.$$

As an example I determine the solutions of the equation

$$x^2 - 2x + 2 = 0. \tag{9.8}$$

Here $p = -2$ and $q = 2$, thus

$$\frac{p^2}{4} - q = 1 - 2 = -1.$$

Thus the radicand is negative and hence the two solutions of Eq. (9.8) are to be calculated according to the second possibility. The result is

$$x_1 = 1 + i$$

and

$$x_2 = 1 - i.$$

As a check, one can now substitute these solutions back into Eq. (9.8); thus one obtains, e.g.

$$x_1^2 - 2x_1 + 2 = (1 + i)^2 - 2(1 + i) + 2 = 1 + 2i + i^2 - 2 - 2i + 2 = 0.$$

Please carry out the test of the second solution x_2 yourself.

Quite similar remarks as for quadratic equations apply also to biquadratic ones, i.e. those of the form

$$x^4 + bx^2 + c = 0,$$

which you learned in the third chapter. There you also saw that such an equation in the real can have up to four solutions, but only if certain numbers are not negative. Exactly this "if"-addition can now be omitted in the complex: Here, *every* biquadratic equation has four solutions, and they can be calculated according to the following procedure:

Complete Solution of Biquadratic Equations
To find the solutions of the biquadratic equation

$$x^4 + bx^2 + c = 0$$

one performs the substitution $u = x^2$ and calculates the solutions u_1 and u_2 of the quadratic equation

$$u^2 + bu + c = 0$$

as shown above. Then calculate the numbers

$$x_{11} = \sqrt{u_1} \quad \text{and} \quad x_{12} = -\sqrt{u_1}$$

as well as

$$x_{21} = \sqrt{u_2} \quad \text{and} \quad x_{22} = -\sqrt{u_2}.$$

The numbers x_{11}, x_{12}, x_{21} and x_{22} are solutions of the biquadratic equation.

To illustrate this method of calculation, I recalculate the last example on biquadratic equations from the third chapter, which had no (real) solution there, with this newly acquired knowledge of complex numbers: It is the equation

$$x^4 + 2x^2 + 4 = 0. \tag{9.9}$$

The substitution $u = x^2$ leads here to the quadratic equation

$$u^2 + 2u + 4 = 0$$

with the two solutions

$$u_{1/2} = -1 \pm \sqrt{1-4} = -1 \pm i\sqrt{3}.$$

Now to get the four solutions of the initial Eq. (9.9), I have to take the two square roots from u_1 and u_2 respectively, and that means first of all to determine the trigonometric form.

I start with $u_1 = -1 + i\sqrt{3}$. The absolute value of this number is $|u_1| = \sqrt{1+3} = 2$, the angle is

$$\varphi = \arctan\left(\frac{\sqrt{3}}{-1}\right) + 180^\circ = -60^\circ + 180^\circ = 120^\circ.$$

Thus I obtain as roots from this and thus as solutions of the biquadratic Eq. (9.9):

$$x_{11} = \sqrt{2} \cdot \left(\cos 60^\circ + i \sin 60^\circ\right) = 0.7071 + i \cdot 1.2247$$

and

$$x_{12} = \sqrt{2} \cdot \left(\cos 240^\circ + i \sin 240^\circ\right) = -0.7071 - i \cdot 1.2247.$$

The two roots from $u_2 = -1 - i\sqrt{3}$ can be calculated in the same way (which leads me elegantly to your last exercise phase) and finally you get

$$x_{21} = \sqrt{2} \cdot \left(\cos 120^\circ + i \sin 120^\circ\right) = -0.7071 + i \cdot 1.2247$$

and

$$x_{22} = \sqrt{2} \cdot \left(\cos 300^\circ + i \sin 300^\circ\right) = 0.7071 - i \cdot 1.2247.$$

And one last time I urge you to recalculate these results, on the one hand so that you can practice these techniques, but also because I unfortunately miscalculate all the time – which is why I became a professor, as a simple mathematician I would have no chance in life.

Exercise 9.6 Determine all the solutions of the following equations:

(a) $x^2 + x + 5/2 = 0$,
(b) $x^4 - x^2 - 2 = 0$.

At the very end of a book like this, it is always a good idea to review the whole process again. At the beginning you had to struggle with basic arithmetic – besides pizza and red wine – you have learned what terms and functions are, you can now solve more complicated equations and deal with inequalities, basic geometric structures are no longer strangers to you, the derivation as well as the integration of functions have become familiar to you, you can even solve entire systems of equations and thus describe linear geometric objects in space and you also learned the basics of probability theory and statistics. Likewise, you are now familiar with dealing with complex numbers, a real stumbling block for most mathematics beginners.

Do not you think that is quite a feat? Well, I do, and you can trust me on this. If you have worked through this book and understood the contents on the whole, you need not be afraid of any kind of mathematics in your studies or in life; and who else can say that about him- or self?

Formulary

10

10.1 Chapter 1: Basics

Arithmetic Laws
An arithmetic operation * on a set M is called **commutative** if for all a and b from M holds:

$$a^* b = b^* a.$$

It is called **associative** if for all a, b and c from M holds:

$$a^* (b^* c) = (a^* b) * c.$$

Addition + as well as multiplication · are both commutative and associative on the set N.

Distributive Law
With multiplication · and addition +, the **distributive law** holds for all numbers, i.e., it holds for any three numbers a, b, c:

$$a \cdot (b + c) = a \cdot b + a \cdot c.$$

Products of Negative Numbers
For any numbers a and b holds:

$$(-a)b = a(-b) = -ab$$

G. Walz et al., *Foundations of Mathematics: A Preparatory Course*,
https://doi.org/10.1007/978-3-662-67809-1_10

and

$$(-a)(-b) = ab.$$

Extending and Shortening Fractions
It applies

$$\frac{p}{q} = \frac{p \cdot a}{q \cdot a} \quad \text{for all } a \neq 0.$$

and

$$\frac{p \cdot b}{q \cdot b} = \frac{p}{q} \quad \text{for all } b \neq 0.$$

Sum of Two Fractions
It applies

$$\frac{p_1}{q_1} + \frac{p_2}{q_2} = \frac{p_1 \cdot \frac{\mathrm{kgV}}{q_1} + p_2 \cdot \frac{\mathrm{kgV}}{q_2}}{\mathrm{kgV}}$$

Dividing Fractions
It applies

$$\frac{p_1}{q_1} : \frac{p_2}{q_2} = \frac{p_1}{q_1} \cdot \frac{q_2}{p_2} = \frac{p_1 \cdot q_2}{q_1 \cdot p_2}.$$

Exponentiation with Natural Numbers
If a *is* an arbitrary number and n is a natural number, then a^n is defined as

$$a^n = \underbrace{a \cdot a \cdots a}_{\text{n times}}.$$

Furthermore, one defines $a^0 = 1$ for $a \neq 0$.

Power Laws
For all nonzero numbers a and b and all integers m and n, the following equations hold:

$$\begin{aligned} a^m \cdot a^n &= a^{m+n} \\ a^m \cdot b^m &= (a \cdot b)^m \\ (a^m)^n &= a^{m \cdot n} \\ a^{-n} &= \frac{1}{a^n}. \end{aligned}$$

Furthermore

$$a^{\frac{1}{2}} = \sqrt{a}$$

and

$$a^{\frac{1}{n}} = \sqrt[n]{a}.$$

For all positive numbers a and all rational exponents p/q holds:

$$a^{\frac{p}{q}} = \left(a^{\frac{1}{q}}\right)^p = (\sqrt[q]{a})^p = (a^p)^{\frac{1}{q}} = \sqrt[q]{a^p}.$$

Binomial Formulas
For all real numbers a and b holds:

$$\begin{aligned} (a+b)^2 &= a^2 + 2ab + b^2 \\ (a-b)^2 &= a^2 - 2ab + b^2 \\ (a+b) \cdot (a-b) &= a^2 - b^2 \end{aligned}$$

Factorial
For any natural number n

$$n! = 1 \cdot 2 \cdot 3 \cdots n.$$

In addition, one sets $0! = 1$.

10.2 Chapter 2: Functions

Functions

A **function** f is a rule that uniquely *assigns to* each element x *of a* set D an element $y = f(x)$ of a set W. We call D the **domain of definition of** f and W the **set of values** of f. In short f: $D \to W$.

Intervals.

If a and b are real numbers with $a \le b$, define the following **intervals** with limits a and b:

$$[a, b] = \{x \in \mathbb{R} \text{ with } a \le x \le b\}$$

(closed interval),

$$(a, b) = \{x \in \mathbb{R} \text{ with } a < x < b\}$$

(open interval),

$$(a, b] = \{x \in \mathbb{R} \text{ with } a < x \le b\}$$

and

$$[a, b) = \{x \in \mathbb{R} \text{ with } a \le x < b\}$$

(half-open intervals).

Concatenation of Functions

The function $f \circ g$: $D \to W$, defined by

$$f \circ g : D \to W, \quad f \circ g(x) = f(g(x))$$

for all $x \in D$, is called a **concatenation** of f and g.

Reverse Function

Let f be a function with domain D. A function f^{-1}, with the property

$$\left(f^{-1} \circ f\right)(x) = x$$

for all $x \in D$, is called an inverse **function** of f. If an inverse function exists for a function f, then f itself is called **invertible**.

Monotone Function

Let f be a function and I a subset of the domain D of definition of f.

We call f **monotonically increasing** on I if holds that if $x_1 < x_2$, then $f(x_1) \leq f(x_2)$.

We call f **monotonically decreasing** on I if holds that if $x_1 < x_2$, then $f(x_1) \geq f(x_2)$.

We call f **strictly monotonically increasing** on I if holds that if $x_1 < x_2$, then $f(x_1) < f(x_2)$.

One calls f **strictly monotonically decreasing** on I if holds that if $x_1 < x_2$, then $f(x_1) > f(x_2)$.

A function is called **strictly monotonic** if it is strictly monotonically increasing or strictly monotonically decreasing.

Strict Monotonicity and Reversibility

Let $f: D \to W$ be a strictly monotone function on all of D with image set $f(D)$. Then there exists an inverse function f^{-1} of f defined on $f(D)$, so f *is* invertible. If f is strictly monotonically increasing, then also f^{-1} is strictly monotonically increasing, and if f is strictly monotonically decreasing, then also f^{-1}.

Polynomials

For $n \in \mathbb{N}_0$ and $a_0, a_1, \ldots, a_n \in \mathbb{R}$ is called

$$p(x) = a_n x^n + a_{n-1} x^{n-1} + \cdots + a_1 x + a_0$$

a **polynomial** of degree at most n with **coefficients** $a_0, a_1, \ldots, a_n$.

Rational Function

Let p and q be two polynomials and let D be a subset of the real numbers that does not contain a zero of q. A function of the form

$$r : D \to \mathbb{R}, \quad r(x) = \frac{p(x)}{q(x)}$$

a **rational function**. A real number that is zero of the denominator but not of the numerator of a given rational function $r(x)$ is called the **pole of** $r(x)$.

Exponential Function to General Base

Let a be a positive real number. The function

$$\exp_a : \mathbb{R} \to \mathbb{R}, \quad \exp_a(x) = a^x$$

is called an exponential function in base a. The function $\exp_a(x) = a^x$ is strictly monotonically increasing on all of $\mathbb{R}$ if $a > 1$, and strictly monotonically decreasing on all of $\mathbb{R}$ if $a < 1$.

Logarithm to Base *a*

Let a and x be positive real numbers and $a \neq 0$. That real number y which satisfies the equation

$$a^y = x$$

is called the **logarithm of** x **to the base** a, denoted $\log_a(x)$.

Conversion of Logarithms to Different Bases

For any positive numbers a and b different from 1 and positive values x the conversion formula applies

$$\log_a(x) = \frac{\log_b(x)}{\log_b(a)}.$$

10.3 Chapter 3: Equations and Inequalities

Transformation of Equations

The following transformations of an equation do not change its solution set and are called **allowed transformations:**

- Addition or subtraction of the same number on both sides of the equation
- Multiplication of both sides by the same non-zero number

Solution of Linear Equation

The linear equation

$$ax + b = 0$$

with $a \neq 0$ has exactly one solution, this is

$$x = -\frac{b}{a}.$$

Solution of Quadratic Equations; ***p–q-Formulas***

The quadratic equation

$$x^2 + px + q = 0$$

has the real solutions

$$x_1 = -\frac{p}{2} + \sqrt{\frac{p^2}{4} - q}$$

and

$$x_2 = -\frac{p}{2} - \sqrt{\frac{p^2}{4} - q},$$

if the expression under the root, the **radicand**, is not negative. In this case the polynomial has the representation

$$p(x) = a(x - x_1)(x - x_2).$$

This is also true if these two zeros are identical, so the radicand in the *p–q formula* is zero. This is called the **decomposition into linear factors**.

Transformation of Inequalities

The following transformations of an inequality do not change its solution set and are called **allowed transformations:**

- Addition or subtraction of the same number on both sides of the inequality
- Multiplication of both sides by the same *positive number*

Multiplying both sides of an inequality by the same negative number does not change its solution set, so it is allowed if you simultaneously invert the inequality sign connecting the two sides, that is, change "<" to ">" and vice versa.

10.4 Chapter 4: Geometry

Heron's Formula

The area of a triangle F with side lengths a, b and c *is* half the product of the incircle radius ρ with the sum of the side lengths: $F = \rho\,(a + b + c)/2$.

Pythagorean Theorem

For right triangles, the sum of the squares of the lengths of the cathets a, b is always equal to the square of the length of the hypotenuse c:

$$a^2 + b^2 = c^2.$$

Euclid's Theorem

For right triangles, the square of the length of a cathetus ag equals the product of the length of the hypotenuse c and the length of the hypotenuse segment p that has a point in common with that cathetus:

$$a^2 = cp.$$

Height Set

For right triangles, the square of the height h on the hypotenuse is equal to the product of the two hypotenuse intercepts p and q:

$$h^2 = p \cdot q.$$

Theorem of Thales

For right triangles with common hypotenuse (of length *c formed* by the vertices A and B) the remaining vertex C always lies on the common circumcircle of these triangles. This circumcircle has the center $S = (A + B)/2$ and the radius $c/2$.

Radians

The conversion formula between degrees and radians of an angle is

$$x = \pi\alpha/180^\circ.$$

Sine, Cosine and Tangent

The following definitions apply to the right triangle:

$$\sin(\alpha) = \frac{\text{opposite cathetus}}{\text{hypotenuse}}$$
$$\cos(\alpha) = \frac{\text{countercathetus}}{\text{hypotenuse}}$$
$$\tan(\alpha) = \frac{\text{countercathetus}}{\text{countercathetus}}$$

Trigonometric Pythagorean Theorem
For any angle α holds:

$$\sin^2(\alpha) + \cos^2(\alpha) = 1.$$

Important Values of the Sine

$$\begin{aligned} \sin(0^\circ) &= 0 \\ \sin(30^\circ) &= \frac{1}{2} \\ \sin(45^\circ) &= \frac{\sqrt{2}}{2} \\ \sin(60^\circ) &= \frac{\sqrt{3}}{2} \\ \sin(90^\circ) &= 1 \end{aligned}$$

Important Values of the Cosine

$$\begin{aligned} \cos(0^\circ) &= 1 \\ \cos(30^\circ) &= \frac{\sqrt{3}}{2} \\ \cos(45^\circ) &= \frac{\sqrt{2}}{2} \\ \cos(60^\circ) &= \frac{1}{2} \\ \cos(90^\circ) &= 0 \end{aligned}$$

Addition Theorems of Sine and Cosine
For all x and y, the following relationships hold:

$$\sin(x+y) = \sin(x)\cos(y) + \cos(x)\sin(y)$$
$$\cos(x+y) = \cos(x)\cos(y) - \sin(x)\sin(y)$$

Cosine Theorem

For any triangle, the square of the length a *of* a side is the difference of the sum of the squares of the lengths b, c *of* the remaining sides with the term $2bc\cos(\alpha)$, where α is the angle enclosed by the sides with lengths b and c:

$$a^2 = b^2 + c^2 - 2\,bc\cos(\alpha).$$

Sine Set

For any triangle with side lengths a, b, and c and corresponding opposite interior angles α, β, and γ holds:

$$a/\sin(\alpha) = b/\sin(\beta) = c/\sin(\gamma).$$

Tangent Set

For any triangle with side lengths a and b and interior angles α and β opposite these sides, the following holds:

$$\tan((\alpha-\beta)/2) \\ = \tan((\alpha+\beta)/2)(a-b)/(a+b).$$

Sum of the Interior Angles of a Quadrilateral

The sum $\alpha + \beta + \gamma + \delta$ of the interior angles α, β, γ, δ of a quadrilateral is always 360°.

Area of a Parallelogram

The area F of a parallelogram with sides a and b and angle α enclosed by these sides is $F = \sin(\alpha)ab$. Specifically, the area F *of* a rectangle ($\alpha = 90°$) is $F = ab$ and the area F of a square ($\alpha = 90°$ and $a = b$): $F = a^2$.

Area of the Trapezoid

The area F of a trapezoid with side lengths a and b *of* the parallel sides and height length h is $F = (a + b)h/2$.

***N*-Corner**

The sum of the interior angles of an *N-corner* is $(N - 2)\ 180°$.

A regular *N-corner* has area $F_N = N\,(a^2/2\,\sin(360°/N))$.

Area and Circumference of the Circle
The unit circle has area $F = \pi$ and circumference $U = 2\pi$. Any circle with radius r has area

$$F = \pi r^2$$

and the scope

$$U = 2\pi r.$$

Circle Equation
A circle with center (x_0, y_0) and radius r has the equation

$$(x - x_0)^2 + (y - y_0)^2 = r^2.$$

10.5 Chapter 5: Linear Algebra

Scalar Product
If $\boldsymbol{x} = (x_1, \ldots, x_n)$ and $\boldsymbol{y} = (y_1, \ldots, y_n)$ are two *n-dimensional* vectors, then set

$$\mathbf{x} \cdot \mathbf{y} = x_1 \cdot y_1 + \cdots + x_n \cdot y_n.$$

This operation is called the scalar product of the two vectors.

Exactly then two vectors are perpendicular to each other, if their scalar product results in zero.

Calculation Rules for the Scalar Product
For $\boldsymbol{x}, \boldsymbol{y}, \boldsymbol{z} \in \mathbb{R}^n$ and $\lambda \in \mathbb{R}$ the following rules hold:

(a) $\boldsymbol{x} \cdot \boldsymbol{y} = \boldsymbol{y} \cdot \boldsymbol{x}$;
(b) $\boldsymbol{x} \cdot (\boldsymbol{y} + \boldsymbol{z}) = \boldsymbol{x} \cdot \boldsymbol{y} + \boldsymbol{x} \cdot \boldsymbol{z}$;
(c) $\lambda \cdot (\boldsymbol{x} \cdot \boldsymbol{y}) = \lambda \cdot \boldsymbol{x} \cdot \boldsymbol{y}$.

Vector Product
If $\boldsymbol{b} = (b_1, \ldots, b_n)$ and $\boldsymbol{c} = (c_1, \ldots, c_n)$ are two three-dimensional vectors, the vector is called

$$\mathbf{b} \times \mathbf{c} = \begin{pmatrix} b_2c_3 - b_3c_2 \\ b_3c_1 - b_1c_3 \\ b_1c_2 - b_2c_1 \end{pmatrix}$$

the vector product or cross product of $\boldsymbol{b}$ and $\boldsymbol{c}$. The vector $\boldsymbol{b} \times \boldsymbol{c}$ is perpendicular to $\boldsymbol{b}$ and to $\boldsymbol{c}$.

Calculation of the Parameter-Free Plane Equation
If $A = (a_1, a_2, a_3)$, $B = (b_1, b_2, b_3)$, and $C = (c_1, c_2, c_3)$ are three points in space that do not all lie on a straight line, then compute the parameter-free equation of the plane that passes through all three points using the following scheme.

(a) One determines

$$\mathbf{b} = \begin{pmatrix} b_1 - a_1 \\ b_2 - a_2 \\ b_3 - a_3 \end{pmatrix}, \quad \mathbf{c} = \begin{pmatrix} c_1 - a_1 \\ c_2 - a_2 \\ c_3 - a_3 \end{pmatrix}.$$

(b) Calculate the normal vector with the cross product

$$\mathbf{n} = \mathbf{b} \times \mathbf{c} = \begin{pmatrix} b_2c_3 - b_3c_2 \\ b_3c_1 - b_1c_3 \\ b_1c_2 - b_2c_1 \end{pmatrix}.$$

(c) For the plane points (x, y, z) one makes the approach

$$\mathbf{n} \cdot \begin{pmatrix} x - a_1 \\ y - a_2 \\ z - a_3 \end{pmatrix} = 0.$$

(d) Calculate the scalar product of c) and simplify the resulting equation until an equation of the form $ax + by + cz =$ dent is obtained. This equation is then the parameter-free equation of the plane.

Linear Dependence and Independence
The vectors $\boldsymbol{x}_1, \ldots, \boldsymbol{x}_m \in \mathbb{R}^n$ are called *linearly dependent* if one of them can be represented as a linear combination of the remaining $m - 1$ vectors. Vectors that are not linearly dependent are called *linearly independent*.

The vectors $\boldsymbol{x}_1, \ldots, \boldsymbol{x}_m \in \mathbb{R}^n$ are linearly independent if and only if from

$$c_1 \cdot \mathbf{x}_1 + c_2 \cdot \mathbf{x}_2 + \cdots + c_m \cdot \mathbf{x}_m = \mathbf{0}$$

always follows:

$$c_1 = c_2 = \cdots = c_m = 0,$$

where 0 is understood to be the zero vector. If, on the other hand, the vectors can be combined to form the zero vector without each of the prefactors $c_1, c_2, \ldots, c_m$ having to become zero, then the vectors are linearly dependent.

If one has m *n-dimensional* vectors and if $m > n$ holds, then these vectors are linearly dependent.

Rules of Calculation for Matrices

If A and B are matrices with m rows and n columns, and λ and μ are real numbers, then the following rules of calculation apply:

(a) $A + B = B + A$;
(b) $\lambda \cdot (A + B) = \lambda \cdot A + \lambda \cdot B$;
(c) $(\lambda + \mu) \cdot A = \lambda \cdot A + \mu \cdot A$.

Matrix Multiplication

There were

$$A = \begin{pmatrix} a_{11} & a_{12} & \cdots & a_{1n} \\ a_{21} & a_{22} & \cdots & a_{2n} \\ \vdots & \vdots & & \vdots \\ a_{m1} & a_{m2} & \cdots & a_{mn} \end{pmatrix}$$

a matrix with m rows and n columns and

$$B = \begin{pmatrix} b_{11} & b_{12} & \cdots & b_{1k} \\ b_{21} & b_{22} & \cdots & b_{2k} \\ \vdots & \vdots & & \vdots \\ b_{n1} & b_{n2} & \cdots & b_{nk} \end{pmatrix}$$

a matrix with n rows and k columns. The product matrix $C = A \cdot B$ then has the form

$$C = \begin{pmatrix} c_{11} & c_{12} & \cdots & c_{1k} \\ c_{21} & c_{22} & \cdots & c_{2k} \\ \vdots & \vdots & & \vdots \\ c_{m1} & c_{n2} & \cdots & c_{mk} \end{pmatrix}.$$

The entries c_{ij} in C are calculated as the scalar product of the *ith* row of A and the *jth* column of B, thus:

$$c_{ij} = a_{i1}b_{1j} + a_{i2}b_{2j} + \cdots + a_{in}b_{nj}.$$

Calculation Rules for the Matrix Product

Let matrices A, B and C *be* given for which the following operations are to be feasible. Then the following rules apply.

(a) $A \cdot (B \cdot C) = A \cdot B \cdot C$.
(b) $A \cdot (B + C) = A \cdot B + A \cdot C$.
(c) $(A + B) \cdot C = A \cdot C + B \cdot C$.
(d) In general, $A \cdot B \neq B \cdot A$.

Inverse Matrix

A matrix $A \in \mathbb{R}^{n \times n}$ is called **invertible**, or **regular**, if there exists a matrix $A^{-1} \in \mathbb{R}^{n \times n}$ with

$$A \cdot A^{-1} = A^{-1} \cdot A = I_n.$$

In this case, A^{-1} is called the **inverse matrix** of A.

Permitted Manipulations of Systems of Equations

The solution set of a linear system of equations is not changed by the following operations.

(a) Multiplication of an equation by a non-zero number.
(b) Adding a multiple of one equation to another.
(c) Swapping two equations.

Gauss Algorithm

Let $A\,\boldsymbol{x} = \boldsymbol{b}$ *be* a linear system of equations with n equations and n unknowns that has a unique solution. We start with the matrix

$$(A\,|\,b) = \begin{pmatrix} a_{11} & \cdots & a_{1n} & b_1 \\ a_{21} & \cdots & a_{2n} & b_2 \\ \vdots & & \vdots & \vdots \\ a_{n1} & \cdots & a_{nn} & b_n \end{pmatrix},$$

which lists the coefficients of the system of equations and its right side. Then you need to perform the following steps.

(a) If $a_{11} = 0$, find a line whose first element is different from zero, swap this line with the first line and rename it. Then $a_{11} \neq 0$ holds.
(b) Subtract a suitable multiple of the first row from the second, third, . . ., last row, so that these rows each start with zero. The new matrix then has only zeros in the first column below a_{11}
(c) Subtract a suitable multiple of the second row from the third, . . ., last row, so that these rows each start with zero and also have a zero in the second place. The new matrix has then in the first column below a_{11} and in the second column below a_{22} only zeros.
(d) Repeat this procedure for the following columns of the matrix until there are only zeros left below the main diagonal elements.
(e) Divide the last row by the last non-zero coefficient so that there is a one at the end of the main diagonal. Then proceed with the appropriate row transformations from bottom to top, so that the matrix has the following form at the end:

$$\begin{pmatrix} 1 & 0 & \cdots & 0 & x_1 \\ 0 & 1 & \cdots & 0 & x_2 \\ \vdots & \ddots & \ddots & \vdots & \vdots \\ 0 & \cdots & 0 & 1 & x_n \end{pmatrix}$$

The last column contains the solution of the linear system of equations.

If this procedure does not lead to the goal, the linear system of equations does not have a unique solution.

If the Gaussian algorithm leads to a line of the form $(0\ 0\ \cdots\ 0\ b)$ with $b \neq 0$, the system of linear equations has no solution.

Inversion of Matrices

Let $A \in \mathbb{R}^{n\times n}$, a matrix whose inverse matrix A^{-1} is to be computed. We combine A and the unit matrix I_n into one matrix by writing the matrix A in the left half and the unit matrix in the right half, and reshape this matrix using the well-known operations from the Gaussian algorithm so that the left half is transformed into the unit matrix I_n. Then the inverse matrix A^{-1} is in the right half of the reshaped large matrix. If it is not possible to produce the unit matrix in the left half, An is not invertible.

10.6 Chapter 6: Differential and Integral Calculus

Differential Quotient
If the limit of the difference quotients of f in x for *his equal to* 0

$$\lim_{h \to 0} \frac{f(x+h) - f(x)}{h}$$

exists, this **differential quotient** of f in x is called $f'(x)$. $f'(x)$ is then the value of the slope of f in x and is called the **derivative of** f at location x.

Tangent of f in x
The linear function t through the point $(x, f(x))$ with slope $f'(x)$ is called the **tangent** of f in x.

Derivation of Important Functions

$$\begin{aligned}
(x^n)' &= n \cdot x^{n-1} \\
(e^x)' &= e^x \\
\sin'(x) &= \cos(x) \\
\cos'(x) &= -\sin(x) \\
\left(\sqrt{x}\right)' &= 1/\left(2\sqrt{x}\right) \\
\ln'(x) &= 1/x
\end{aligned}$$

Summation Rule
If f and g are derivable functions, then

$$(f+g)'(x) = f'(x) + g'(x).$$

Product Rule
If f and g are derivable functions, then

$$(f \cdot g)'(x) = f'(x)\,g(x) + g'(x)f(x).$$

Chain Rule
If f and g are derivable functions, then

$$(f \circ g)'(x) = g'(x) f'(g(x)).$$

Quotient Rule

If f and g are derivable functions, then

$$\left(\frac{f(x)}{g(x)}\right)' = \frac{f'(x)g(x) - g'(x)f(x)}{(g(x))^2}.$$

Monotony

A derivable function f is monotonically increasing if and only if $f'(x) \geq 0$. It is monotonically decreasing if and only if $f'(x) \leq 0$.

Strict Monotony

A derivable function f with $f'(x) > 0$ for all x is strictly monotonically increasing. A derivable function f with $f'(x) < 0$ for all x is strictly monotonically decreasing.

Inverse Derivative

If f is a derivable function with $f'(x) > 0$ or $f'(x) < 0$ for all x, then its inverse function f^{-1} is derivable and the following holds

$$(f^{-1})'(\overline{x}) = \frac{1}{f'(x)},$$

Where $\overline{x} = f(x)$.

Necessary Criterion for Local Extremes

If a derivable function f in x_0 has a local extremum, then $f'(x_0) = 0$.

Sufficient Criterion for Local Extremes

If f is a twice derivable function and if $f'(x_0) = 0$, then f has a local extremum in x_0 if, in addition, $f''(x_0) \neq 0$ holds. If $f''(x_0) < 0$, then x_0 is a local maximum, if $f''(x_0) > 0$, then x_0 is a local minimum.

Convexity and Sign of the Second Derivative

The graph of twice derivable functions f is convex exactly if the second derivative f'' becomes nonnegative for all x: $f''(x) \geq 0$.

Turning Points

A point x_0 is called an **inflection point of** f if f' has a local extremum in x_0. The point $(x_0, f(x_0))$ is then called an **inflection point of** the graph of f. Turning points at which the first derivative also vanishes are called **saddle points.**

Necessary and Sufficient Criterion for Turning Points

If a twice derivable function f has an inflection point in x_0, then $f''(x_0) = 0$. If f is a three times derivable function and x_0 is a point with $f''(x_0) = 0$, then f has an inflection point in x_0 if moreover $f'''(x_0) \neq 0$ holds. If $f'(x_0) = 0$, then x_0 is a saddle point.

Integrability and Definite Integral

A function f is said to be **integrable** on $[a, b]$ if for each subdivision of $[a, b]$ into n subintervals $I_{j,n} = [x_{j,n}, x_{j+1,n}]$ whose lengths $x_{j+1,n} - x_{j,n}$ go towards 0 for $n \to \infty$ and for each choice of points $\widetilde{x}_{j,n} \in I_{j,n}$, where $j \in \{0, \ldots, n-1\}$, the limit of sequence

$$\begin{aligned} \mathcal{F}_n &= f(\widetilde{x}_{0,n})(x_{1,n} - x_{0,n}) + \cdots \\ &\quad + f(\widetilde{x}_{n-1,n})(x_{n,n} - x_{n-1,n}) \end{aligned}$$

exists for $n \to \infty$ and is equal in each case. In this case the limit is called

$$\begin{aligned} \int_a^b f(x)\,dx &= \lim_{n \to \infty} \mathcal{F}_n \\ &= \lim_{n \to \infty} \sum_{j=0}^{n-1} f(\widetilde{x}_{j,n})\,(x_{j+1,n} - x_{j,n}) \end{aligned}$$

(read "integral from a to b over f to $d\,x$") the **definite integral of** f **with respect to** $[a, b]$.

Properties of the Definite Integral

$$\int_a^b f(x)\,dx = \int_a^c f(x)\,dx + \int_c^b f(x)\,dx$$

$$\int_a^b f(x)\,dx = -\int_b^a f(x)\,dx$$

$$\int_a^a f(x)\,dx = 0$$

$$\int_a^b (\lambda \cdot f)(x)\,dx = \lambda \int_a^b f(x)\,dx$$

$$\int_a^b (f+g)(x)dx = \int_a^b f(x)\,dx + \int_a^b g(x)\,dx.$$

Important Stem Functions

$$\begin{aligned}
\int x^n\,dx &= \frac{x^{n+1}}{n+1} + C \\
\int e^x\,dx &= e^x + C \\
\int a^x\,dx &= \frac{a^x}{\ln(a)} + C \\
\int \sin(x)\,dx &= -\cos(x) + C \\
\int \cos(x)\,dx &= \sin(x) + C \\
\int \frac{1}{x}\,dx &= \ln(|x|) + C
\end{aligned}$$

Derivability of the Integral Function of Continuous Functions
For any continuous function f on $[a, b]$ is

$$F_a(t) = \int_a^t f(x)\,dx, \quad t \in [a,b]$$

a derivable function whose first derivative coincides with f:

$$F_a'(t) = f(t), \quad t \in [a,b].$$

Master Functions Differ Only by Additive Constants
Two primitive functions F and G *of* the same function f differ only by an additive constant c: $F(x) = G(x) + c$ for all x.

Main Theorem of Differential and Integral Calculus

If f is a continuous function on $[a, b]$ and F is an arbitrary primitive function of f, then the definite integral of f *with respect to* $[a, b]$ can be calculated as follows: $\int_a^b f(x)\,dx = F(b) - F(a)$.

Partial Integration

If f and g are functions derivable on $[a, b]$, the following is true

$$\int_a^b f'(x)\,g(x)\,dx$$

$$= f(x)\,g(x)|_a^b - \int_a^b f(x)\,g'(x)\,dx.$$

Substitution Rule

If f is continuous on $[a, b]$ and g is derivable on $[a, b]$, then

$$\int_a^b g'(x) f(g(x))\,dx = \int_{g(a)}^{g(b)} f(\overline{x})\,d\overline{x}.$$

10.7 Chapter 7: Probability Calculation

Experiment with Finitely Many Equally Probable Outcomes

If a trial has n distinct outcomes, all of which are equally likely, then each outcome occurs with probability $1/n$.

Binomial Coefficient

If m and l are two natural numbers with $l \leq m$, the **binomial coefficient** is

$$\binom{m}{l},$$

pronounced "m over l", defined as

$$\binom{m}{l} = \frac{m!}{l! \cdot (m-l)!}.$$

Selection with Backspacing with Attention to the Sequence

To select k balls from a set of n balls with laying back and with respect to the order, there are n^k possibilities.

Selection with Backspacing Without Regard to the Sequence

To select k balls from a set of n balls with reclining, but without paying attention to the order, there is

$$\binom{n+k-1}{k}$$

Possibilities.

Selection Without Backspacing with Consideration of the Sequence

To select k balls from a set of n balls without putting them back, but with respect to the order, there is

$$\frac{n!}{(n-k)!} = n \cdot (n-1) \cdot (n-2) \cdots (n-k+1)$$

Possibilities. Special case $k = n$: There are $n!$ possibilities to arrange *n-distinguishable* elements of a set.

Selection Without Backspacing Without Consideration of the Sequence

To select k balls from a set of n balls without laying back and without paying attention to the order, there is

$$\binom{n}{k} = \frac{n!}{k!(n-k)!}$$

Possibilities.

Axiomatic Definition of Probability

On the set of all results of a random experiment, define a function p that assigns a real number to each result. This function, or its values, is called a **probability** if it has the following properties:

(1) For any event A, $0 \leq p(A) \leq 1$ holds.
(2) The **safe** event Ω has the value 1: $p(\Omega) = 1$.
(3) For each pair of incompatible events A and B, *the* following applies

$$p(A \cup B) = p(A) + p(B).$$

(4) Property (3) also holds for the sum of any number of events, including infinitely many, if they are pairwise incompatible.

Rules for Calculating Probabilities
In addition to the above axioms, the following rules apply to probabilities:

(5) The impossible event has probability zero: $p(\varnothing) = 0$.
(6) If $\overline{A}$ is the counter-event to A, then $p(\overline{A}) = 1 - p(A)$.
(7) If an event A *is* part of another event B, that is, if the occurrence of A implies that of B *in* all cases, then $p(A) \leq p(B)$ holds.
(8) For any events A and B, *the* following holds true

$$p(A \cup B) = p(A) + p(B) - p(A \cap B).$$

Conditional Probability
If A and B are two events of the same random trial with $p(B) > 0$, then $p(A|B)$, defined by

$$p(A \mid B) = \frac{p(A \cap B)}{p(B)}$$

the **conditional probability of** A *with* **respect to** B.

10.8 Chapter 8: Descriptive Statistics

Statistical Scales

(a) Nominal scale: Characteristic expressions are distinguishable, but only qualitative properties are named to distinguish them, usually verbally.
(b) Ordinal scale or ranking scale: The proficiencies allow a comparison, so that a sorting is possible.
(c) Metric scale: The expressions are real numbers and the differences between the expressions describe the difference of the respective characteristic expressions.

Absolute and Relative Frequencies

If n characteristic carriers – i.e. n data records – and the characteristic values $A_1, \ldots, A_j$ are given, the absolute frequency n_i describes the number of characteristic carriers that have the value A_i. The following always applies

$$n_1 + \cdots + n_j = n.$$

The relative frequency of characteristic expression A_i is calculated by

$$h_i = \frac{n_i}{n}.$$

It always applies

$$h_1 + \cdots + h_j = \frac{n_1 + \cdots + n_j}{n} = 1.$$

Cumulative and Relative Cumulative Frequencies

If n characteristic carriers – i.e. n data sets – of a metrically scaled characteristic with the characteristic values $A_1, \ldots, A_j$ are given and if $A_1 < A_2 < \cdots < A_j$, the cumulative frequency K_i describes the number of characteristic carriers that have a value less than or equal to A_i. The following applies

$$K_i = n_1 + \cdots + n_i,$$

where $n_1, \ldots, n_j$ are the absolute frequencies of the characteristic values.

The relative cumulative frequency k_i of the characteristic expression A_i is calculated by

$$k_i = \frac{n_1}{n} + \cdots + \frac{n_i}{n} = \frac{K_i}{n},$$

that is, as the sum of the relative frequencies $h_1 = n_1/n$ to $h_i = n_i/n$.

It is always true that $K_j = n$ and $k_j = 1$, i.e.: the last cumulative frequency is equal to the number of feature carriers, the last relative cumulative frequency is equal to 1.

Modal Value

If the characteristic values of a characteristic are given, the characteristic value that occurs most frequently is a modal value of the characteristic. A characteristic can have several modal values.

Arithmetic Mean

If n values $x_1, \ldots, x_n$ of a metrically scaled characteristic are given, their arithmetic mean $\overline{x}$ is calculated according to the formula

$$\overline{x} = \frac{x_1 + x_2 + \cdots + x_n}{n}.$$

If the individual expressions of a characteristic are not known, but only a class division is given, the arithmetic mean is calculated according to the formula:

$$\overline{x} = \frac{n_1 \cdot x_1 + n_2 \cdot x_2 + \cdots + n_k \cdot x_k}{n}.$$

Here, k denotes the number of classes, $x_1, \ldots, x_k$ the respective class midpoints of the classes, $n_1, \ldots, n_k$ the absolute frequencies of the feature expressions in the respective class, and $n = n_1 + \cdots + n_k$ the total number of these expressions.

Median

If n characteristic values of a metrically scaled or ordinally scaled characteristic are available in ordered form, i.e. $x_1 \leq x_2 \leq \cdots \leq x_n$, the median $\widetilde{x}$ is determined according to the following formula:

$$\widetilde{x} = x_{\frac{n+1}{2}}, \quad \text{if } n \text{ is odd}$$

respectively

$$\widetilde{x} = \frac{x_{\frac{n}{2}} + x_{\frac{n}{2}+1}}{2}, \quad \text{falls } n \text{ ungerade ist.}$$

Span

The range of expressions of a characteristic is the difference between the largest and smallest expression of the characteristic.

Mean Absolute Deviation

If n characteristics $x_1, \ldots, x_n$ of a metrically scaled characteristic are given, the size is denoted by

$$\frac{|x_1 - \overline{x}| + |x_2 - \overline{x}| + \cdots + |x_n - \overline{x}|}{n}$$

as the mean absolute deviation with respect to the arithmetic mean $\overline{x}$, and the quantity

$$\frac{|x_1 - \tilde{x}| + |x_2 - \tilde{x}| + \cdots + |x_n - \tilde{x}|}{n}$$

as the mean absolute deviation with respect to the median $\tilde{x}$.

Dispersion and Standard Deviation
If n expressions $x_1, \ldots, x_n$ of a metrically scaled characteristic are given, the size is denoted by

$$s^2 = \frac{(x_1 - \overline{x})^2 + (x_2 - \overline{x})^2 + \cdots + (x_n - \overline{x})^2}{n - 1}$$

as variance or dispersion. The size

$$s = \sqrt{\frac{(x_1 - \overline{x})^2 + (x_2 - \overline{x})^2 + \cdots + (x_n - \overline{x})^2}{n - 1}}$$

is called the standard deviation.

10.9 Chapter 9: Complex Numbers

Imaginary Unit
The **imaginary unit** is defined by

$$i^2 = -1.$$

Complex Numbers
If i *is* the imaginary unit and a and b are arbitrary real numbers, a number of the form $a + ib$ is called a **complex number**. The set of all complex numbers is called $\mathbb{C}$. The number a is called the **real part**, the number b the **imaginary part of** the complex number $a + ib$.

Reciprocal of a Complex Number
For any complex number $z = a + ib \neq 0$ is

$$\frac{1}{z} = \frac{1}{a + ib} = \frac{a}{a^2 + b^2} - i \cdot \frac{b}{a^2 + b^2}.$$

Absolute Value
The **absolute value of** the complex number $z = a + ib$ is the real number

$$|z| = \sqrt{a^2 + b^2}.$$

Angle of a Complex Number

The angle φ that a complex number $z = a + ib$ makes with the positive real axis can be calculated as follows:

$$\varphi = \begin{cases} \arctan\left(\frac{b}{a}\right), & \text{falls } a > 0 \text{ and } b \geq 0, \\ \arctan\left(\frac{b}{a}\right) + 360^\circ, & \text{falls } a > 0 \text{ and } b < 0, \\ \arctan\left(\frac{b}{a}\right) + 180^\circ, & \text{falls } a < 0, \\ 90^\circ, & \text{falls } a = 0 \text{ and } b > 0 \\ 270^\circ, & \text{falls } a = 0 \text{ and } b < 0, \\ 0^\circ, & \text{falls } a = 0 \text{ and } b = 0. \end{cases}$$

Trigonometric Form of a Complex Number

The trigonometric form of the complex number z is the representation

$$z = |z| \cdot (\cos\varphi + i \sin\varphi).$$

Multiplication and Division of Complex Numbers in Trigonometric Form

For

$$z_{1,2} = |z_{1,2}| \cdot \left(\cos\varphi_{1,2} + i \sin\varphi_{1,2}\right)$$

$$z_1 \cdot z_2 = |z_1| \cdot |z_2| \cdot (\cos(\varphi_1 + \varphi_2) + i \sin(\varphi_1 + \varphi_2))$$

and – if $z_2 \neq 0$ –

$$\frac{z_1}{z_2} = \frac{|z_1|}{|z_2|} \cdot (\cos(\varphi_1 - \varphi_2) + i \sin(\varphi_1 - \varphi_2)).$$

Exponentiation of Complex Numbers
Is

$$z = |z| \cdot (\cos\varphi + i\sin\varphi)$$

is a complex number and $n \in \mathbb{N}$, then

$$z^n = |z|^n \cdot (\cos n\varphi + i\sin n\varphi).$$

Roots of Complex Numbers
Let z be any nonzero complex number in the trigonometric form

$$z = |z| \cdot (\cos\varphi + i\sin\varphi)$$

and n is any natural number.

There are exactly n *nth* roots of z, denoted $\zeta_0, \zeta_1, \ldots, \zeta_{n-1}$. If one sets

$$\varphi_k = \frac{\varphi + k \cdot 360^\circ}{n}$$

for $k = 0, 1, \ldots, n - 1$, then

$$\zeta_k = \sqrt[n]{|z|} \cdot (\cos(\varphi_k) + i \cdot \sin(\varphi_k))$$

for $k = 0, 1, \ldots, n - 1$.

Complete Solution of Quadratic Equations
The quadratic equation

$$x^2 + px + q = 0$$

has the solutions

$$x_{1/2} = -\frac{p}{2} \pm \sqrt{\frac{p^2}{4} - q},$$

if the radicand $p^2/4 - q$ is not negative. Otherwise the solutions are the complex numbers

$$x_{1/2} = -\frac{p}{2} \pm i \cdot \sqrt{-\left(\frac{p^2}{4} - q\right)}.$$

Complete Solution of Biquadratic Equations
To find the solutions of the biquadratic equation

$$ax^4 + bx^2 + c = 0$$

one performs the substitution $u = x^2$ and calculates the solutions u_1 and u_2 of the quadratic equation

$$au^2 + bu + c = 0.$$

Then the numbers are

$$x_{11} = \sqrt{u_1} \quad \text{and} \quad x_{12} = -\sqrt{u_1}$$

and

$$x_{21} = \sqrt{u_2} \quad \text{and} \quad x_{22} = -\sqrt{u_2}.$$

the solutions of the biquadratic equation.

Appendix: Solutions of the Exercises

For your self-control, the solutions of the exercises are given here; detailed solutions can be found on the publisher's website.

Exercises Chap. 1

Solution of Exercise 1.1 It is

$$a + b + c = a + (b + c) = (b + c) + a = (c + b) + a = c + b + a$$

Solution of Exercise 1.2

(a) $(3 - 6)(-2 - 3) = (-3)(-5) = 15$
(b) $-(8 - 3)(-2 + 5) = (-5)\,3 = -15$
(c) $(9 - 3 - 8)(3 - 1 - 7)(-2 + 1) = (-2)(-5)(-1) = -10$

Solution of Exercise 1.3

$$\frac{231}{22} = \frac{21 \cdot 11}{2 \cdot 11} = \frac{21}{2}$$

$$\frac{52}{28} = \frac{4 \cdot 13}{4 \cdot 7} = \frac{13}{7}$$

$$\frac{16}{64} = \frac{1}{4}$$

G. Walz et al., *Foundations of Mathematics: A Preparatory Course*,
https://doi.org/10.1007/978-3-662-67809-1

Solution of Exercise 1.4

$\frac{9}{4}\cdot\frac{6}{3}=\frac{9}{2}$

$\frac{13}{7}\cdot\frac{21}{26}=\frac{3}{2}$

$\frac{52}{76}\cdot\frac{19}{13}=1$

Solution of Exercise 1.5

$\frac{3}{4}:\frac{7}{8}=\frac{6}{7}$

$\frac{52}{76}:\frac{13}{19}=1$

$\frac{143}{11}:\frac{130}{22}=\frac{11}{5}$

Solution of Exercise 1.6

(a) $(((2-3)\cdot 4)-2)\cdot(-2)=12$
(b) $2-((1-4)\cdot(3-2)+4)=1$

Solution of Exercise 1.7

(a) $\left(\frac{1}{2}\right)^3\cdot\left(\frac{2}{3}\right)^5\cdot\left(\frac{3}{4}\right)^2=\frac{1}{108}$
(b) $(a^2b^3c^{-1})^2\cdot(c^2)^2=a^4b^6c^2$

Solution of Exercise 1.8

(a) $\sqrt[4]{81}=3$
(b) $\sqrt[3]{2^2\cdot 2}=2$
(c) $\sqrt[5]{(a^5)^2}=a^2$

Solution of Exercise 1.9

(a) $4^{1/3}\cdot 2^{1/3}\cdot 16^{1/4}=4$
(b) $120^{1/2}\cdot 900^{1/4}=60$
(c) $\sqrt{0.16}=0.4$

Solution of Exercise 1.10

(a) $(a^2+b-c)(a+bc)(a^2b)=a^5b+a^3b^2-a^3bc+a^4b^2c+ab^{23}\mathrm{c}-a^2b^2c^2$
(b) $(1+x+x^2+x^3+x^4)(1-x)=1-x^5$

Solution of Exercise 1.11

(a) $34xyz^2-17x^3yz+51y^4=17y\,(2xz^2-x^3z+3y^3)$

(b) $2a^2bc - 4ab^2c + 8abc^2 = 2abc\,(a - 2b + 4c)$

Solution of Exercise 1.12

(a) $\frac{4a^3(bc)^2}{(2abc - 4ab)a^2} = \frac{2bc^2}{c-2}$

(b) $\frac{\left(3x^2yz\right)^3}{9(xy^2z)^3} = 3\left(\frac{x}{y}\right)^3$

Solution of Exercise 1.13

(a) $\sum\limits_{n=1}^{33} 3n$

(b) $\sum\limits_{n=0}^{10} (2n+1)$

Solution of Exercise 1.14

$\sum\limits_{i=-1}^{1} \sum\limits_{j=-2i}^{i} i^2 \cdot j = -2$

Solution of Exercise 1.15

(a) $8! = 40.320$

(b) $9 \cdot 10 \cdot 11 = \frac{11!}{8!}$

Exercises Chap. 2

Solution of Exercise 2.1

$f([-1,2]) = \left[-1, -\frac{1}{4}\right]$

Solution of Exercise 2.2

(a) $f\colon \mathbb{N} \to \mathbb{N}, f(x) = x^2$ is a function because the square of a natural number is again a natural number.

(b) $g\colon [-1,1] \to \mathbb{R},\ g(x) = 1/x$ is not a function because no value exists for $x = 0$.

(c) $h\colon \mathbb{Q} \to \mathbb{Q},\ h(x) = 2x^4 - 5/3x^3 + 1/2x$ is a function because the values are always rational numbers.

Solution of Exercise 2.3 The concatenation is possible in both cases; in the first case the result is

$(g \circ f)(x) = (\sqrt{x} + 1)^2$,
in the second

$(g \circ f)(x) = \sqrt{\frac{1+2x^2}{1+x^2}}$.

Solution of Exercise 2.4 The function f is monotonically increasing, the functions g and h are monotonically decreasing.

Solution of Exercise 2.5 Since all three functions are strictly monotonic, the inverse functions are defined; they are as follows.

$$f^{-1} : \mathbb{R} \to \mathbb{R}, \qquad f^{-1}(x) = 10x + 170,$$
$$g^{-1} : \mathbb{R} \to \mathbb{R}, \qquad g^{-1}(x) = 170 - 17x;$$
$$h^{-1} : \left[\frac{1}{17}, 1\right] \to \mathbb{R}, \quad h^{-1}(x) = \frac{1}{x}.$$

Solution of Exercise 2.6 f and h are not polynomials, $g(x) = x + 1$ is a polynomial.

Solution of Exercise 2.7 The function value $p(0) = 117$ is positive, while, for example, the values $p(-1000)$ and $p(1000)$ are negative; so there must be a zero in between.

Solution of Exercise 2.8 f is not a rational function because of the term x^x, g is a rational function with maximum domain $\mathbb{R} \setminus \{1\}$.

Solution of Exercise 2.9 The values can be taken from the following table:

a	$x = -2$	$x = 0$	$x = 1$	$x = 2$	$x = 10$
0.5	4	1	0.5	0.25	≈ 0.00097
1	1	1	1	1	1
2	0.25	1	2	4	1024
4	0.0625	1	4	16	1,048,576

Solution of Exercise 2.10 She should choose bank A because she will receive about 9540.15 Euros there.

Solution of Exercise 2.11

$\left(\frac{1}{2}\right)^{\frac{10{,}000}{5776}} \approx 0.301$ Gramm

Solution of Exercise 2.12

$\log_{10}(0.001) = -3, \quad \log_7\left(\sqrt[4]{7^3}\right) = \log_7\left(7^{\frac{3}{4}}\right) = \frac{3}{4}.$

Solution of Exercise 2.13

(a) after a 1000 years
(b) after 1160.96 years

Exercises Chap. 3

Solution of Exercise 3.1 It is the equation

$12 + 6x = 60$
to solve, it follows $x = 8$. So you can invite eight guests if you are starving yourself, or seven if you are also eating a pizza.

Solution of Exercise 3.2

(a) $x = 1$
(b) $x = -2$
(c) $x = \frac{1}{9}$

Solution of Exercise 3.3

(a) $x_1 = x_2 = -\sqrt{2}$
(b) $x_1 = \sqrt{3}, x_2 = -3$

Solution of Exercise 3.4

(a) The equation is linear, and the solution is $x_1 = 5/2$.
(b) The equation is general, so it is solved by any real number x, but one must not insert the zeros of the denominators. The solution set is therefore $\mathbb{L} = \mathbb{R} \setminus \{-2, 1\}$.

Solution of Exercise 3.5

(a) $r_1(x) = \frac{x(x+12)(x-5)}{3(x-4)(x-5)} = \frac{x(x+12)}{3(x-4)}$
(b) $r_2(x) = \frac{x^2-4}{2x-4} = \frac{(x-2)(x+2)}{2(x-2)} = \frac{x+2}{2}$

Solution of Exercise 3.6 Simplified equation:

$x^2 - 2x - 1 = 0.$
Solutions: $x_{1/2} = 1 \pm \sqrt{2}.$

Solution of Exercise 3.7
(a) Unsolvable (already the substituted equation has no solution).
(b) Two solutions: $x_{2,1} = -3$, $x_{2,2} = 3$.
(c) Four solutions: $x_{1,1} = -2$, $x_{1,2} = 2$, $x_{2,1} = -3$, $x_{2,2} = 3$.

Solution of Exercise 3.8
(a) Two solutions: $x_1 = -2$, $x_2 = 2$.
(b) Two solutions: $x_1 = 1$, $x_2 = 1/2$.
(c) The solution is $x_1 = 7$, and the *apparent solution* is $x_2 = 2$.

Solution of Exercise 3.9
(a) $x = 0$
(b) $x = 2$
(c) $x \approx 5.419$

Solution of Exercise 3.10 About 352.25 years.

Solution of Exercise 3.11
(a) $\mathbb{L} = \{x \in \mathbb{R} \mid x > 6\}$
(b) $\mathbb{L} = \{x \in \mathbb{R} \mid x > -9/2\}$
(c) $\mathbb{L} = \{x \in \mathbb{R} \mid x < -1 \text{ or } x > 2\}$

Exercises Chap. 4

Solution of Exercise 4.1 The area F of the triangles in a), b) and c) is given by: $F = 1/2 \cdot 4 \cdot 2 = 4$. In d), use Heron's formula to calculate $F = 4$.

Solution of Exercise 4.2 The remaining cathetus has the length $b = 3$.

Solution of Exercise 4.3 The cathetes have lengths $a = -1 + \sqrt{3}$ and $b = 3 - \sqrt{3}$, while the length of the hypotenuse is $c = 2(\sqrt{3} - 1)$.

Solution of Exercise 4.4 For the length of the height h on the side with length b we have according to the Pythagorean theorem: $h = \sqrt{a^2 - b^2/4}$. Inserting into the formula for the area leads to $F = bh/2 = b\sqrt{4a^2 - b^2}/4$.

Solution of Exercise 4.5 The remaining quantities are $q = 1$, $h = 1$ and $a = \sqrt{2}$.

Solution of Exercise 4.6 The two remaining interior angles are $\alpha = 60°$ and $\beta = 30°$. The hypotenuse length is $c = 6$. The area of the triangle is $F \approx 7.7942$.

Solution of Exercise 4.7 The possible heights are $h_c = 4\sin(30°) = 2$ and $h_a = 5\sin(30°) = 5/2$. In both cases, the area of the triangle is calculated as $F = 5$.

Solution of Exercise 4.8 (a) $\sin(x) = 1$ exactly when $x = \pi/2 + 2k\pi$, where k is any integer, (b) $\cos(x) = -1$ exactly when $x = \pi + 2k\pi$, where k is any integer.

Solution of Exercise 4.9 All real numbers $x \neq \pi/2 + k\pi$, where k is an integer, because $\cos(x) \neq 0$ holds true for these x.

Solution of Exercise 4.10 According to the second addition theorem, we have:

$$\cos(x) = \cos\left(\frac{x+y}{2} + \frac{x-y}{2}\right) = \cos\left(\frac{x+y}{2}\right)\cos\left(\frac{x-y}{2}\right) - \sin\left(\frac{x+y}{2}\right)\sin\left(\frac{x-y}{2}\right),$$

$$\cos(y) = \cos\left(\frac{x+y}{2} + \frac{y-x}{2}\right) = \cos\left(\frac{x+y}{2}\right)\cos\left(\frac{y-x}{2}\right) - \sin\left(\frac{x+y}{2}\right)\sin\left(\frac{y-x}{2}\right).$$

Due to the symmetry properties of the cosine and sine functions, we have

$$\cos\left(\tfrac{y-x}{2}\right) = \cos\left(\tfrac{x-y}{2}\right) \quad \text{and} \quad -\sin\left(\tfrac{y-x}{2}\right) = \sin\left(\tfrac{x-y}{2}\right),$$

from which the assertion.

Solution of Exercise 4.11 The cosine theorem gives $\cos(\alpha) = \cos(\beta) = 0.75$. From this it follows that $\alpha = \beta \approx 41.41°$ and $\gamma \approx 97.18°$.

Solution of Exercise 4.12 No! Consider, for example, the given lengths $a = 1$, $b = 2$ and $c = 3$. If we assume that there is a triangle with these side lengths, then according to the triangle inequality $a + b > c$ should apply. But this is *not* the case for the given lengths.

Solution of Exercise 4.13 One calculates $\sin(\beta) = 4\sin(10°) \approx 0.6946$. Thus: $\beta \approx 43.99°$. Thus: $\gamma \approx 126.01°$. Finally: $c \approx 2\sin(126.01°)/\sin(10°) \approx 9.3167$. If $\alpha = 40°$, then $\sin(\beta) = 4\sin(40°) \approx 2.5712$. But this cannot be because of $|\sin(x)| \leq 1$ – no such plane triangle exists.

Solution of Exercise 4.14 Given two side lengths b and c *of* a triangle and the angle α enclosing these two sides, we first have $F = b\,c\,\sin(\alpha)/2$. Using the relation $\sin(\alpha) = a/2r$ then gives $F = abc/4r$, where r is the radius of the circumcircle.

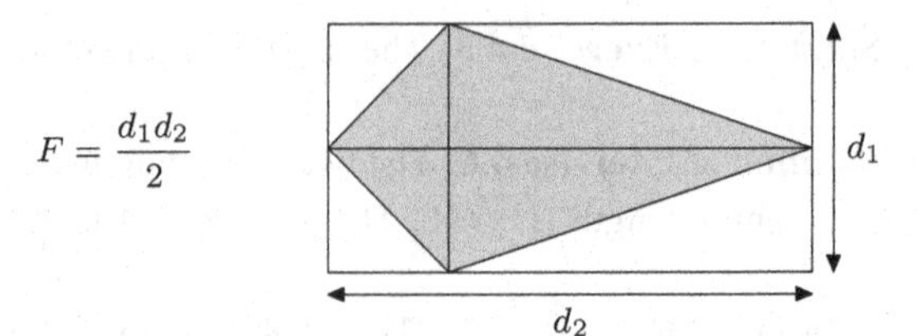

Fig. A.1 The area F of a kite (*grey area*) with diagonals of length d_1 and d_2 is half the area of a rectangle with sides d_1 and d_2

Solution of Exercise 4.15 It holds that $\alpha = 180° - \beta = 60°$ and therefore the length of the diagonal d in the rhomboid is $d = \sqrt{a^2 + b^2 - 2ab\cos(\alpha)} = \sqrt{19}$. The area of the parallelogram is $F = \sin(\alpha)\,ab = 15\sqrt{3}/2$.

Solution of Exercise 4.16 The solution is shown in Fig. A.1.

Solution of Exercise 4.17 For the length of the height h on the side with length a the following holds true:$h = \sin(\alpha)c = \sin(45^\circ)\sqrt{2} = 1$, and thus we get $F = (a + b)\, h/2 = 5$.

Solution of Exercise 4.18 From the area formula for the trapezoid it follows that the length of the height is $h = \sqrt{6}/3$. Thus the lengths a and b *of* the parallel sides of the trapezoid are calculated as $a = h = \sqrt{6}/3$ and $b = 2h = (2\sqrt{6})/3$. Since the trapezoid is isosceles, the remaining side length c is determined by using the Pythagorean theorem $c = \sqrt{30}/6$. From $\sin(\alpha) = h/c$ follows that $\alpha \approx 63.43°$ and because of the isosceles nature of the trapezoid we have for the the remaining angle $\beta \approx 116.57°$.

Solution of Exercise 4.19

(a) From $F_{3600} = 450\, b^2 \sin((0.1)°)/\sin^2((0.05)°)$, calculate $F_{3600} \approx 1{,}031{,}323.769\, b^2$
(b) The sum of the lengths of the sides of the regular N-corner is $U_N = N\, b = 2\, N \sin(180°/N)$, since $a = 1$. For $N = 3600$, we get $U_{3600} = 7200 \sin((0.05)°) \approx 6.28318451$.

Solution of Exercise 4.20 The area of each pizza portion is 200π cm$^2 \approx 628$ cm^2 and the corresponding boundary has length 10π cm ≈ 31.42 cm. You accept the pizza maker's "offer", of course, because according to it the total area of the circular ring pizza is $(2500 - 625)\pi$ cm^2 – everyone is then entitled to 234.38π cm$^2 \approx 736$ cm^2 pizza. However, the boundary of each portion has then the length ≈ 58.90 cm.

Solution of Exercise 4.21 The approach $\pi 10^2 2/3 = \pi(2b)\, b$ leads to the minor axis radius $b = 10\sqrt{3}/3 \approx 5.77$ cm and the major axis radius $r = 20\sqrt{3}/3 \approx 11.55$ cm of the Burger-ellipse.

Exercises Chap. 5

Solution of Exercise 5.1

$$\begin{pmatrix} 1 \\ -2 \\ 5 \end{pmatrix} - 4 \cdot \begin{pmatrix} -2 \\ 1 \\ 4 \end{pmatrix} + 2 \cdot \begin{pmatrix} 12 \\ 0 \\ 1 \end{pmatrix} = \begin{pmatrix} 33 \\ -6 \\ -9 \end{pmatrix},$$

$$-\begin{pmatrix} 3 \\ -4 \end{pmatrix} + 3 \cdot \left[\begin{pmatrix} 2 \\ 1 \end{pmatrix} - 2 \cdot \begin{pmatrix} -1 \\ 4 \end{pmatrix} \right] = \begin{pmatrix} 9 \\ -17 \end{pmatrix}.$$

Solution of Exercise 5.2

(a) $\begin{pmatrix} 3 \\ -4 \\ 1 \end{pmatrix} \cdot \begin{pmatrix} 1 \\ 0 \\ -12 \end{pmatrix} = -9, \begin{pmatrix} 2 \\ 1 \\ -1 \end{pmatrix} \cdot \begin{pmatrix} -1 \\ 5 \\ 6 \end{pmatrix} = -3.$

(b) $\begin{pmatrix} -1 \\ 2 \end{pmatrix} \cdot \left(\begin{pmatrix} 3 \\ 4 \end{pmatrix} + \begin{pmatrix} -2 \\ 5 \end{pmatrix} \right) = \begin{pmatrix} -1 \\ 2 \end{pmatrix} \cdot \begin{pmatrix} 1 \\ 9 \end{pmatrix} = 17$ and

$\begin{pmatrix} -1 \\ 2 \end{pmatrix} \cdot \left(\begin{pmatrix} 3 \\ 4 \end{pmatrix} + \begin{pmatrix} -2 \\ 5 \end{pmatrix} \right) =$

$\begin{pmatrix} -1 \\ 2 \end{pmatrix} \cdot \begin{pmatrix} 3 \\ 4 \end{pmatrix} + \begin{pmatrix} -1 \\ 2 \end{pmatrix} \cdot \begin{pmatrix} -2 \\ 5 \end{pmatrix} = 5 + 12 = 17.$

Solution of Exercise 5.3

(a) $\begin{pmatrix} 1 \\ 2 \\ 3 \end{pmatrix}, \begin{pmatrix} 2 \\ 4 \\ 7 \end{pmatrix}$ are linearly independent.

(b) $\begin{pmatrix} -2 \\ 5 \end{pmatrix}, \begin{pmatrix} 12 \\ 17 \end{pmatrix}, \begin{pmatrix} 234 \\ 1 \end{pmatrix}$ are linearly dependent.

(c) $\begin{pmatrix} 2 \\ -1 \\ 2 \end{pmatrix}, \begin{pmatrix} -4 \\ 2 \\ -4 \end{pmatrix}$ are linearly dependent.

Solution of Exercise 5.4

(a) $2 \cdot \begin{pmatrix} 1 & 0 & -2 \\ -3 & 5 & 1 \\ 4 & 2 & -1 \end{pmatrix} + 3 \cdot \begin{pmatrix} -6 & 3 & 4 \\ 1 & 2 & -2 \\ -3 & -4 & 0 \end{pmatrix} = \begin{pmatrix} -16 & 9 & 8 \\ -3 & 16 & -4 \\ -1 & -8 & -2 \end{pmatrix}.$

$$\text{(b)}\quad -3\cdot\begin{pmatrix}-1 & 5 & 2\\ -2 & 1 & 0\\ 3 & 1 & -2\end{pmatrix}+\begin{pmatrix}2 & -5 & -2\\ 2 & -1 & 0\\ -3 & -1 & 3\end{pmatrix}=\begin{pmatrix}5 & -20 & -8\\ 8 & -4 & 0\\ -12 & -4 & 9\end{pmatrix}.$$

Solution of Exercise 5.5

$$2\cdot\begin{pmatrix}1 & 4\\ -2 & 6\end{pmatrix}-3\cdot\left[\begin{pmatrix}0 & 2\\ 3 & -1\end{pmatrix}+4\cdot\begin{pmatrix}1 & 0\\ -1 & -2\end{pmatrix}\right]=\begin{pmatrix}-10 & 2\\ -1 & 39\end{pmatrix}.$$

Solution of Exercise 5.6

$$\text{(a)}\quad\begin{pmatrix}1 & 0 & -2\\ -3 & 5 & 1\\ 4 & 2 & -1\end{pmatrix}\cdot\begin{pmatrix}-6 & 3 & 4\\ 1 & 2 & -2\\ -3 & -4 & 0\end{pmatrix}=\begin{pmatrix}0 & 11 & 4\\ 20 & -3 & -22\\ -19 & 20 & 12\end{pmatrix}.$$

$$\text{(b)}\quad\begin{pmatrix}-1 & 5 & 2\\ -2 & 1 & 0\end{pmatrix}\cdot\begin{pmatrix}2 & -5 & -2 & 1\\ 2 & -1 & 0 & 2\\ -3 & -1 & 3 & 0\end{pmatrix}=\begin{pmatrix}2 & -2 & 8 & 9\\ -2 & 9 & 4 & 0\end{pmatrix}.$$

Solution of Exercise 5.7 The result matrix is

$$\begin{pmatrix}-17 & 38 & -5\\ -11 & 25 & 26\\ -38 & 35 & 7\end{pmatrix}.$$

It does not matter whether you first process the matrix operations within the square brackets and then multiply the matrices or multiply them out immediately and then combine them.

Solution of Exercise 5.8 If $A=\begin{pmatrix}1 & 2\\ 2 & 4\end{pmatrix}$ had an inverse $A^{-1}=\begin{pmatrix}a & b\\ c & d\end{pmatrix}$, then it should hold:

$$\begin{pmatrix}1 & 0\\ 0 & 1\end{pmatrix}=A\cdot A^{-1}=\begin{pmatrix}1 & 2\\ 2 & 4\end{pmatrix}\begin{pmatrix}a & b\\ c & d\end{pmatrix}=\begin{pmatrix}a+2c & b+2d\\ 2a+4c & 2b+4d\end{pmatrix}.$$

So $a + 2c = 1$ and at the same time $2a + 4c = 0$. Because of $2a + 4c = 2(a + 2c)$ this is impossible.

Solution of Exercise 5.9 In the first system of equations, $x = 1$, $y = -1$, $z = 2$. In the second, $x = -1$, $y = 7$, $z = -1/2$, $u = 3$.

Solution of Exercise 5.10 The system of equations is unsolvable.

Solution of Exercise 5.11 The first system of equations has the solutions

$x = z + 1, \quad y = z - 1, \quad z \in \mathbb{R}.$

The second system of equations has the solutions

$x = u, \quad y = 2u, \quad z = -2u, \quad u \in \mathbb{R}.$

Solution of Exercise 5.12 The vectors are linearly independent.

Solution of Exercise 5.13

$$A^{-1} = \begin{pmatrix} 1 & 1 & -4 \\ 2 & 2 & -7 \\ 0 & 1 & -4 \end{pmatrix}.$$

Solution of Exercise 5.14

$$\text{(a)}\ \overrightarrow{AB} = \begin{pmatrix} -3 \\ -3 \\ 2 \end{pmatrix}, \overrightarrow{CD} = \begin{pmatrix} 1 \\ -1 \end{pmatrix}.$$

$$\text{(b)}\ \overrightarrow{\mathbf{0}A} = \begin{pmatrix} 3 \\ 6 \\ -1 \end{pmatrix}, \overrightarrow{\mathbf{0}B} = \begin{pmatrix} 4 \\ 7 \\ 9 \end{pmatrix}.$$

Solution of Exercise 5.15

$$\mathbf{a} + 2\mathbf{b} - \mathbf{c} = \begin{pmatrix} 3 \\ 3 \end{pmatrix}, \quad 3\mathbf{a} + \mathbf{b} + \mathbf{c} = \begin{pmatrix} -1 \\ 8 \end{pmatrix}.$$

The first calculation is entered in the graphical representation (Fig. A.2).

Solution of Exercise 5.16 The equation is

$$\begin{pmatrix} x \\ y \\ z \end{pmatrix} = \begin{pmatrix} 2 \\ 1 \\ 3 \end{pmatrix} + \lambda \cdot \begin{pmatrix} -3 \\ -1 \\ 2 \end{pmatrix} + \mu \cdot (-1\ 0\ -2) \quad \text{with } \lambda, \mu \mathbb{R}.$$

Solution of Exercise 5.17

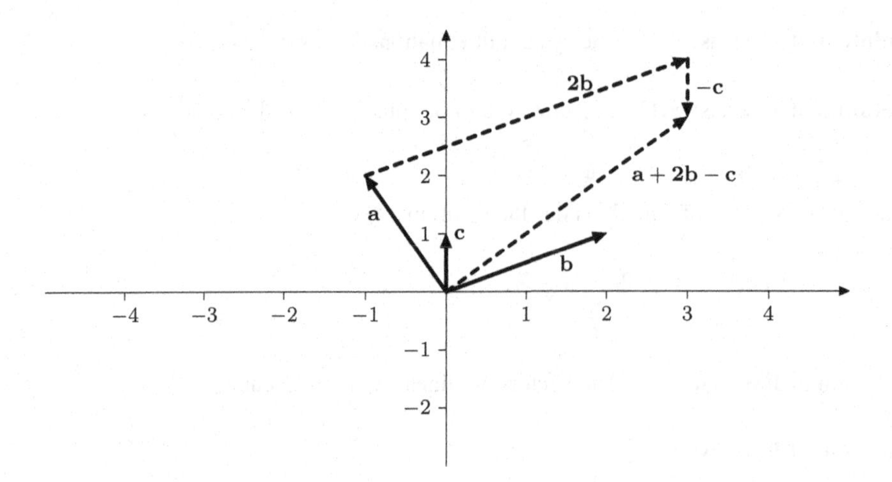

Fig. A.2 Vectors a, b, c and $a + 2b - c$

$$\begin{pmatrix} 2 \\ -1 \\ 4 \end{pmatrix} \times \begin{pmatrix} -3 \\ 1 \\ 2 \end{pmatrix} = \begin{pmatrix} -6 \\ -16 \\ -1 \end{pmatrix}$$

and

$$\begin{pmatrix} -6 \\ -16 \\ -1 \end{pmatrix} \cdot \begin{pmatrix} 2 \\ -1 \\ 4 \end{pmatrix} = \begin{pmatrix} -6 \\ -16 \\ -1 \end{pmatrix} \cdot \begin{pmatrix} -3 \\ 1 \\ 2 \end{pmatrix} = 0.$$

Solution of Exercise 5.18 The equation is $3x + 6y + 9z = 12$ or, if you divide by 3 on both sides, $x + 2y + 3z = 4$.

Exercises Chap. 6

Solution of Exercise 6.1

(a) $f'(x) = 2x$, thus: $f'(3) = 6$ and $f'(-2) = -4$
(b) $x = -1$

Solution of Exercise 6.2

$(x + h)^3 - x^3 = x^3 + 3x^2h + 3xh^2 + h^3 - x^3 = h(3x^2 + 3xh + h^2)$

and thus it follows for the limit value of the quotient of the differences that

$\lim_{h \to 0} 3x^2 + 3xh + h^2 = 3x^2.$

Solution of Exercise 6.3

$(x + h)^2 + (x + h) - x^2 - x = 2hx + h^2 + h = h(2x + h + 1)$
and thus it follows for the limit of the difference quotient, i.e. the differential quotient, that:

$\lim_{h \to 0} 2x + h + 1 = 2x + 1.$

Solution of Exercise 6.4

(a) $f(x) = |x - 2|$ is not derivable at the position $x = 2$, therefore f is not differentiable
(b) $f(x) = 1/2((x - 2)^3 + |x - 2|^3)$ is differentiable and in particular differentiable in $x = 2$: $f'(2) = 0$.

Solution of Exercise 6.5

$(3 \cos(x) - 2 \sin(x))' = 3\cos'(x) - 2\sin'(x) = -3 \sin(x) - 2 \cos(x).$

Solution of Exercise 6.6

(a) $f'(x) = -3x^2 - 4x + 1/8$
(b) $f_t'(x) = 4t^2x^3 - 4t^4x$
(c) $f_x'(t) = -8x^2t^3 + 2(x^4 + 1)t$

Solution of Exercise 6.7

(a) $f'(x) = -5x^4 + 16x^3 - 6x^2$
(b) $f'(x) = (2x + 1)\cos(x) - (x^2 + x)\sin(x)$
(c) $f'(x) = x(2 + x)e^x$
(d) $f'(x) = (\cos(x) - \sin(x))e^x$
(e) $f'(x) = -2\sin(x)\cos(x)$

Solution of Exercise 6.8

(a) $f'(x) = 16(2x - 4)(x^2 - 4x + 2)^{15}$
(b) $f'(x) = (2x - 1)e^{x^2 - x}$
(c) $f'(x) = (n/m)\, x^{(n/m) - 1}$, where $x > 0$
(d) $f'(x) = 2x(x^2 - 3)(x^2 + 1)^{-3}$

Solution of Exercise 6.9

(a) $f'(x) = (-2x^2 + 16x + 1)/(2x^2 + 1)^2$, *x arbitrary*
(b) $f'(x) = (-2x^2 - 16x - 1)/(2x^2 - 1)^2$, $x \neq \sqrt{2}/2$ and $x \neq -\sqrt{2}/2$

(c) $f(x) = (\cos(x)x - \sin(x))/x^2$, $x \neq 0$
(d) $f(x) = (2x(x^2 - 3))/(x^2 + 1)^3$, x arbitrary, this is the same (derivative) function as in Exercise 6.8 d)
(e) $f(x) = (x^2 - 3x - 1)/e^x$, x arbitrary

Solution of Exercise 6.10
(a) $p'(x) = 16x^3 - 12x, p''(x) = 48x^2 - 12, p^{(3)}(x) = 96x, p^{(4)}(x) = 96, p^{(i)}(x) = 0, i \geq 5$
(b) $f_t'(x) = 3t^3x^2 - 2tx + t, f_t''(x) = 6t^3x - 2t, f_t'''(x) = 6t^3, f_t^{(i)}(x) = 0, i \geq 4$

Solution of Exercise 6.11

$\left(e^{x^2-x}\right)''' = \left(8x^3 - 12x^2 + 18x - 7\right)e^{x^2-x}$ and

$\left(e^{x^2-x}\right)'''' = \left(16x^4 - 32x^3 + 72x^2 - 56x + 25\right)e^{x^2-x}$

Solution of Exercise 6.12 $f'(x) = -|x|$ is not differentiable in $x = 0$.

Solution of Exercise 6.13 $\left(e^{x^2-x}\right)' = 0$ has only the solution $x = 1/2$, the value of the second derivative $\left(e^{x^2-x}\right)'' = (4x^2 - 4x + 3)e^{x^2-x}$ in $x = 1/2$ is > 0, $T = (1/2|e-^{1/4})$ is the global minimum of the graph of e^{x^2-x}

Solution of Exercise 6.14
(a) $f'(x) = (1/3)x^3, f''(x) = x^2, f'''(x) = 2x$, test of inflection points $f'(x) = 0$ yields the candidate $x = 0$, however, $f'''(0) = 0$, $x = 0$ is not an extremum of f', hence not an inflection point of f.
(b) $\left(e^{x^2-x}\right)'' = (4x^2 - 4x + 3)e^{x^2-x} = 0$ leads to quadratic equation $x^2 - x + 3/4 = 0$, which has no real solutions – so there are no candidates for inflection points.

Solution of Exercise 6.15 The domain of definition of f_t is $\mathbb{R} \setminus \{0\}$. The zeros of f_t are $x = -t$ and $x = t$. f_t has no local extrema and points of inflection. The graph of f_t is symmetric to the y-axis. $a_{f_t}(x) = -t$ is horizontal asymptote of f_t.

Solution of Exercise 6.16 It is a square area with side length $\sqrt{2} \approx 1.41$ km.

Solution of Exercise 6.17

(a) $\int_1^3 x^2\,dx = 26/3$

(b) $\int_1^t x^2\,dx = t^3/3 - 1/3 = 7/3$ and $t > 1$ imply $t = 2$

Solution of Exercise 6.18

(a) Oriented area: $\int_{-1}^{1} x^3 - x\,dx = 0,$

absolute area: $\int_{-1}^{1} |\,x^3 - x\,|\,dx = \int_{-1}^{0} x^3 - x\,dx - \int_0^1 x^3 - x\,dx = 1/2$

(b) $\int_{-1}^{1} 4x^4 - 6x^2 + e^8\,dx = -12/5 + 2e^8, \int_{-1}^{1} t^3x^3 - tx^2 + tx + t^3\,dx = 2t^3 - (2/3)t$

Solution of Exercise 6.19

$\int_0^t -x^2 + 2x\,dx = -t^3/3 + t^2 = F_0(t),$

first derivative F_0' of F_0 coincides with integrand: $F_0'(t) = -t^2 + 2t$.

Solution of Exercise 6.20

(a) $H'(x) = -x^2 + 2x = h(x)$

(b) Since $\left(e^{x^2-x}\right)' = (2x-1)\,e^{x^2-x}$ holds, it follows that $F(x) = e^{x^2-x} + c$ are all primitive functions of f (c is an arbitrary constant here).

Solution of Exercise 6.21

(a) $\int_0^2 -x^2 + 2x\,dx = [-x^3/3 + x^2]_0^2 = 4/3$

(b) $\int_1^2 (2x-1)\,e^{x^2-x}\,dx = e^{x^2-x}\Big|_1^2 = e^2 - 1$

(c) $\int_1^2 (1-x^2)/(1+x^2)^2\,dx = [x/(1+x^2)]_1^2 = -1/10$

(d) From Exercise 6.7 b) it follows: $F(x) = \cos(x)\,e^x$ is the primitive function of

(e) $f(x) = (\cos(x) - \sin(x))\,e^x$. Therefore:

$\int_0^{\pi} (\cos(x) - \sin(x))\,e^x\,dx = \cos(x)e^x\Big|_0^{\pi} = -(e^{\pi} + 1).$

(f) $F(x) = -1/x$ is primitive function of $f(x) = 1/x^2$. Thus:

$\int_1^2 1/x^2\,dx = 1/x\Big|_1^2 = 1/2.$

Solution of Exercise 6.22

$$\int^t \sin(x)\,x\,dx + c = \sin(t) - t\cos(t) + c \quad \text{und}$$

$$\int^t x^2 e^x\,dx = (t^2 - 2t + 2)\,e^t + c,$$

where c is an arbitrary constant.

Solution of Exercise 6.23

$$\int^t 3x^2 \left(x^3 + 4\right)^{88} dx + c = \int^{t^3+4} \overline{x}^{88}\,d\overline{x} + c = \left(t^3 + 4\right)^{89}/89 + c$$

$$\int^t (2x - 1)\,e^{x^2 - x}\,dx + c = \int^{t^2 - t} e^{\overline{x}}\,d\overline{x} + c = e^{t^2 - t} + c,$$

where c is an arbitrary constant.

Solution of Exercise 6.24 Defining $g(x) = x^2 + 1$, $g'(x) = 2x$ holds true. Applying the substitution rule then gives:

$$\tfrac{1}{2}\int_1^2 2x/(x^2 + 1)\,dx = \tfrac{1}{2}\int_{g(1)}^{g(2)} 1/\overline{x}\,d\overline{x} = \tfrac{1}{2}(\ln(5) - \ln(2)).$$

Exercises Chap. 7

Solution of Exercise 7.1 There are $32^3 = 32{,}768$ possibilities, so the probability we are looking for is

$$\frac{1}{32{,}768}.$$

Solution of Exercise 7.2 Because of $n = 7$ and $l = 5$ there is

$$\binom{11}{5} = 462$$

possibilities, only one of which results in the desired word. The searched probability is therefore

$$\tfrac{1}{462} \approx 0.00216.$$

Solution of Exercise 7.3 The probability we are looking for is

$$\frac{(7-5)!}{7!} = \frac{1}{3\cdot4\cdot5\cdot6\cdot7} = \frac{1}{2520} \approx 0.0003968.$$

Solution of Exercise 7.4

(a) The coach has
$$\binom{15}{11} = \frac{15!}{11!\cdot4!} = 1365$$
possibilities.

(b) The new coach has
$$\binom{7}{5}\cdot\binom{6}{5}\cdot\binom{2}{1} = \frac{7!\cdot6!\cdot2!}{5!\cdot2!\cdot5!\cdot1!\cdot1!} = 252$$
possibilities.

Solution of Exercise 7.5

(a) There are
$$3^{11} = 177,147$$
different tips.

(b) With "5 out of 25" the probability for a main hit is equal to
$$\binom{25}{5} = \frac{1}{53,130},$$
but on 4/20, she is not.
$$\binom{20}{4} = \frac{1}{4845},$$
so much higher.

(c) In total, passengers have
$$\frac{7!}{(7-4)!} = 840$$
possibilities.

(d) The organizer has
$$\binom{12}{4}\cdot\binom{8}{4} = \frac{12!\cdot8!}{4!\cdot8!\cdot4!\cdot4!} = 34,650$$
possibilities.

Solution of Exercise 7.6 Adding up the individual frequencies gives $n = 797$. This means that

$$h_{797}(A_1) = \frac{140}{797} \approx 0.176,$$
$$h_{797}(A_2) = \frac{393}{797} \approx 0.493,$$
$$h_{797}(A_3) = \frac{277}{797} \approx 0.348.$$

Solution of Exercise 7.7 There are a total of 166 winnings to be drawn, so the probability sought is

$p(G) = \frac{166}{500} = 0.332.$
Alternatively, you can calculate as follows: The probability of drawing a blank is

$\frac{334}{500} = 0.668,$
and since this is just the probability of *not* winning, it follows that

$p(G) = 1 - 0.668 = 0.332.$

Solution of Exercise 7.8 The incompatibility of the three events can be seen immediately. The individual probabilities are $p\ (A) = 1/6$, $p\ (B) = 1/3$, $p\ (C) = 1/6$. Thus.

$p(A) + p(B) = 1/2,$
$p(B) + p(C) = 1/2,$
$p(A) + p(C) = 1/3.$
This is also obtained by calculating $p(A \cup B)$, $p(B \cup C)$, and $p(A \cup C)$.

Solution of Exercise 7.9 Let E = "student is learning English" and L = "student is learning Latin". Then

$p(E) = \frac{35}{50} = 0.7$ and $p(L) = \frac{25}{50} = 0.5.$
Since each student learns at least one language, $p(E \cup L) = 1$. It follows:

(a) $1 = p(E) + p(L) - p(E \cap L)$, so $p(E \cap L) = 0.2$.
(b) p ("English only") $= p(E) - p(E \cap L) = 0.5$.
(c) p ("one language only") $= p$ ("English only") $+ p$ ("Latin only") $= 0.5 + 0.3 = 0.8$.

Solution of Exercise 7.10 If A and B are incompatible, then $A \cap B$ is umpossible, so $p(A \cap B) = 0$. Thus (7.4) agrees with (7.3).

Solution of Exercise 7.11 Let B be the event "Part comes from company B" and OK *be* the event "Part is intact". Then, after specifying $p(OK \cap B) = 0.23$, it follows that

$p(OK \mid B) = \frac{p(OK \cap B)}{p(B)} = \frac{0.23}{0.25} = 0.92,$
92% of the parts supplied by B are therefore intact.

Exercises Chap. 8

Solution of Exercise 8.1

(a) The values 5.7; 5.8; 5.9; 6.0; 6.1; 6.2 occur.
(b) The table should look something like this:

Table of screw diameters

Sequence number	Diameter	af	rf	cf	rcf
1	5.7	1	0.04	1	0.04
2	5.8	3	0.12	4	0.16
3	5.9	6	0.24	10	0.4
4	6.0	9	0.36	19	0.76
5	6.1	4	0.16	23	0.92
6	6.2	2	0.08	25	1.0

(c) Bar chart of absolute frequencies:

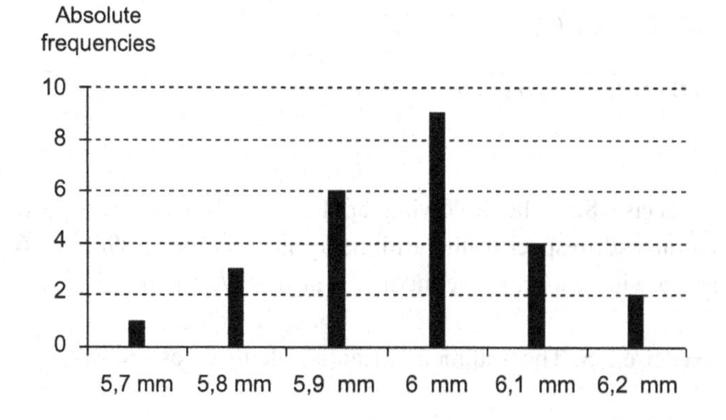

Solution of Exercise 8.2

Classes of body sizes					
Sequence Number	Height	af	rf	kf	rcf
1	160 to <165	3	0.1	3	0.1
2	165 to <170	6	0.2	9	0.3
3	170 to <175	6	0.2	15	0.5
4	175 to <180	6	0.2	21	0.7
5	180 to <185	5	0.167	26	0.867
6	185–190	4	0.133	30	1

The associated bar chart of absolute frequencies has the following form:

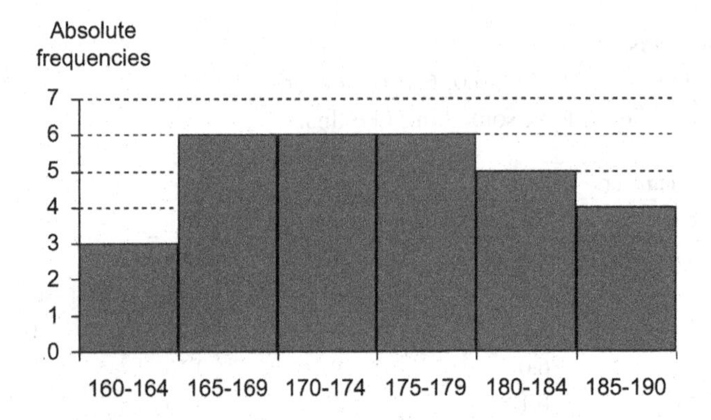

Solution of Exercise 8.3 It applies:

$$\bar{x} = \frac{149.3}{25} = 5.972; \quad \tilde{x} = 6.0;$$
$$s^2 = \frac{0.3704}{24} = 0.0038; s = \sqrt{0.0038} = 0.0616.$$

Solution of Exercise 8.4 The following applies: $\bar{x} = 174.77$; $\tilde{x} = 174.5$. The mean absolute deviation with respect to the arithmetic mean is then 189/30 = 6.3. The mean absolute deviation with respect to the median is also 189/30 = 6.3.

Solution of Exercise 8.5 The arithmetic mean is calculated as follows:

$$\frac{3 \cdot 162.5 + 6 \cdot 167.5 + 6 \cdot 172.5 + 6 \cdot 177.5 + 5 \cdot 182.5 + 4 \cdot 187.5}{30} = \frac{5255}{30} = 175.17.$$

Exercises Chap. 9

Solution of Exercise 9.1 All three expressions are complex numbers.

Solution of Exercise 9.2

$$z_1 \cdot z_2 = 8 - 9i,$$
$$\frac{z_1}{z_2 \cdot z_3} = -\frac{9}{50} - \frac{37}{50}i,$$
$$\frac{z_1 + z_2}{z_2 - z_3} = \frac{1+3i}{-2+i} = \frac{1}{5} - \frac{7}{5}i.$$

Solution of Exercise 9.3

(a) $1 - i = 1.4142 \cdot (0.7071 - 0.7071\, i)$,
(b) $-5 - 3\, i = 5.8309 \cdot (-0.8575 - 0.5145\, i)$.

Solution of Exercise 9.4

(a) $(-1 + 2\, i)^5 = -41 - 38i$,
(b) $\left(\frac{1}{\sqrt{2}} + \frac{i}{\sqrt{2}}\right)^8 = 1$.

Solution of Exercise 9.5

(a) The two second roots from $-2 + 3i$ are
$0.8960 + 1.6741i$ and $-0.8960 - 1.6741i$.

(b) The three third roots of 8 are 2, $-1 + i\sqrt{3}$ and $-1 - i\sqrt{3}$.

Solution of Exercise 9.6

(a) $x_{1/2} = -1/2 \pm i \cdot 3/2$.
(b) The four solutions are i, $-i$, $\sqrt{2}$ and $-\sqrt{2}$.

Bibliography

Fritzsche K (1995) Mathematik für Einsteiger. Spektrum Akademischer Verlag, Heidelberg

Knorrenschild M (2004) Vorkurs Mathematik. Fachbuchverlag Leipzig

Reinhard F, Soeder H (1998a) Grundlagen Algebra und Geometrie. dtv-Atlas Mathematik, Bd. 1. Deutscher Taschenbuch Verlag, München

Reinhard F, Soeder H (1998b) Analysis und Angewandte Mathematik. dtv-Atlas Mathematik, Bd. 2. Deutscher Taschenbuch Verlag, München

Rießinger T (2004) Übungsaufgaben zur Mathematik für Ingenieure. Springer-Verlag, Berlin Heidelberg

Rießinger T (2005) Mathematik für Ingenieure. Springer-Verlag, Berlin Heidelberg

Rommelfanger H (2002/2004) Mathematik für Wirtschaftswissenschaftler, Bd. 1, 2. Elsevier, München

Schwarze J (2005) Lineare Algebra, Lineare Optimierung und Graphentheorie. Mathematik für Wirtschaftswissenschaftler, Bd. 3. Verlag Neue Wirtschafts-Briefe

Walz G (Hrsg.) (2002) Lexikon der Mathematik in 6 Bänden. Spektrum Akademischer Verlag, Heidelberg

Walz G (2016) Mathematik für Hochschule und Duales Studium. Springer Verlag, Heidelberg

G. Walz et al., *Foundations of Mathematics: A Preparatory Course*,
https://doi.org/10.1007/978-3-662-67809-1

Index

G. Walz et al., *Foundations of Mathematics: A Preparatory Course*,
https://doi.org/10.1007/978-3-662-67809-1

Printed in the United States
by Baker & Taylor Publisher Services